To: Dr. Nicolas Humphrey
From: Tetsuro Matsuzawa

Thanks a lot!

T. Matsuzawa

May 18th, 2011

at Cambridge

Primatology Monographs

Series Editors

Tetsuro Matsuzawa
Inuyama, Japan

Juichi Yamagiwa
Kyoto, Japan

For further volumes:
http://www.springer.com/series/8796

Tetsuro Matsuzawa · Tatyana Humle
Yukimaru Sugiyama

Editors

The Chimpanzees of Bossou and Nimba

Editors
Tetsuro Matsuzawa, Ph.D.
Professor
Director, Primate Research Institute
Kyoto University
41-2 Kanrin, Inuyama
Aichi 484-8506, Japan
matsuzaw@pri.kyoto-u.ac.jp

Yukimaru Sugiyama, Ph.D.
Professor Emeritus
Primate Research Institute
Kyoto University
41-2 Kanrin, Inuyama
Aichi 484-8506, Japan
hqvg62yd@qc.commufa.jp

Tatyana Humle, Ph.D.
School of Anthropology and Conservation
The Marlowe Building
University of Kent
Canterbury CT2 7NR, UK
t.humle@kent.ac.uk

ISSN 2190-5967   e-ISSN 2190-5975
ISBN 978-4-431-53920-9   e-ISBN 978-4-431-53921-6
DOI 10.1007/978-4-431-53921-6
Springer Tokyo Dordrecht Heidelberg London New York

Library of Congress Control Number: 2011921514

Cover
Front cover: From top, clockwise: An infant female chimpanzee (Joya) watching her mother (Jire) cracking nuts with a pair of stone tools (photo by Etsuko Nogami). Tree nursery for the green corridor project (photo by Tatyana Humle). World Natural Heritage site, the Nimba Mountains, a natural boundary between Guinea, Liberia and Côte d'Ivoire (photo by Kathelijne Koops). A group of chimpanzees at Bossou crossing a road in front of villagers (photo by Tetsuro Matsuzawa).
Spine: An infant male chimpanzee (Flanle) clinging to his mother. Notice that he has a sixth finger (photo by Kathelijne Koops).
Back cover: From top, clockwise: Jire's family in a tree: the mother, Jire (*right*), rests while her daughter, Joya (*middle*), and her older son, Jeje (*left*), play together (photo by Tatyana Humle). Jire and her daughter, Joya, eating leaves in a cultivated field (photo by Tatyana Humle). Fresh oil palm fruits and dried nuts placed in the outdoor laboratory for the observation of nut cracking (photo by Tetsuro Matsuzawa). Thunderstorm at night in Bossou (photo by Akihiro Hirata, Mainichi Newspapers).

Printed on acid-free paper

Springer is part of Springer Science+Business Media (www.springer.com)

Foreword

From an evolutionary perspective, chimpanzees (*Pan troglodytes*) are our closest living relatives. Nearly 50 years of research on chimpanzees in their natural habitat has revealed many remarkable facets of their social and physical cognitive abilities and their capacity to survive in a range of habitat types. Wild chimpanzees primarily inhabit evergreen forest, but some populations also persist in deciduous woodland and grassland biotopes interspersed with gallery forest. Wild chimpanzees can be found today in 21 countries in Africa lying between 13°N and 7°S of the equator. Despite this wide distribution, our current understanding of their behavior comes from only six long-term field sites – Gombe and Mahale (Tanzania), Kibale and Budongo (Uganda), Taï (Côte d'Ivoire), and Bossou (Guinea) – and a few other newer sites including Nimba (Guinea/Côte d'Ivoire/Liberia), Fongoli (Senegal), Gashaka (Nigeria), Goualougo (Republic of Congo), Kalinzu and Semliki (Uganda), and Ugalla (Tanzania).

According to the International Union for the Conservation of Nature and Natural Resources Red List (IUCN 2009), chimpanzees in their natural habitat are in danger of extinction. Their populations have declined by more than 66% in the past 30 years, from around 600,000 to fewer than 200,000 individuals (Kormos et al. 2003). This tendency is all the more concerning as chimpanzees are extremely vulnerable to demographic decline and are unable to recover as rapidly as other species. Female chimpanzees indeed typically only give birth to a single offspring every 5–6 years. In addition, the majority of wild chimpanzees (e.g., more than 90% in Guinea, West Africa) (Kormos et al. 2003) live outside protected areas and are thus extremely vulnerable to human anthropic pressures.

In this regard, Bossou is an exceptional site because of its close proximity with human settlements and activities. This site harbors a single chimpanzee community that has cohabited alongside the local Manon people for many generations. Research at Bossou began in the 1960s with Adrian Kortlandt, a Dutch primatologist. More systematic research focused on this unique chimpanzee community began in 1976 with Prof. Sugiyama of Kyoto University, Japan. The research presence of Kyoto University in the region led to a convention for scientific cooperation between Guinea and Japan through the Direction National de la Recherche Scientifique et Technologique (DNRST) and Kyoto University Primate Research Institute (KUPRI). This collaboration gave me the amazing opportunity to be granted a fellowship from the Japanese government in 1985. Under the supervision of Prof. Sugiyama, I studied Japanese monkeys (*Macaca fuscata*) in Japan for 7 years and received an M.A. and a Ph.D. degree in primatology from KUPRI.

Ever since 1986, when Prof. Matsuzawa joined the Bossou field site, research and cooperation between Guinea and Japan have considerably gained in strength. The Institut de Recherche Environnementale de Bossou (IREB) was created, in 1995, as a symbol of collaboration between Guinea and Japan. The research facility of IREB as it stands today was completed in 2001. This facility greatly improved the in situ living conditions and research activities. The desire of Prof. Matsuzawa to bridge oriental and occidental primatology by bringing together students from different parts of the world ultimately led to the emergence of KUPRI-International. This initiative began in 1995. Dr. Tatyana Humle was the first non-Japanese student to pursue her Ph.D. research on chimpanzees at Bossou and surrounding areas. In the following years, students from countries as widespread as Portugal, Hungary, The Netherlands, the UK, the USA, Brazil, and France joined the team of researchers and students from Kyoto University. Together they have been implementing a detailed research program on chimpanzees at Bossou, the Nimba Mountains (6 km distant on the border with Liberia and Côte d'Ivoire), Diécké (50 km away), and surrounding areas, including some in Liberia, in collaboration with the DNRST and IREB and national authorities of neighboring countries. In parallel to this research program, this team in collaboration with IREB and local people has been promoting environmental education in the region and has developed a reforestation program known as the Green Corridor Project. This project, which was initiated by Prof. Matsuzawa in 1997, aims to connect the semi-isolated Bossou chimpanzee community with those chimpanzees inhabiting the Nimba Mountains region, a World Heritage Site characterized by its exceptional biodiversity and landscape.

In November 2006, an international symposium took place in Conakry in celebration of 30 years of research at Bossou. This meeting marked my renewed involvement in chimpanzee research and conservation at Bossou and the Nimba Mountains. In April 2009, I was appointed the new director of IREB. While pursuing research and conservation on chimpanzees in the region, in collaboration with KUPRI-International, our prime objective will be to prepare and train Guinean students to become the researchers and conservationists of tomorrow.

This book describes the achievements of the KUPRI international team, which has made significant contributions to our current understanding of behavior, ecology, sociology, culture, and cognition in chimpanzees, as well as that of conservation issues related to health, great ape–human conflicts, and traditional beliefs, and also local perceptions and practices. Although this book focuses on one region, on one great ape species and its habitat, the lessons learned and knowledge gained should serve to help promote conservation of all great apes living in their natural habitat. Aside from Ebola outbreaks, human activity constitutes today the greatest threat to the great apes. However, Bossou embodies ways that humans and wildlife can persist together in close proximity. Our goal in collaboration with the local people is to help preserve this harmony.

Aly Gaspard Soumah, Ph.D.
Director of the Institut de Recherche Environnementale de Bossou (IREB)

Preface

This book is about the wild chimpanzees of Bossou and Nimba, West Africa, as well as those of the surrounding areas. Bossou chimpanzees are renowned for their use of a pair of mobile stones as hammer and anvil to crack open oil-palm nuts. They use a set of folded broad leaves to drink water from tree hollows. They modify sticks to fish for algae floating on the surface of ponds. They manufacture pestles from palm fronds to pound the center of the palm crown to mash the palm heart. These examples of tool use and manufacture are unique to this chimpanzee community. Just like human communities, each chimpanzee community has its own set of cultural traditions. This book aims to illuminate the unique way of life of the wild chimpanzees dwelling in Bossou and its surrounding areas.

Bossou is the name of a village that is located in the southeastern corner of the Republic of Guinea. This village is situated about 1,000 km from the capital, Conakry. The last nationwide chimpanzee census in Guinea suggests that the country may be home to several thousand chimpanzees. However, to anyone travelling to Bossou by land, it quickly becomes clear that the natural habitat is highly disturbed by human activity such as logging, cultivation, cattle farming, and so forth. Bossou is a rare exception in the coexistence between humans and chimpanzees. Bossou chimpanzees can therefore easily be observed. The community comprised about 20 chimpanzees for more than four decades at least up until very recently.

The coexistence between humans and chimpanzees at Bossou is made possible by the local Manon people, who in the majority respect chimpanzees as their totem and consider them the reincarnation of their ancestors. The villagers have protected this community and parts of its core habitat for many generations. That is why Bossou chimpanzees continue to thrive alongside a human-dominated habitat that supports thousands of people. The habitat of Bossou chimpanzees is a mosaic of forest, savanna, and cultivated fields. It is isolated by savanna from the larger forests of the Nimba Mountains.

Bossou is located only about 10 km from the main ridge of the Nimba Mountains. Nimba is a UNESCO-designated world heritage site renowned for its rich fauna and flora and its unique biodiversity, which has attracted scientists ever since colonial times. Nimba is at the center of the Upper Guinean forest hotspot and a landmark in West Africa because it borders three countries: Guinea, Côte d'Ivoire, and Liberia. The chimpanzee population of the Nimba Mountains is estimated to be in the hundreds.

This book is the outcome of a collective effort of scholars who have endeavored to understand and conserve the chimpanzees of Bossou and Nimba. Yukimaru Sugiyama began his study of Bossou chimpanzees in 1976. He was later joined by Tetsuro Matsuzawa in 1986 and then followed by other Japanese researchers. Tatyana Humle joined as the first non-Japanese researcher in 1995 and was followed by an international team including Guinean scholars. According to our records as of March 2010, Yukimaru Sugiyama visited Bossou 22 times for a total of 57 months, Tetsuro Matsuzawa 21 times for 40 months, Gen Yamakoshi 13 times for 43 months, Gaku Ohashi 12 times for 42 months, Tatyana Humle 11 times for 43 months, and so forth. In that sense, this book is the product of 48 researchers from nine countries (Japan, Guinea, USA, UK, France, Hungary, Mexico, Brazil, and Portugal) who cumulatively visited Bossou 181 times for a total of 538 months. These 44.8 observation-years always involved the collaboration of local Manon field assistants. Without their dedication and help, we would have not been able to comprehensively appreciate the unique way of life of the wild chimpanzees of Bossou.

We celebrated the 30-year anniversary of the Bossou project in November 2006 in Conakry and then in Bossou. The idea of this book was born at this anniversary meeting, but much extra effort was needed to finally realize this endeavor. During the process, we lost a number of important Guinean collaborators: Mr. Gouanou Goumy, our first guide; Mr. Tino Zogbila, our second guide; Mrs. Nyonko Traore, our first cook; Mr. Soh Pleta Bonimy, an NGO collaborator; and Dr. Koulibaly Bakary, an administration officer. The book is dedicated to them, as a tribute to their lifetime commitment to promote the study and conservation of chimpanzees in Bossou and Nimba. Although they have passed away, the pleasant times and memories we have shared together still remain engraved in our hearts.

We have also lost beloved chimpanzees: Kai, Nina, Pru, Poni, Jokro, Veve, Jimato, and Jodoamon, among others. There was a flu-like epidemic among Bossou chimpanzees at the end of 2003 which led to the death of five chimpanzees—the most tragic event in the site's history. Since then, the number of chimpanzees at Bossou has not recovered. One of our conservation initiatives, the Green Corridor project, was initiated in 1997. This project aims to plant trees in the savanna to connect the isolated habitats of Bossou with the forest of the Nimba Mountains. This initiative is progressing well but will require further effort and investment before completion. In spite of our conservation efforts and initiatives, conflict between humans and chimpanzees has worsened in recent years. We think that this might be due to the negative impact of researchers' habituation of wild chimpanzees in conjunction with a steady increase in the size of the local human population. The population of Bossou was about 1,000 for a long time but it now probably exceeds 3,000. This increase concords with the influx of refugees fleeing civil war in neighboring countries, especially Liberia. This book provides us an opportunity to reflect, and to assess what we have done in our attempts to understand chimpanzees. We sincerely hope that the collaboration among all the people concerned about chimpanzees at Bossou and the surrounding areas continues to develop. Such cooperation among all stakeholders is crucial in continuing to identify and implement realistic and practical solutions which will ensure an enduring peaceful coexistence between humans and chimpanzees, our closest evolutionary relatives.

In the final part of the preface, we would like to mention the people and the organizations who have contributed to the Bossou–Nimba project. This project has mainly been financed over the years by the Ministry of Science, Technology, Education, Culture and Sports of Japan (for example, MEXT 12002009, 16002001, 20002001 in recent years). We are also immensely grateful to the Japan Society for the Promotion of Science (JSPS) core-to-core program HOPE, the Ministry of the Environment (Japan), the U.S. Fish and Wildlife Services, Conservation International, the Houston Zoo (USA), the Matsushita International Foundation, the Leakey Foundation (USA), the Wenner–Gren Foundation, the Lucie Burgers Foundation (the Netherlands), the Schure–Beijerinck–Popping Foundation (the Netherlands), the International Primatological Society, the National Institutes of Health (USA), the IUCN/SSC Primate Specialist Group, the Royal Society (UK), the University of Cambridge (UK), and the University of Stirling (UK).

We also sincerely appreciate the collective efforts of many Guinean collaborators and counterparts. The Bossou–Nimba project is based on two conventions: one between KUPRI (Kyoto University Primate Research Institute) and DNRST (Direction Nationale de la Recherche Scientifique et Technologique), and another between KUPRI and IREB (Institut de Recherche Environnementale de Bossou). IREB is a unique institute for the promotion of environmental research, and is located on-site in Bossou. Since its establishment in situ in 2001, this institution has truly promoted the collaboration between Guinean and foreign researchers. We would also like to acknowledge and express our thanks for the collaboration of the Ivorian Government for granting us the permission to work in the Nimba Mountains in Yealé and Gouéla, on the Ivorian side of the massif, and the Government of Liberia for permission to conduct surveys around Yekepa.

The following people have all participated as core researchers in the Bossou–Nimba project: Yukimaru Sugiyama, Jérémy Koman, Aly Gaspard Soumah, Tetsuro Matsuzawa, Osamu Sakura, Takao Fushimi, Rikako Tonooka, Gen Yamakoshi, Hiroyuki Takemoto, Makoto Shimada, Tatyana Humle, Masako Myowa-Yamakoshi, Satoshi Hirata, Dora Biro, Naruki Morimura, Shiho Fujita, Gaku Ohashi, Claudia Sousa, Nicolas Granier, Misato Hayashi, Laura Martinez, Asami Kabasawa, Joël Gamys, Kathelijne Koops, Miho Ito, Shigeo Kobayashi, Kazunari Ushida, Kimberley Hockings, Ryo Hasegawa, Susana Carvalho, Makan Kourouma, and Michiko Fujisawa.

Finally, we deeply appreciate and are most grateful for the collaboration of the local people of Bossou, Nimba, and Diécké, especially our field assistants in Bossou: Gouanou Goumy, Tino Zogbila, Pascal Goumy, Paquilé Chérif, Jiles Doré, Marcel Doré, Boniface Zogbila, Henry Gbéregbé, and Cé Goumy; our Green Corridor assistants: Buna Zogbila and Rémy Traoré; our field assistants in Seringbara: Kassié Doré, Fromo Doré, and Fokayé Zogbila; and our field assistants in Yealé: David Droh, Anatole Gogo, Philibert Pahon, Ferdinand Tonga, Anthony Gopou, Alexis Wonseu, and Pascal Gondo.

We are also extremely grateful to the Japanese Embassy in Guinea, especially the current ambassador, Hiroshi Sumimoto, and his predecessors, as well as the French Embassy and the British Embassy in Guinea.

In closing, we would like to express our gratitude to our publisher, Springer Japan. We especially thank Ms. Aiko Hiraguchi for her editorial work. Without her continuous support and encouragement, this book would never have seen the light of day. Thanks are also due to all the editorial staff at Springer who participated in the publishing process. This book is the product of the collaboration of a large and diverse team of people. We really hope that this volume will provide stimulating reading to all those interested in chimpanzees, our closest evolutionary neighbors, by illuminating their past, present, and future.

Kyoto University, Japan   Tetsuro Matsuzawa
University of Kent, UK   Tatyana Humle
Kyoto University, Japan   Yukimaru Sugiyama

Contents

Part VII   Conservation

Contributors

Dora Biro (*Chapters 17 and 25*)
Department of Zoology, University of Oxford, South Parks Road, Oxford OX1 3PS, UK

Blandine Bril (*Chapter 20*)
École des Hautes Études en Sciences Sociales, 54 Bd Raspail, 75006 Paris, France

Susana Carvalho (*Chapters 15 and 30*)
Leverhulme Centre for Human Evolutionary Studies, Department of Biological Anthropology, University of Cambridge, Henry Wellcome Building, Fitzwilliam Street, Cambridge CB2 1QH, UK

Gilles Dietrich (*Chapter 20*)
Université Paris Descartes, 1 rue Lacretelle, 75015 Paris, France

Shiho Fujita (*Chapters 3 and 36*)
Department of Veterinary Medicine, Faculty of Agriculture, Yamaguchi University,
Yoshida 1677-1, Yamaguchi 753-8515, Japan

Nicolas Granier (*Chapters 29, 37, 39 and Appendix G*)
Behavioral Biology Unit, Department of Environmental Sciences, University of Liège, Quai Van Beneden, 22, 4020 Liège, Belgium

Misato Hayashi (*Chapters 18 and 19*)
Primate Research Institute, Kyoto University, 41-2 Kanrin, Inuyama, Aichi 484-8506, Japan

Satoshi Hirata (*Chapters 14, 19 and 20*)
Great Ape Research Institute, Hayashibara Biochemical Laboratories, Inc., 952-2 Nu, Tamano, Okayama 706-0316, Japan

Kimberley Jane Hockings (*Chapters 22 and 23*)
Department of Anthropology, Faculty of Social and Human Sciences,
New University of Lisbon, Avenida de Berna, 26-C, 1069-061 Lisbon, Portugal
and
Department of Psychology, Stirling University, Stirling, UK

Tatyana Humle (*Chapters 1, 2, 6, 9, 11, 12, 27, 32, 37, 38, 40 and Appendix B*)
School of Anthropology and Conservation, The Marlowe Building, University of
Kent, Canterbury CT2 7NR, UK

Noriko Inoue-Nakamura (*Chapter 18*)
Showa Women's University, 1-7 Taishido, Setagaya-ku, Tokyo 154-8533, Japan

Asami Kabasawa (*Chapter 5*)
Graduate School of Asian and African Area Studies, Kyoto University,
46 Shimoadachi-cho, Yoshida, Sakyo-ku, Kyoto 606-8501, Japan

Kathelijne Koops (*Chapter 28*)
Department of Biological Anthropology, Leverhulme Centre for
Human Evolutionary Studies, University of Cambridge, Fitzwilliam Street,
Cambridge CB2 1QH, UK

Rebecca Kormos (*Chapter 40*)
1170 Grizzly Peak Blvd, Berkeley, CA 94708, USA

Makan Kourouma (*Chapter 37*)
Institut de Recherche Environnementale de Bossou, Bossou, Republic of Guinea

Laura Martinez (*Chapter 39 and Appendix G*)
Research Institute of EcoScience, Ewha Womans University, B 365, Science
Building,
11-1 Daehyun-Dong, Seodaemun-Gu, Seoul 120-750, Republic of Korea

Tetsuro Matsuzawa (*Chapters 1, 7, 11, 12, 13, 16, 21 and 37*)
Primate Research Institute, Kyoto University, 41-2 Kanrin, Inuyama,
Aichi 484-8506, Japan

Yuu Mizuno (*Chapter 14*)
Department of Children, Faculty of Children Studies, Chubu-Gakuin University,
30-1 Nakaoita, Kakamihara, Gifu 504-0837, Japan

Masako Myowa-Yamakoshi (*Chapter 24*)
Graduate School of Education, Kyoto University, Yoshida-honmachi, Sakyo,
Kyoto 606-8501, Japan

Michio Nakamura (*Chapter 26*)
Wildlife Research Center of Kyoto University, 2-24 Tanaka-Sekiden-cho, Sakyo,
Kyoto 606-8203, Japan

Gaku Ohashi (*Chapters 31, 37, Appendices A and D*)
Japan Monkey Centre, Inuyama, Aichi 484-0081, Japan
and
Primate Research Institute, Kyoto University, 41-2 Kanrin, Inuyama,
Aichi 484-8506, Japan

Makoto K. Shimada (*Chapter 34 and Appendix E*)
Primate Research Institute, Kyoto University, 41-2 Kanrin, Inuyama,
Aichi 484-8506, Japan
and
National Institute of Genetics, Mishima 411-8540, Japan
and
Institute of Comprehensive Medical Science, Fujita Health University,
1-98 Dengakugakubo, Kutsukake-cho, Toyoake, Aichi 470-1192, Japan

Aly Gaspard Soumah (*Chapter 37*)
Institut de Recherche Environnementale de Bossou, Bossou, Republic of Guinea

Cláudia Sousa (*Chapter 8*)
Department of Anthropology, Faculty of Social and Human Sciences,
New University of Lisbon, Avenida de Berna, 26-C, 1069-061 Lisbon, Portugal
and
Centre for Research in Anthropology (CRIA), Avenida das Forças Armadas,
Edifício ISCTE, 1049-029 Lisbon, Portugal

Yukimaru Sugiyama (*Chapter 3*)
Primate Research Institute, Kyoto University, 41-2 Kanrin, Inuyama,
Aichi 484-8506, Japan

Hiroyuki Takemoto (*Chapter 33 and Appendix C*)
Primate Research Institute, Kyoto University, 41-2 Kanrin, Inuyama,
Aichi 484-8506, Japan

Kazunari Ushida (*Chapter 35 and Appendix F*)
Laboratory of Animal Science, Graduate School of Life and Environmental
Sciences, Kyoto Prefectural University, Shimogamo, Kyoto 606-8522, Japan

Gen Yamakoshi (*Chapters 4, 10, 11, 12 and 24*)
Graduate School of Asian and African Area Studies, Kyoto University,
46 Shimoadachi-cho, Yoshida, Sakyo-ku, Kyoto 606-8501, Japan

Shinya Yamamoto (*Chapter 12*)
Primate Research Institute, Kyoto University, 41-2 Kanrin, Inuyama,
Aichi 484-8506, Japan
and
Great Ape Research Institute, Hayashibara Biochemical Laboratories, Inc., 952-2
Nu, Tamano, Okayama 706-0316, Japan

About the Accompanying DVD Compilation

This book offers a unique DVD compilation of video clips of Bossou chimpanzees (*Pan troglodytes verus*) using tools and performing various other behaviors referred to in this volume. This exceptional visual ethogram includes examples of algae scooping and pestle pounding, as well as water drinking with folded leaves from tree holes. These behaviors represent tool-use signatures of the Bossou chimpanzee community and have therefore never been recorded elsewhere. Video footage also describes stone-tool selectivity and transport, as well as metatool use, i.e., the use of a third stone as a wedge to balance an unstable anvil stone in the process of oil palm nut cracking.

The viewer may also watch clips of the coula nut experiment, which has yielded invaluable insights into cultural transmission among chimpanzees, as well as the variety of ant-dipping techniques targeted at army ants (*Dorylus* sp.) displayed by members of this community. In addition, the DVD contains unique clips portraying examples of deception, of fruit sharing among adults, and of palm-wine drinking with leaves, in addition to education-by-master apprenticeship in action with regard to nut

Fig. 1 Video footage of nut cracking of oil palm nuts (*Elaeis guineensis*) being collected in the outdoor laboratory on the top of the Hill of Gban in the core area of the Bossou chimpanzee community (photograph by Dora Biro)

cracking and water drinking. Cross-sectional clips permit the viewer to appreciate the sequential stages involved in the ontogeny of these tool-use behaviors.

Included is also a range of video footage depicting the unique features of the mother–offspring bond and the complex array of communicative and playful behaviors that shape social interactions among members of this community. Some clips also illustrate their nesting, feeding, and processing skills.

This DVD additionally provides glimpses into the coexistence that exists between humans and chimpanzees at Bossou. Bossou chimpanzees have indeed evolved several behavioral adaptations to crossing roads with minimal risk and to other anthropogenic modifications and threats to their habitat, e.g., crop raiding. In combination with the array of chapters and themes addressed in this volume, this DVD is a perfect complement further illustrating the intelligence and behavioral flexibility of this unique community of chimpanzees located in southeastern Guinea, West Africa.

We are extremely grateful to Miho Nakamura for putting together this DVD compilation based on video archive contributions from ANC Productions Inc., Japan, Tetsuro Matsuzawa, Tatyana Humle, Gaku Ohashi, Kimberly Hockings, and Gen Yamakoshi.

Color Plates

1. An infant female chimpanzee (Joya) gazing at human observers. Photo by Kathelijne Koops.
2. Two chimpanzees (Yolo and his elderly mother, Yo) vocalizing in a tree. Photo by Pascal Goumy.
3. A young male chimpanzee (Jeje) performing algae scooping. Photo by Henry Didier Camara.
4. A young male chimpanzee (Peley) sitting on a tree trunk. Photo by Henry Didier Camara.
5. Two chimpanzees (Tua and Yo) screaming. Photo by Pascal Goumy.
6. An adult male chimpanzee (Yolo) grooming the back of a female chimpanzee (Fanle), who is cracking nuts. Photo by Jules Gondo Doré.
7. A young mother (Fanle) cracking nuts while holding her son (Flanle). Photo by Jules Gondo Doré)
8. Male chimpanzees (*left to right*: Yolo, Foaf, and Peley) cracking nuts in the outdoor laboratory. Photo by Henry Didier Camara.
9. Nimba Mountains in the mist. Photo by Kathelijne Koops.
10. A young male chimpanzee (Jeje) crossing a road under the observation of a researcher and a local guide. Photo by Henry Didier Camara.
11. An infant male chimpanzee (Flanle) held by his mother (Fanle) in a tree. Photo by Kathelijne Koops.
12. A mother–infant chimpanzee pair (Jire and Joya) looking back while crossing a narrow path. Photo by Etsuko Nogami.

T. Matsuzawa et al. (eds.), *The Chimpanzees of Bossou and Nimba*,
DOI 10.1007/978-4-431-53921-6_1, © Springer 2011

1

2

3

4

5

11

12

Part I
Introduction

Chapter 1
Bossou: 33 Years

Tetsuro Matsuzawa and Tatyana Humle

1.1 Outline of Chimpanzee Research at Bossou

A small group of chimpanzees inhabits the forested hills surrounding the village of Bossou in southeastern Guinea, West Africa. Over the years, cultural primatology has driven much of the research on this unique group of wild chimpanzees.

The village of Bossou is located west of the Nimba Mountains, the only World Natural Heritage site (UNESCO/MAB) in Guinea (see Chaps. 27, 28, 29, 39, and 40). The highest peak in the Nimba Mountains, which is 1,752 m above sea level, is a landmark in West Africa. This region of Guinea is known as Forest Guinea (Guinée Forestière) and belongs to the Upper Guinea forest ecosystem that extends from southern Guinea into Sierra Leone and eastward from Liberia to Western Togo. The Upper Guinea forests constitute a major African biodiversity hotspot (Myers et al. 2000). From these forests originate major rivers including the Niger, the Gambia, and the Senegal. The region of Bossou and the Nimba Mountains is truly located at the center of the Upper Guinea Forest network, at the crossroads between Guinea, Côte d'Ivoire, and Liberia.

The Republic of Guinea (République de Guinée in French), formally known as French Guinea, or also referred to today as Guinea-Conakry or simply Guinea, was a French colony that acquired its independence in 1958. The country's current population is estimated to be a little more than ten million inhabitants (CIA 2008). The 1997 census estimated the population at about seven million. The country has therefore experienced significant population increase and consequent mounting demographic pressure during the last decade. Guinea extends across 245,857 km² (94,926 square miles), which corresponds in size to the state of Michigan in the USA or to about two-thirds of the surface area of Japan. The population density is

T. Matsuzawa (✉)
Primate Research Institute, Kyoto University, 41-2 Kanrin, Inuyama, Aichi 484-8506, Japan
e-mail: matsuzaw@pri.kyoto-u.ac.jp

T. Humle
School of Anthropology and Conservation, The Marlowe Building, University of Kent, Canterbury CT2 7NR, UK

T. Matsuzawa et al. (eds.), *The Chimpanzees of Bossou and Nimba*,
DOI 10.1007/978-4-431-53921-6_2, © Springer 2011

about 38 individuals per square kilometer. Guinea is curve shaped, limited on its western front by the Atlantic Ocean; the country extends eastward inland before tipping south toward Sierra Leone and Liberia on its eastern front. Guinea borders to the north with Guinea-Bissau and Senegal, to the northeast with Mali, to the southeast with Côte d'Ivoire, and to the south with Liberia and Sierra Leone.

During the French colonization and thereafter, French and Dutch scholars carried out early studies of the fauna and flora, including chimpanzees, in and around the Nimba Mountains (Kortlandt 1986; Lamotte and Roy 2003) (see Chaps. 4 and 39). In November 1976, Yukimaru Sugiyama of the Kyoto University Primate Research Institute carried out his first long-term field study of Bossou chimpanzees (Sugiyama and Koman 1979a, b). At the time, the country was still governed by Ahmed Sékou Touré, the first post-independence president of the country. Sugiyama carried out field surveys for 4–7 months three times during the first 10 years. In those days, Guinea primarily maintained international relations with Eastern Bloc countries and had a closed economy, rendering it difficult to conduct in situ fieldwork.

In February 1986, Tetsuro Matsuzawa began field research at Bossou alongside Yukimaru Sugiyama. Since then, many researchers from the Kyoto University Primate Research Institute (KUPRI) have contributed in a joint effort to study the chimpanzees of Bossou and the Nimba Mountains and to promote the conservation of their habitat (Matsuzawa 2006a, b, c). In July 1995, Tatyana Humle joined the KUPRI team as the first non-Japanese scientist. Since then, the research team has grown to become increasingly international.

At present, the international team of researchers, known as KUPRI-International, continues to contribute to the long-term research of the Bossou–Nimba chimpanzees. The research is based on a formal convention between KUPRI and two Guinean authorities: the Direction Nationale de la Recherche Scientifique et Technologique (DNRST) and Institut de Recherche Environnementale de Bossou (IREB). The two institutions represent the official counterparts of the scientific collaboration between KUPRI and the Guinean government.

In December 2008, the president Lansana Conte, who had been in power since 1984, passed away. Captain Moussa Dadis Camara subsequently took over until December 2009, after falling victim to an assassination attempt. This event forced him to abandon his position as president. Even though the situation of the country remains unstable, the collaboration between KUPRI and the Guinean authorities has remained strong.

The chimpanzees of Bossou have numbered around 20 individuals for decades since the advent of research in 1976 (see Chaps. 3 and 4). This community possesses quite remarkable life history (see Chap. 3), genetic (see Chap. 34), and physiological (see Chap. 35) features. It is also well known for its use of a variety of different tools (see Chaps. 6 and 16), including a stone hammer and a stone anvil to crack open oil-palm nuts (Biro et al. 2003; Matsuzawa 1994) (Fig. 1.1). At Bossou, nut-cracking has not only been studied from a developmental perspective (see Chaps. 18, 21, and 24), complemented by studies in captivity (see Chaps. 19 and 20), but also from an archeological (see Chaps. 7 and 15) and cultural (see Chap. 17) perspective, supplemented by field studies in surrounding areas, including

Fig. 1.1 Chimpanzees at Bossou cracking nuts with stone-tool (photograph by Etsuko Nogami)

the Nimba Mountains (see Chaps. 27–29) and Diécké (see Chap. 30). Other unique examples of tool use in the repertoire of the Bossou chimpanzees include use of leaves for drinking water (see Chap. 8), ant-dipping (Humle and Matsuzawa 2002; Humle et al. 2009; see Chap. 9), pestle pounding (Yamakoshi and Sugiyama 1995; see Chap. 10), and algae scooping (Matsuzawa et al. 1996; see Chap. 11), as well as recent innovations (e.g., ant-fishing: Yamamoto et al. 2008; see Chap. 12) or more rarely observed behavioral patterns including playing with a log doll (see Chap. 13) and animal toying (Hirata et al. 2001; see Chap. 14).

Bossou chimpanzees also exhibit culturally specific social behaviors (Nakamura and Ohashi 2003; see Chap. 26) and fascinating ecological adaptations (see Chap. 33), including their notable close relationship with the local human population. Because chimpanzees represent a totem animal for many local Manon families, especially the founding family of the village, they are protected and typically tolerated by the local people; although both compete for overlapping resources. Bossou chimpanzees indeed regularly frequent human areas and raid crops (see Chaps. 22 and 23). In that sense, Bossou embodies a truly remarkable example of coexistence between humans and chimpanzees.

1.2   The 30-Year Anniversary Symposium in 2006

This book was inspired by an international symposium celebrating the 30th anniversary of chimpanzee research at Bossou. This symposium, entitled "Research and Conservation of the African Great Apes: The 30th Anniversary of the Bossou-Nimba Project" and organized by KUPRI in association with its two Guinean institutional counterparts (DNRST and IREB), took place between the 27th and the 29th of November, 2006, in Conakry, the capital city of Guinea (Fig. 1.2). Dr. Tamba Tagbino, vice-director of the DNRST, presented a summary of the collaboration between Japan and Guinea concerning chimpanzee research at Bossou, the Nimba Mountains, and surrounding areas. There were three plenary talks. The first one was given by Yukimaru Sugiyama, who summarized 30 years of chimpanzee research at Bossou. The second talk was delivered by Tetsuro Matsuzawa, who spoke about the current program of Bossou–Nimba research and the Green Corridor Project. The third talk was presented by William McGrew of Cambridge University, UK, who highlighted the unique contribution of the Bossou long-term research to primatology worldwide.

The sessions that followed focused on the various studies of chimpanzees at Bossou and Nimba by the KUPRI-International team. The speakers (with their affiliations in 2006) included Gen Yamakoshi (Kyoto University, Japan), Tatyana Humle (University of Wisconsin, USA), Dora Biro (Oxford University, UK), Claudia Sousa (Lisbon New University, Portugal), Gaku Ohashi (KUPRI, Japan), Kathelijne Koops (Cambridge University, UK), Kimberly Hockings (Stirling University, UK), Kazunari Ushida (Kyoto Prefectural University, Japan), Asami Kabasawa (Kyoto University, Japan), Nicolas Granier (University of Liege, Belgium), and Susana Carvalho (Lisbon University of Technology, Portugal).

The symposium also included four invited talks on other research topics in Guinea: one on baboons in Guinea by Marie-Claude Huynen (Liege University,

Fig. 1.2 Group photograph of invited speakers at the international symposium held in Conakry in November 2006 (photograph by KUPRI-International)

Belgium), another two summarizing conservation efforts for chimpanzees in Guinea by Christine Sagno (Directrice Nationale des Eaux et Forêts, Guinea) represented by Marthe Sany Gbansara and Sédibinet Sidibe (Directeur du Centre National d'Observation et de Suivi Environnemental, Guinea), and finally one on the tree nursery for the Green Corridor Project in Bossou and Nimba by Makan Kourouma (Directeur of IREB, Guinea).

There were four other guest speakers from different parts of Africa, addressing the following topics: conservation efforts, especially environmental education in Guinea and the Gambia by Janis Carter (Gambia), chimpanzees in eastern Congo by Augustin Basabose (Democratic Republic of Congo), chimpanzees of the Mahale Mountains in Tanzania by Michio Nakamura (Japan), and chimpanzees in Liberia by Joel Gamys (Conservation International, Liberia). There were about 80 participants in the symposium (see Fig. 1.2).

The film festival on the third day was open to the public. Four films were shown: "A hard nut to crack," made by ANC/NHK, illuminating the developmental changes in stone-tool use in Bossou chimpanzees, "Jokro: The death of an infant chimpanzee" by ANC focusing on a chimpanzee mother carrying the mummified body of her dead infant, "Return to the Great Apes Planet: the Chimpanzees of Bossou" by TF1/WLP/Ushuaia, and "Le Pacte de Bossou" by France3/Gaia-Video-Concept, documenting the unique coexistence of humans and chimpanzees at Bossou. All films were in French, and after each presentation Tatyana Humle answered questions from the audience.

1.3   Support from the Local Manon Community

Without the support from the local community, it would be impossible for us to continue the long-term research at Bossou and the Nimba Mountains. After the symposium in Conakry, some of the participants traveled to Bossou, 1,050 km from the capital. A ceremony was held in the village on December 3 to commemorate the project's 30th anniversary. The Manon people of Bossou were the hosts of the ceremony and were joined by Guinean and foreign researchers and our local assistants.

The Manon of Bossou have a religious belief that chimpanzees represent the reincarnation of their ancestors, inhabiting the sacred forest of Mont Gban situated behind the village. The ceremony of Mont Gban entailed the appearance of forest devils and traditional dancing accompanied by drumming (Fig. 1.3).

One of the key people from the local community was our first guide, Mr. Gouano Goumi (1945?–2006), who passed away in December 2006, soon after the 30th anniversary ceremony in Bossou. Another key member of the village was Mr. Tino Zogbila (1945?–2005), our second guide. Some members of their families continue to work with us as local assistants.

Our local research assistants currently include Pascal Goumi, Paquillé Cherif, Bonifas Zogbila, Jiles Zogbila, and Henry Gberegbe at Bossou; Henri Kassié Doré,

Fig. 1.3 Forest devils' appearance with traditional drumming in the village of Bossou during the celebration of Gban hill on the occasion of the 30th Anniversary of Bossou Research (photographs by Tatyana Humle)

Fig. 1.4 The facilities of Kyoto University Primate Research Institute (KUPRI) and the Institut de Recherche Environnementale de Bossou (IREB) located beside the village of Bossou (photograph by Tatyana Humle)

Fromo Doré, and Fokayé Zogbila at Seringbara; and David Droh, Anatole Gogo, Filbert Pahon, Anthony Gopu, and Alexi Wanseu in the Nimba Mountains. The creation of IREB in 2001 was also a landmark in the Bossou–Nimba research (Fig. 1.4).

1.4   Current Situation of the Bossou Chimpanzees

There are currently 13 chimpanzees at Bossou (as of May 2010). Bossou chimpanzees are quite unique as they are not afraid of humans and coexist peacefully alongside the local villagers. This situation in part reflects the attitude of the local Manon people, in addition to the long-term habituation by researchers.

Bossou is home to about 2,500 villagers. The chimpanzees live in the secondary forests of the hills surrounding the village (see Chap. 2), and they cross roads to move from one part of the forest to another (Hockings et al. 2006; see Chap. 23). During the past 30 years, the number of chimpanzees in the Bossou community comprised approximately 20 individuals (range, 16–22). However, a flu-like epidemic occurred in November 2003, which resulted in the loss of five chimpanzees within a short time period (Matsuzawa et al. 2004; see Chap. 32) and revealed the strength of the mother–offspring bond (see Chap. 25). This tragedy alerted us all the more to the threat of disease(s) to the future survival of chimpanzees and to the necessity for a health monitoring program (see Chap. 36).

No losses have occurred since then. The group of 13 has remained stable during the past 5 years. We have applied strict rules to protect this precious community; for example, we always keep our distance from the chimpanzees, limit the number of tourists, and wear masks during observations (see Chap. 32).

Bossou is located only 4 km from the border with Liberia and 8 km from the border with Côte d'Ivoire. These two neighboring countries were recently shaken by civil war, and many refugees have fled into the Bossou area since the early 1990s. Bossou chimpanzees also occasionally range into Liberia (see Chap. 31).

The KUPRI-International researchers, in collaboration with IREB, the villagers, and local non-governmental organizations (NGOs), initiated a reforestation program called the Green Corridor Project (Projet Corridor Vert) in 1997 (Hirata et al. 1998a, b; Matsuzawa and Kourouma 2008; see Chap. 37). This project aims to plant trees to enlarge the forests of Bossou and create a corridor across the savanna that separates the hills of Bossou from the Nimba Mountains.

In Nimba, there are other chimpanzee communities. These chimpanzees build ground nests, a unique cultural feature of the Nimba chimpanzees (Koops 2005; see Chap. 28). The Green Corridor aims to connect the fragmented forests of Bossou to the large primary forest of Nimba to secure the exchange of individuals between adjacent communities. Bossou chimpanzees have increasingly been utilizing the forests of the Nimba Mountains.

The KUPRI-International team, in collaboration with its Guinean counterparts, DNRST and IREB, will pursue its monitoring effort of the Bossou chimpanzees and will continue to promote conservation efforts in the area and the region (see Chaps. 5, 39, and 40). For further information, please visit the following website for further information: http://www.greenpassage.org.

Acknowledgments The research at Bossou is a collaboration between the Kyoto University Primate Research Institute and Guinean researchers belonging to the DNRST and IREB. The 2006 symposium was financially supported by Kyoto University, KUPRI, and the JSPS-HOPE program. The long-running study is financially supported by MEXT grants (#16002001, 20002001), etc. Special thanks are offered to these organizations for their support.

Part II
History and Ecology

Chapter 2
Location and Ecology

Tatyana Humle

2.1 Geographic Location

2.1.1 Guinea

The Republic of Guinea, located on the west coast of Africa, lies between 7°05′–12°51′ N and 7°30′–15°10′ W and covers a surface area of 245,857 km² (CIA 2008). Guinea borders six countries including Guinea Bissau, Senegal, Mali, Côte d'Ivoire, Liberia, and Sierra Leone with the Atlantic Ocean to the west (Fig. 2.1). Guinea is politically divided into 34 prefectures, including a total of 345 sub-prefectures. The country can also naturally be divided into four regions: Guinée Maritime (36,208 km²), the Fouta Djallon (or Moyenne Guinée; 63,608 km²), Haute Guinée (96,667 km²), and Guinée Forestière (49,375 km²). Each one of these four regions differs remarkably in its vegetation, climate, topography, and geology. From Guinée Maritime to the capital city Conakry on the Atlantic coast, the terrain rises up to the highlands of the Fouta Djallon, a mountainous region located in the center of the country. The highest point in the Fouta Djallon is Mali (1,538 m). To the east of the Fouta Djallon are the relatively flat plains of Haute Guinée, where the average elevation is only about 300 m. To the south of Haute Guinée lies the region of Guinée Forestière with the highest points located in the Nimba Mountains (1,752 m), Pic de Fon (1,656 m), Pic de Tibe (1,504 m), and Mont Ziama (1,387 m). Because of its high elevation, Guinea is a vast water catchment and the source of many of the major rivers of West Africa, including the Gambia, the Senegal, and the Niger. Guinea straddles three main climatic and vegetation zones. A transitional woodland–grassland mosaic extends across the center of the country, and dry Sudanian savanna vegetation zones dominate the northeast (White 1983). Mangroves shape the northern coastline. A large part of the surface area of the country is covered

T. Humle (✉)
School of Anthropology and Conservation, The Marlowe Building, University of Kent, Canterbury CT2 7NR, UK
e-mail: t.humle@kent.ac.uk

T. Matsuzawa et al. (eds.), *The Chimpanzees of Bossou and Nimba*,
DOI 10.1007/978-4-431-53921-6_3, © Springer 2011

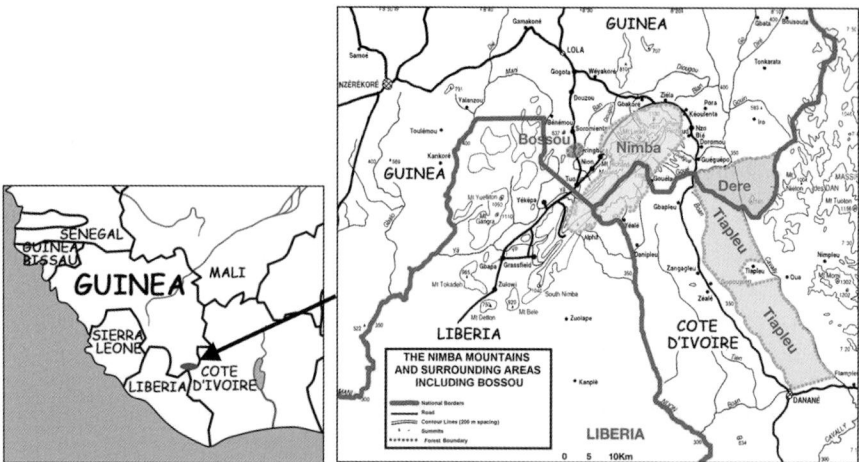

Fig. 2.1 Maps indicating the location of the Nimba Mountains Biosphere Reserve with its three core areas in Guinea, West Africa: Bossou, the Nimba massif, and Déré, and the delimitation of the transboundary priority site for the conservation of chimpanzees decided upon in 2002, which includes the Bossou, Nimba, Déré, and Tiapleu ecosystems (Kormos et al. 2003)

in agricultural and fallow lands, as well as villages and roads. The rainforest in the south, especially dominant in the Guinée Forestière region, forms part of the Upper Guinea Forest block (Sayer et al. 1992).

2.1.2 *Guinée Forestière: The Forest Region of Guinea*

The forest region of Guinea covers approximately 20% of the surface area of the country and harbors approximately 20% of the Guinean human population (Kormos et al. 2003b). This region has served at least since the 1990s as a refuge for thousands of refugees from Liberia, Sierra Leone, and Côte d'Ivoire during the periods of instability or civil unrest experienced by each one of these neighboring countries, respectively. About 9% of the surface area of the region is legally protected with status ranging from Classified Forests ($n=40$) to Biosphere Reserves ($n=2$), including a World Heritage site, the Nimba Mountains, gazetted in 1981 for Guinea and in 1982 for Côte d'Ivoire (Lamotte 1998a, b). Nevertheless, habitat encroachment and forest loss, including in protected areas, are of increasing concern in a region known for its high biodiversity, high levels of floral and faunal endemism, a relatively high prevalence of chimpanzees (*Pan troglodytes verus*), and harboring the last remaining populations of forest elephants (*Loxodonta africana*) in the country in the Biosphere Reserve of Ziama (see Chaps. 30, 31, 39, and 40). This region with its high population growth is also increasingly drawing the attention of extractive industries, including logging and mining, especially for iron ore in the regions of the Pic de Fon and Nimba and for bauxite in the northwest, while already serving as a hub for large-scale oil-palm (*Elaeis guineensis*) plantations in the region of Diécké (see Chap. 30).

2.1.3 Bossou

Bossou is a sub-prefecture in the prefecture of Lola, located in the forest region of the country, about 1,050 km from the capital city, Conakry (Fig. 2.1). The village of Bossou is 550 m above sea level and is home to 2,500 inhabitants (Hasegawa, personal communication). A small community of wild chimpanzees lives in the forest surrounding the village (latitude 7°38′71.7 N and longitude 8°29′38.9 W), located about 6 km from the foothills of the Nimba Mountains, which span the border with Côte d'Ivoire and Liberia, a massif that also harbors chimpanzees (Fig. 2.1; see Chaps. 27–29 for more details on Nimba chimpanzees). The Guinean portion of the massif was declared a Biosphere Reserve in 1980, comprising a large portion of the Nimba range (with the exception of an area designated for mining) and the Bossou and Déré ecosystems (Fig. 2.1). Bossou is one of six long-term wild chimpanzee research sites in Africa and is one of the only two long-term sites focused on the West African subspecies *P. t. verus*. Bossou presents, therefore, a long history of research and conservation on wild chimpanzees (see Chap. 4).

2.2 The Ecological Setting

2.2.1 Human–Chimpanzee Coexistence

Bossou is home to the Manon people, an ethnic group now dispersed among several villages in that southeastern region of Guinea, northern Côte d'Ivoire, and Liberia. Bossou provides a rare example of a site where wild chimpanzees and local people have been living harmoniously, sharing the resources of the same forest. This fragile, yet mostly peaceful, coexistence stems from the beliefs of many Manon families, who hold chimpanzees as one of their animal totems, that the chimpanzee represents the reincarnation of their ancestors (Kortlandt 1986; see Chap. 4). This totemization of the chimpanzee by the villagers explains why this species of great apes has survived so close to the village for so many generations.

2.2.2 The Habitat and Home Range of the Bossou Chimpanzees

The village of Bossou is surrounded by small hills 70–150 m high that are covered in primary and secondary forest (Sugiyama and Koman 1979a; Yamakoshi 1998) (Fig. 2.2). Terrestrial herbaceous vegetation, including Zingiberaceae and Marantaceae species, prevails throughout primary and secondary forest areas. The village of Bossou is surrounded by hills that constitute the core area of the Bossou chimpanzees (Fig. 2.2). Primary forest covers less than 1 km^2 and is concentrated on one of the four main hills located near the village. This hill, which is known as Gban and holds a

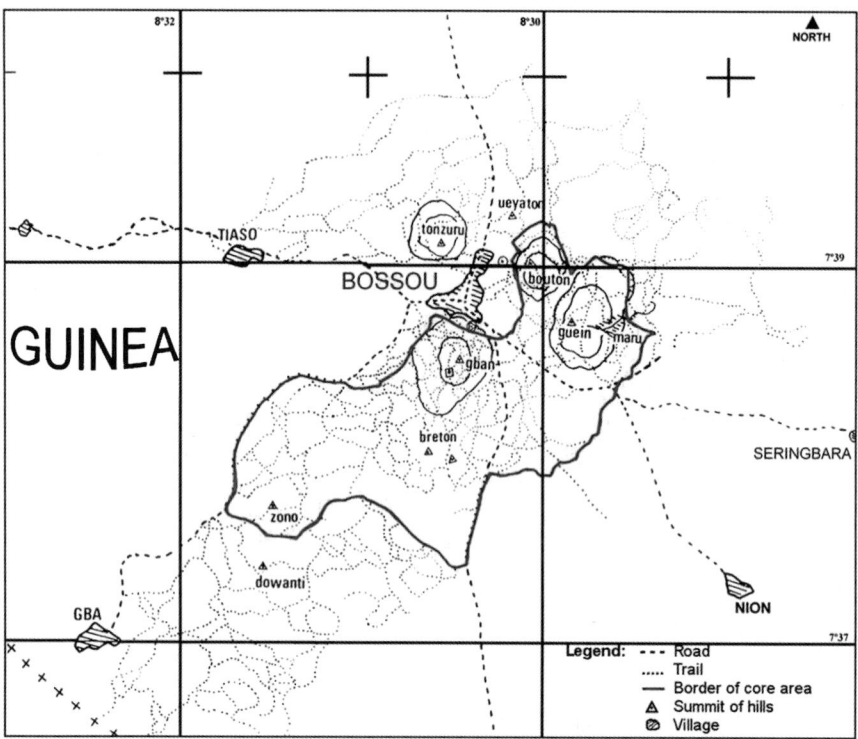

Fig. 2.2 Map of Bossou and surrounding hills indicating the core area (outlined here in *grey* and in *red* on online figure) utilized by the chimpanzees

sacred traditional value for the Manon people of Bossou (see Chaps. 1 and 4), is dominated by species such as *Antiaris africana*, *Ceiba pendandra*, *Ficus mucuso*, *Cola cordifolia*, and *Ricinodendron heudelotii*. The secondary forest, the most dominant vegetation type in the core area, is exemplified by species such as the oil palm, the umbrella tree (*Musanga cecropioides*), the bush pineapple (*Myrianthus arboreus*), and other species including *Carapa procera*, *Spondias mombin*, *Albizia zygia*, and *Alchornea cordifolia* (for a more comprehensive listing of tree species, see Appendix B). At the foot of those hills, cultivated or abandoned fields and secondary, riverine, and scrub forests form a patchy mosaic for about 6 km in all directions.

The Bossou chimpanzees mostly confine their daily activities within this core area of about 6 km^2, although they sometimes travel to adjacent forests using the few remaining gallery forest corridors, thus extending their home range to around 15–20 km^2 (Sugiyama 1984; Ohashi 2006a). The nearest currently known chimpanzee populations have their ranges in the Nimba Mountains, about 6 km west of Bossou (see Fig. 2.1; see Chaps. 27–29 for further information). Since 1976, the chimpanzee community inhabiting the forest surrounding the village of Bossou has been habituated to observers without provisioning. All members of this community can be identified individually (see Chaps. 1, 3, and 4 for further details).

2.2.3   The Climate

The massif of Nimba is located at the crossroads of distinctly seasonal climatic influences typical of West Africa: basking for most of the year in the moist air of equatorial monsoons, then being swept by the Harmattan winds originating from the Sahel region in the dry season. Indeed, Bossou has two seasons: a short dry season that extends from November to February, and a long rainy season that extends from March to October. Monthly precipitation may vary from 0 mm in the dry season to more than 700 mm in the rainy season (cf. Yamakoshi 1998; Takemoto 2002; Humle 2003a; Hockings 2007). Typically, July, August, and September are the months with the greatest rainfall, while January and February are the driest months. Temperatures can be as low as 12°C and as high as 43°C in the dry season, although the daily temperature range is less dramatic in the rainy season, with fewer extremes in minima and maxima temperatures (see Appendix C).

2.3   Feeding Ecology

2.3.1   Dietary Diversity

The feeding behavior of chimpanzees varies seasonally and is greatly influenced by fruit availability and habitat type. Chimpanzees are omnivorous and have a very diverse diet, which provides them with all the nutrients they require for their survival. On a daily basis, chimpanzees travel from one food patch to the next, mainly searching for fruit and leaves. Indeed, the pulp of fruits comprises the largest portion of the wild chimpanzee diet. As the feeding repertoire of chimpanzees at different sites is being compiled and expanded, it is becoming apparent that there are emerging differences in species eaten across sites that cannot be explained by differences in their biotic environment (Nishida et al. 1983; McGrew 1992). These differences also relate to food processing techniques (Nishida et al. 1983) and to the use of plants for self-medication purposes (Huffman and Wrangham 1994).

At Bossou, chimpanzees consume more than 200 plant species (see Appendix B), representing approximately 30% of available species in the habitat and more than 246 plant parts (Sugiyama and Koman 1987, 1992). Bossou chimpanzees may spend between 45.6% and 76.7% of their monthly feeding time consuming fruit (Yamakoshi 1998; Takemoto 2002; Hockings 2007; Humle, unpublished data) (Fig. 2.3). Leaves and pith (mainly from oil-palm fronds and terrestrial herbaceous vegetation) are the two next most important foods for the chimpanzees at this site. Indeed, Bossou chimpanzees may spend between 5.1% and 31.2% of their monthly time, respectively, feeding on these two food items (Yamakoshi 1998; Takemoto 2002; Hockings 2007; Humle, unpublished data). Seeds and the pith of herbaceous plants also comprise a nonnegligible portion of their diet. Takemoto (2002) noted that cultivars comprise 6.4% of the annual diet of Bossou chimpanzees (Fig. 2.4;

Fig. 2.3 Yolo (13-year-old male) feeds on the fruit of the bush pineapple, *Myrianthus arboreus*, which grows primarily in secondary forest areas (photograph by Gaku Ohashi)

Fig. 2.4 Jeje, an adolescent male, feeding on maize raided from a cultivated field at Bossou (photograph by Tatyana Humle)

see Chap. 22 for further information). Bossou chimpanzees also eat flowers, bark, roots, and tubers, the gum of *Albizia* sp. (Ushida et al. 2006), and insects; including adult termites (*Pseudacanthotermes* and *Macrotermes* sp.) and ants (*Dorylus* sp., *Camponotus* sp., *Oecophylla longinoda*), and the eggs and larvae of ants, bees, and several species of beetle, including those of the Raphia coleopteran (*Rhynchophorus quadrangulus*). Ushida et al. (2006) showed that 20–30 g gum exudate of *Albizia zygia* can provide the chimpanzees with sufficient amounts of calcium, manganese, magnesium, and potassium to fulfill their daily mineral requirements (see Chap. 35). Other food items consumed more infrequently include dead wood, soil from *Pseudacanthotermes* termite mounds, algae (*Spirogyra* sp.), mushrooms, honey, bird eggs, and mammals such as the tree pangolin (*Manis tricuspis*) (Sugiyama and Koman 1992). Hunting for animal prey at this site is relatively rare compared to other sites where chimpanzees have been studied, probably because of the paucity of other mammalian species in the habitat.

2.3.2 Seasonality, Tool-Use, and Activity Budget

Fruit availability at Bossou tends to peak during the dry season, especially during the month of December (Yamakoshi 1998; Hockings 2007). However, the rainy season, especially the months of May, June, and July, tends to correspond to a period of lower fruit abundance and diversity for the Bossou chimpanzees (Takemoto 2002; Yamakoshi 1998; Hockings 2007). Yamakoshi (1998) showed that, when fruits are scarce, Bossou chimpanzees effectively increase their tool-use activities, especially nut-cracking and pestle pounding (see Chaps. 6, 7, and 10), to gain access to otherwise inaccessible food resources and to boost their energy intake. In addition, during such times of fruit scarcity, when food resources are more patchily distributed and rarer, Takemoto (2002) demonstrated that Bossou chimpanzees spend less time feeding and moving and thus decrease their energy expenditure. Dietary diversity does not, however, necessarily decrease during the rainy season, as findings indicate interannual variation in patterns of dietary diversity (Takemoto 2002; Hockings 2007).

2.3.3 Role of Fallback Foods and Cultivars

During periods of fruit scarcity, Bossou chimpanzees depend heavily on human-impacted habitats, including secondary forest, scrub forest, orchards, and cultivated fields (Yamakoshi 1998, 2005; Hockings 2007). Such habitats provide the chimpanzees with numerous important natural fallback plant foods, including the oil palm, the umbrella tree, terrestrial herbaceous vegetation of the Zingeberaceae and Marantaceae families, and the bark of a variety of vine and tree species (Takemoto 2002), or cultivars, those either available all year round such as cassava (*Manihot esculenta*), papaya (*Carica papaya*), or banana (*Musa* sp.), or on a more seasonal basis, such as oranges (*Citrus aurantifolia*), mandarins (*C. reticulata*), or mangoes (*Mangifera indica*) (see Chap. 22).

The oil palm is the most valuable and important fallback food for Bossou chimpanzees because it provides them with year-round food resources, including the rich mesocarp of the fruit, the oily nut kernel, the petiole of young palm fronds, the base of immature flowers, the pith of mature leaves, and the sugary and nutritious palm heart (Yamakoshi and Sugiyama 1995; Humle and Matsuzawa 2004). In addition, the oil palm is a highly preferred nesting species for Bossou chimpanzees (Humle 2003a).

At some chimpanzee study sites, staple fallback foods such as figs (*Ficus* sp.), characterized by their aseasonal fruiting patterns, may constitute up to 100% of the diet during periods of low fruit diversity and abundance (Wrangham et al. 1996; Tweheyo and Lye 2003; Marshall and Wrangham 2007). Although figs are a preferred food for Bossou chimpanzees (Takemoto 2002), the contribution of figs to the diet of Bossou chimpanzees is not as significant as at other study sites for reasons of their apparent lower density and biomass (Yamakoshi 1998). In addition, as already noted, cultivars play a significant role in the dietary repertoire of the Bossou chimpanzees; although their seasonal proportion in the diet is variable (see Chap. 22). In some cases, cultivar consumption correlates with the low availability of natural fruits, while some cultivars constitute highly preferred foods sought by the chimpanzees in spite of the availability of natural fruits (see Chap. 22).

2.4   Conclusion and Summary

With the exception of Taï in Côte d'Ivoire (Boesch and Boesch-Achermann 2000), Bossou in Guinea is the only chimpanzee site located in West Africa that has contributed more than three decades of understanding of the socioecology, life history, and cognition of chimpanzees. Bossou is uniquely situated at the crossroads with two other West African countries, Côte d'Ivoire and Liberia. The Bossou chimpanzee community is unique for reasons of its close proximity to humans and its long history of tolerated cohabitation with humans. Bossou has importantly demonstrated that chimpanzees and humans can coexist. However, this coexistence does present some major disadvantages, including (1) the absence of recent immigrations into the community from neighboring communities in the Nimba Mountains, as these nonhabituated chimpanzees may be shy to venture into habitats with high human presence (see Chaps. 27, 28, 31, 34, and 37), (2) decreased tolerance of chimpanzee crop-raiding as local people increasingly economically depend on farming for their survival (see Chaps. 4 and 22), and (3) increased risk of disease transmission (see Chaps. 32, 35, and 36).

The chimpanzees' significant reliance on and preference for human cultigens, including the oil palm and a whole range of cultivars, does create specific socioecological and demographic conditions not observed elsewhere (see Chap. 3). The habitat of the Bossou community is rather heterogeneous, ranging from open grassland savanna to cultivated fields to primary forest. Bossou chimpanzees use all vegetation types at their disposal.

Finally, the Bossou community of chimpanzees importantly reveals that chimpanzees may thrive in human-impacted habitats and may even preferentially seek resources favored and selected for by humans [a similar pattern is emerging from data on the Sumatran orangutan (*Pongo abelii*); see Hockings and Humle 2009]. Such a pattern is not surprising considering the remarkable socioecological behavioral plasticity of the chimpanzees, their propensity for socially biased learning, and their omnivorous diet.

Acknowledgments I would like to thank the Ministère de l'Enseignement Supérieur et de la Recherche Scientifique, in particular the Direction Nationale de la Recherche Scientifique et Technologique (DNRST) and the Institut de Recherche Environnementale de Bossou (IREB), for granting us over the years the permission to carry out research at Bossou and in the Nimba Mountains. I am particularly grateful to Yukimaru Sugiyama and Tetsuro Matsuzawa, the founders of the research on chimpanzees at Bossou and Nimba, for their continual support and advice, and Jeremy Koman, Paquillé Chérif, Gen Yamakoshi, Hiroyuki Takemoto, and Kim Hockings, for their essential contributions to our understanding of the ecology of Bossou chimpanzees. Finally, I am forever grateful to the local villagers and all our local assistants at Bossou and Nimba for all their hard work and their invaluable contributions and collaboration.

Chapter 3
The Demography and Reproductive Parameters of Bossou Chimpanzees

Yukimaru Sugiyama and Shiho Fujita

3.1 Data Collection

In February 1969, a Dutch research team recorded 18 chimpanzees (*Pan troglodytes verus*) at the summit of the hill of Gban in the Bossou chimpanzee community's core area (Albrecht and Dunnett 1971; see Chap. 2 for geographical location and Chap. 4 for research history). Seven years later, when our long-term study at Bossou started in November 1976, the community numbered 21 chimpanzees (Sugiyama and Koman 1979a). Since December 1976, we have individually identified all members of the community (see Chap. 1).

Between 1976 and 1985, Sugiyama was the only primatologist conducting research at Bossou. During this period, he carried out three expeditions: the first one lasted 7 months from November 1976 to May 1977, the second one 4 months from December 1979 to March 1980 (Sugiyama 1981a), and the third one 4 months from December 1982 to March 1983. Data gathered during this first phase of the long-term research were therefore based on 15 months of in situ observations and split into three periods, with a 20-months interval without on-site research presence.

Sugiyama undertook his fourth expedition in December 1985. Matsuzawa joined him in February 1986 when Sugiyama was compelled to leave Bossou owing to a severe bout of malaria. Matsuzawa carried on the fourth expedition until March 1986. This expedition marked the beginning of the second phase of the long-term chimpanzee research at Bossou. Osamu Sakura, the third researcher to join the team, accompanied Sugiyama and Matsuzawa during the fifth expedition

Y. Sugiyama (✉)
Primate Research Institute, Kyoto University, 41-2 Kanrin, Inuyama, Aichi 484-8506, Japan
e-mail: hqvg62yd@qc.commufa.jp

S. Fujita
Department of Veterinary Medicine, Faculty of Agriculture, Yamaguchi University,
Yoshida 1677-1, Yamaguchi 753-8515, Japan
e-mail: fujita@yamaguchi-u.ac.jp

T. Matsuzawa et al. (eds.), *The Chimpanzees of Bossou and Nimba*,
DOI 10.1007/978-4-431-53921-6_4, © Springer 2011

(from September 1987 until March 1988). Since the sixth expedition in November 1988, the research team has been growing, and field missions have been organized annually.

The first full year-round continuous observation of the Bossou chimpanzees was carried out by Gen Yamakoshi from January 1994 to January 1995 (Yamakoshi 1999). It was also the first time we welcomed a foreign student: Tatyana Humle, in July 1995. Intervals of time without any in situ researcher presence then progressively decreased, until we finally established year-round continuous observations of the Bossou chimpanzees. During the course of the past 33 years, a total of 50 researchers, excluding visitors, have worked at Bossou (see Appendix G).

In his initial effort in 1976–1977, Sugiyama named each individual chimpanzee with a local Manon word or name (see Appendix A). Sugiyama estimated the age of each individual chimpanzee based on his experience with captive chimpanzees at the Arnhem Zoo in the Netherlands. In addition, published literature on the Gombe chimpanzees in Tanzania (Goodall 1968) and direct observations of chimpanzees of the Mahale Mountains of Tanzania with the assistance of researchers, Koshi Norikoshi and Shigeo Uehara, helped fine-tune these initial age estimates.

The age estimation of chimpanzees during the following years became more precise as the number of observers and the observation periods increased (see Appendix A). The following summary of population dynamics and reproductive parameters is the result of the collective efforts of all the researchers involved in chimpanzee research at Bossou during the past three decades.

3.2   Demography of the Bossou Community

One unique characteristic of the Bossou community has been its demographic stability. From 1976 until 2003, that is, 26 years, the same seven adult females prevailed in the community: Kai, Nina, Fana, Jire, Yo, Velu, and Pama (The last one, Pama, was not fully mature before 1977). Tua, the long-term alpha male of the community, was present during the first observations of the group in 1976 as the second ranking male among three adult males.

Figure 3.1 shows the number of individuals and age composition of the Bossou community since 1976. With the exception of 1984 and 1986 when the group numbered fewer than 18 individuals, the population remained stable at about 20 individuals until 2003. During this period, group size ranged from 18 to 23 individuals (Sugiyama 2004). The first decline in population size in 1984 coincided with the death of Sékou Touré, the President of the Republic of Guinea, and the coup d'état that ensued. The political and economical system subsequently drastically changed and became increasingly westernized, and villagers expanded their agricultural activities. The second decline in 1992 arose after a significant increase in the human population in and around Bossou resulting from an influx of Liberian refugees fleeing civil war in their home country. The presence of refugees had a significant negative impact on the forest, as cultivation and the gathering of construction materials for

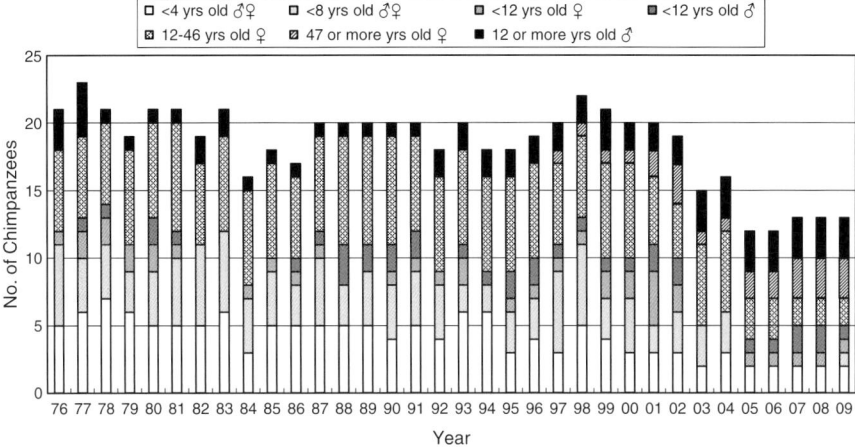

Fig. 3.1 Age-class composition of the Bossou community between 1976 and 2009

building houses soared. Nevertheless, after these two events, the Bossou chimpanzee community gradually recovered to its original size of about 20.

However, in November, 2003, an epidemic of a flu-like disease spread among members of the community and its size shriveled to 12 individuals (Matsuzawa et al. 2004; see Chaps. 25 and 32 for further details). We lost 5 chimpanzees: two old females, two infants, and one 10-year-old adolescent male. We found all the corpses except the one of an old female named Nina. Nina was estimated to be more than 50 years old at the time. Therefore, we strongly believe that she died, rather than emigrated to one of the neighboring communities in the Nimba Mountains or Liberia. The 2003 epidemic thus brought about the death of 5 chimpanzees and significantly impacted this small community.

In 2009, the Bossou community numbered 13 chimpanzees. The villagers of Bossou are primarily Manon and believe that chimpanzees are the reincarnation of their ancestors. The local Manon people of Bossou have thus helped protect this chimpanzee community, which numbered about 20 individuals for many generations. Demographic data suggest that the carrying capacity of the environment of Bossou and its surrounding area might be about 20 individuals for chimpanzees if the environment is minimally impacted and relatively stable (see Chap. 4 for further details on the historical context).

3.3 Ranging Behavior and Party Size

Just like other chimpanzee communities, the Bossou community has a fission-fusion social organization. Several individuals travel in parties in search of scattered food patches. The core area of the Bossou community is about 6 km² including several

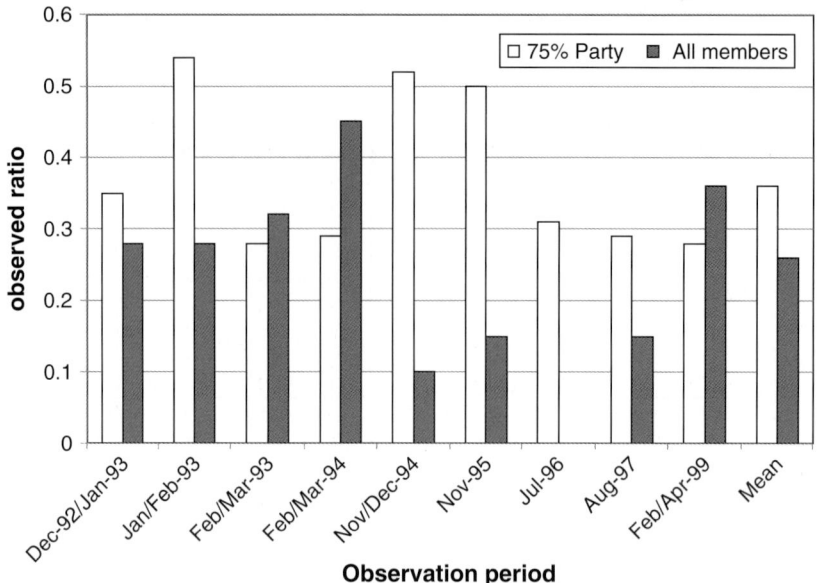

Fig. 3.2 Party size between 1993 and 1999. Only independent foragers of 7 years of age or older were counted. Largest party size during each observation day is shown in this figure. Community size consisted of 18–20 individuals. Between 75 and 99% of community members formed a party in 36% of observation days and all members in 25.8% of observation days

small hills (see Chap. 2), and community members can easily communicate with each other via loud calls, i.e., pant hoots. Figure 3.2 shows party size between 1993 and 1999.

As party membership is dynamic, observers cannot always determine with confidence all party members. Figure 3.2 presents the largest party observed in a day. Since the first phase of the research, we noted that the members of the Bossou chimpanzee community gather in large parties (Sugiyama 1984). In Figure 3.2, only independent individuals (≥7 years old) were recorded and the community size fluctuated between 18 and 20 individuals. Between 75% and 99% of community members aged 7 years or older gathered in a single party on 36.0% of observation days, and all members did so on 25.8% of observation days. Bossou chimpanzees are therefore highly gregarious and frequently travel, rest, and forage together as a single party.

During foraging and resting, the dominance relationships between adult males were always clear. However, in the past 33 years, the community never contained more than four adult males, thus potentially favoring a more stable hierarchy among adult males. In contrast, dominance relationships among adult females have rarely been recognized. Agonistic interactions between females are seldom observed, and Bossou females groom and affiliate with one another more than in other chimpanzee communities (Sugiyama 1988).

3.4   Reproductive Maturation, Birth, and Paternity

Swelling of the perigenital area begins to develop in female chimpanzees when they initiate their ovarian cycle. Sugiyama (1984, 1989b, 1994a, 2004) has reported many of the unique life history, demographic, and reproductive parameters characterizing the Bossou community. To date, of the five females (Pili, Juru, Nto, Vuavua, and Fotaiu) for whom data concerning first swelling were recorded, four of them started swelling at 7 years of age and the other at 8 years of age (mean±SD, 7.6±0.4; range, 7.1–8.3). After reaching sexual maturity, these females did not conceive immediately and showed irregular menstrual cycles for about 2 years. Although chimpanzee females usually disappear around sexual maturity, five females (Kie, Pili, Vuavua, Fotaiu, and Fanle) gave birth for the first time at Bossou, that is, in their natal group. Only three nulliparous females (Vube, Nto, and Juru) disappeared and possibly emigrated when they were 8–10 years old (mean±SD, 9.0±1.0, $n=3$). Bossou females first give birth (primiparity) on average at the age of 10.6 years (range, 9.5–13 years, $n=6$).

Seven young females remained in Bossou until the age of 12. Among these seven females, five gave birth before the age of 11. The remaining two were already members of the community in 1976 when their ages were estimated.

At Bossou, age-specific birth rate is relatively constant after the first birth; it peaks between 19 and 28 years of age, then gradually decreases (Fig. 3.3). The oldest female to give birth was 51 years old (estimated) and the second oldest was 41 years old. Mean birth rate of females between 9 and 46 years of age is 0.15 per year. If cases where infants died before 4 years of age are excluded, mean birth

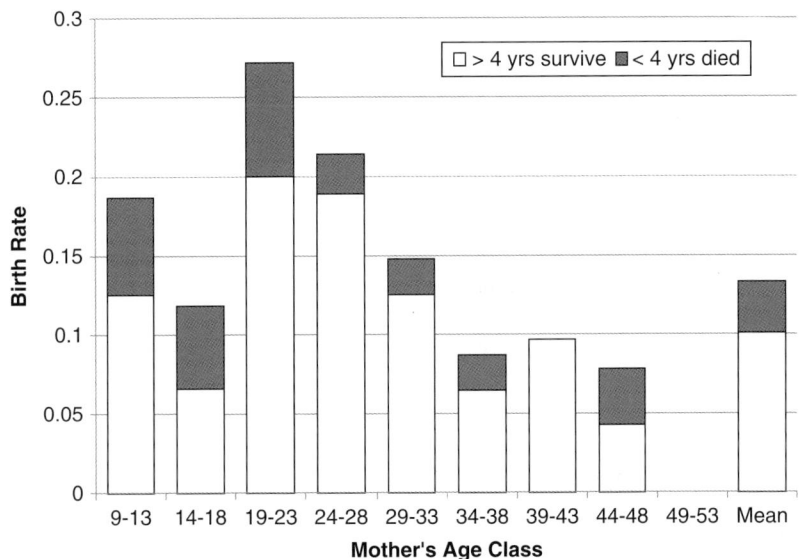

Fig. 3.3 Age-specific birth rate (1976–2008). Mothers were classed into age intervals of 5 years. *Bars* detail whether infants died before 4 years of age (*black*) or survived more than 4 years (*white*)

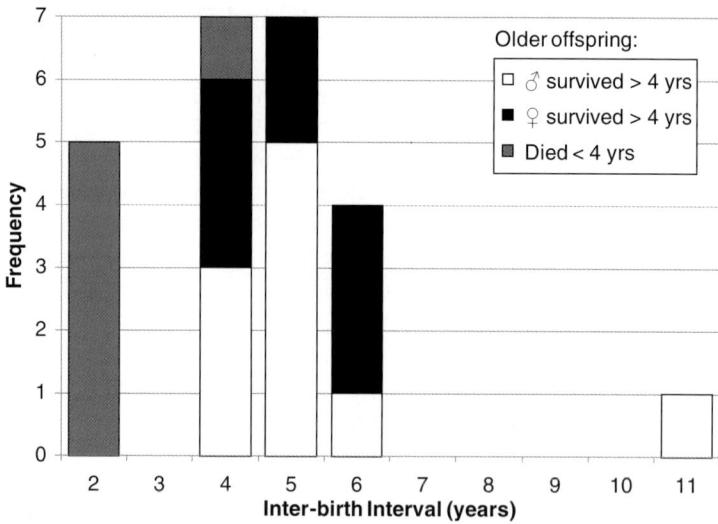

Fig. 3.4 Inter-birth interval (IBI). The graph distinguishes IBI for cases when an older brother/sister died before the age of 4 years or survived more than 4 years. Sample size is 24 mother and offspring pairs

rate is 0.11. The mean inter-birth interval for Bossou female chimpanzees is 4.5 years, or 5.2 years if cases in which the previous infant died before 4 years of age are excluded (Fig. 3.4).

Until 2001, all life history parameters indicated that Bossou chimpanzees have a higher reproductive rate than wild chimpanzees in other communities. However, consequent to the 2003 epidemic and the disappearance of a young adult female and her infant in 2004, lifetime reproductive success based on data gathered until 2007 is within the range of other populations.

In 1993, we determined the paternity of all young individuals of the Bossou community through DNA fingerprinting using individually identified hairs, feces, and food wadges (food remnants spat out during foraging) (Sugiyama et al. 1993b). We thus identified the paternity of all young individuals except Vui, son of Velu, born in 1986. Vui's biological father was not a member of the community, because Tua was the only adult male in the community between 1984 and 1988. We therefore speculate that Velu might have visited the peripheral area of the community's home range, perhaps toward the foothills of the Nimba Mountains or the Liberian border, and mated with a male from another community.

3.5 Death or Dispersal

Between 1976 and 2009, we recorded 37 births at Bossou. The two most recent births were in 2007 and 2009, respectively. Among the remaining 35 previous births, 26 infants (74%) remained in the Bossou community more than 4 years. Nine infants

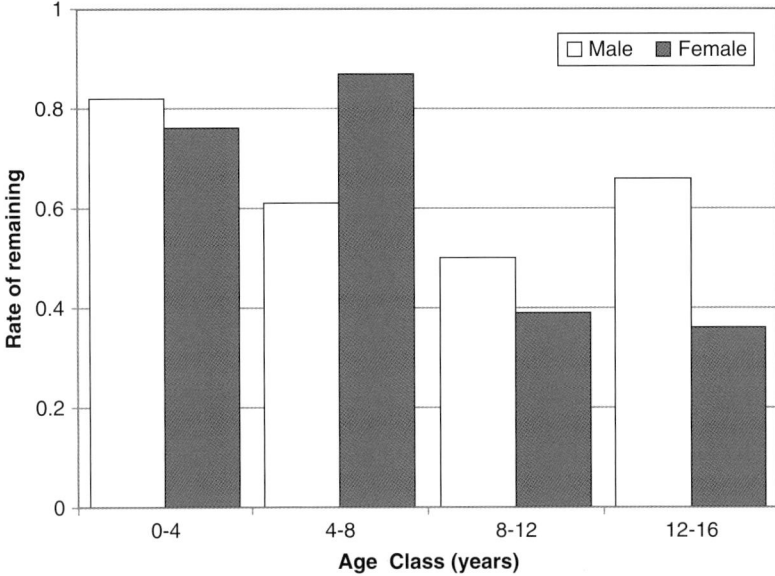

Fig. 3.5 Rate of remaining of young chimpanzees in Bossou. Sample size is 160 male-years and 156 female-years. Beyond 14 years of age, only two males of known ages and one female whose age was estimated remained in Bossou

(26%) either died or emigrated with their mother before the age of 4 (Fig. 3.5). Of these nine infants, seven were confirmed dead or presumed dead (in three cases, two mothers were witnessed carrying the mummified body of their offspring for weeks after their infant's death; see Chaps. 13, 25 and 32). The other two infants disappeared with their mothers, who were all primiparous young adult females, just shortly before being weaned or just after weaning.

Therefore, with the exception of the potential cases of emigration, among 33 Bossou-born young, 26 (about 79%) survived more than 4 years and seven died before the age of 4. Thus, infant mortality rate before 4 years of age is about 0.21, which is lower than the average reported from other communities (Emery-Thompson et al. 2007).

The probability that individuals stayed in the community dramatically decreased after 8 years of age in females and 7 years of age in males. Most young chimpanzees of both sexes disappeared before reaching 15 years of age. Indeed only two Bossou-born males, Foaf and Yolo, remained in the community beyond the age of 14 (see Appendix A). All the females born in Bossou disappeared by the age of 15, either before giving birth or after having given birth to their first infant. This pattern partly supports the idea that male-biased philopatry characterizes chimpanzee society.

In 33 years, we discovered the dead bodies of only five chimpanzees: Npei (6.5 years old at death, found in 1988; Matsuzawa et al. 1990) and four of the victims of the 2003 flu-like epidemic (Kai, female, aged 53 years; Poni, male, aged

10 years and 9 months; Veve, female, aged 2 years and 7 months; and Jimato, male, aged 1 year and 1 month). All the corpses were examined and the bone remains were preserved.

Over the years, we have recorded at least four cases of adolescent and young adult males disappearing from Bossou: Vui (13 years old in 1999 when he disappeared), Na (11 years old in 1996), Pru (11 years old in 1991), and Jieza (10 years old in 1988). This high rate of disappearance of young males is unique to the Bossou community. We have inferred six additional cases of disappearance of young males since 1976 (Sugiyama 1989) between the intermittent observation periods that took place during the first phase of research at Bossou: Vuna (6 years old in 1983 when he disappeared), Yana (5 years old in 1983), Jima (8 years old in 1980), Yiri (6 years old in 1980), Vu (8 years old in 1980), and Non (8 years old in 1977). Sugiyama (2004) argued that some of these disappearances of juvenile and adolescent males may be the result of emigrations.

This claim is partly supported by two observations of male visitors to the Bossou community during the first decade of research: Safi (old adult male, stayed in the Bossou community for 20 days from January 21 until February 9 in 1977) and Sakai (full adult male, stayed in Bossou at least from December 1982 until March 1983) (Sugiyama 1981a,b, 1984). The two adult males temporarily immigrated into the Bossou community. Details on their cases are provided below.

3.6 Immigration

Although Bossou females may emigrate, we have not yet ever been able to confirm their presence in neighboring communities. One certainty, however, is that there has been no female immigration into Bossou in the past 33 years. The absence of female immigration may be a by-product of human habituation and the close proximity of the chimpanzees' core-area to the village. Indeed, naïve females born into unhabituated communities may be too shy to immigrate into a community habituated to human observers.

In contrast, as already mentioned, two cases of temporary immigration by adult males have been recorded (Sugiyama 1981a). The first male, named Safi, was very old with scars on his face and cuts on his ears and nose. When he was first seen in January 1977, he directly walked into the center of a party containing all members of the Bossou community. He tightly embraced the then-alpha male of the community, Bafu, while they grinned at each other. Other Bossou members surrounding them were screaming and barking with grin faces. The commotion continued for 4 h until dusk. During the next 3 days, all the chimpanzees stayed in the center of their core area, thus traveling very little. During these 3 days, all members of the community, particularly Bafu and Safi, were very excited. On the 4th day, their excitement gradually abated and Safi stayed in the community and foraged alongside Bafu and other members of the community for 19 days. He was calm and partook in allogrooming. On the 20th day, he disappeared without a trace (Sugiyama and Koman 1979a).

Based on this incident, we could unfortunately not determine whether Safi was native to the Bossou community as no genetic sample was collected.

The second male, Sakai, was estimated to be slightly less than 20 years old when he was first seen in December 1982 after the absence of Sugiyama for more than a year. He nurtured a calm relationship with other members of the community and displayed no behavior indicating that he was a stranger to other community members. He remained at least 4 months in the community (throughout Sugiyama's study period), perhaps more than a year, and disappeared in the absence of researchers on site.

Both these temporary male immigrations occurred in the early days of research at Bossou when observers kept their distance from the chimpanzees. The gradual habituation of the Bossou chimpanzees to human observers that has occurred since then may discourage foreign male chimpanzees, as well as females, from immigrating into the community. Indeed, even if foreign males or females approached the core-area, it is likely they would go unnoticed if they fled immediately upon sighting students, researchers, or local assistants with the chimpanzees.

3.7 Longevity

The issue of menopause remains debated in chimpanzees (Emery-Thompson et al. 2007); however, the oldest female, Kai, lived 27 years after her last birth (Fig. 3.6). The second oldest one, Nina, lived for 10 years after her last birth. They both succumbed to the flu-like epidemic at the estimated ages of 53 and 49 years, respectively. The estimated ages in 2009 of the other older females still remaining

Fig. 3.6 Kai, the oldest member of the Bossou community, who died in 2003 at the estimated age of 53. Kai had not given birth for 27 years at the time of her death (photograph by Gaku Ohashi)

in the community, that is, Fana, Jire, Velu, Yo, and Pama, are 53, 51, 50, 48 and 42 years, respectively (see Appendix A). The low levels of interfemale competition, as a consequence of the absence of female immigration into the community, may explain their notable longevity in the wild. However, we still lack sufficient data on longevity across other field sites to assess whether Bossou females tend to live longer than chimpanzee females elsewhere.

The oldest male, Tua, is estimated to be 52 years old (end of 2009). He is still alive although he has lost his alpha-male position and has declined in dominance rank, ranking third in 2009. All these senior chimpanzees were present in Bossou in 1976.

3.8 Comparison of Demographic Parameters with Other Chimpanzee Populations

Research on wild chimpanzee populations since the 1960s across different study sites in Africa has revealed a great deal of variation in reproductive parameters (Boesch and Boesch-Achermann 2000; Goodall 1986; Nishida et al. 1990; Sugiyama 1994a; Wallis 1997; see Table 3.1). Even within a single population, female reproductive potential may fluctuate seasonally (Nishida 1990a; Nishida et al. 1990; Wallis 1997, 2002). These observations suggest that environmental factors may constrain the reproductive potential of wild chimpanzees. For instance, abundance and fluctuation of food availability in a habitat can influence female reproductive potential (Bronson 1989). Although the carrying capacity of the habitat at Bossou does not appear to exceed much more than 20 individuals, the diversity of habitat types, as well as the chimpanzees' crop-raiding habits, may be favoring precocious reproductive development among females of this community (see Table 3.1).

However, in addition to ecological factors, social factors may also affect a female's fertility and fecundity and consequently her life history. The pattern of early menarche and primiparity among Bossou females is similar to that of captive chimpanzees in good condition (see Table 3.1). Age at first birth can indeed potentially be socially influenced as most nulliparous female chimpanzees do not start breeding until they transfer from their natal group and join another community. When a female immigrates into a new group, she often faces harassment from other females and occasionally fails to immigrate altogether (Nishida et al. 1990; Boesch and Boesch-Achermann 2000), which can result in a delay in her age at first parturition. However, age at primiparity was still younger at Bossou when compared with that of females from other study communities who gave birth in their natal group. Therefore, early sexual maturity rather than emigration stress most likely explains the early age at first birth at Bossou.

The chimpanzee population size at Bossou has for nearly three decades remained stable at about 20 individuals. Although this community displays a precocious age at primiparity and a higher survival rate before the age of 4 years compared with other communities, most young individuals born within the community either die or disperse post weaning. The flu-like epidemic that occurred at the end of 2003 resulted in a significant loss of community members, which is now among one of

Table 3.1 Reproductive parameters and success of wild chimpanzees across five study sites and of captive chimpanzees across three facilities

	Age at first swelling (menarche) (years)	Age at first birth (years)	Survival to primiparity	IBI[a]	LRS[b]	Reference
Research site						
Bossou 2009	8	10.6	0.39	5.2	2.94	This study
Bossou 2001	7.6	10.9	0.58	5.1	4.23	Sugiyama (2004)
Taï	–	13.7	0.22	5.9	1.46	Boesch and Boesch-Achermann (2000)
Gombe	10.8	13.3	0.46[c]	5.5	3.32	Goodall (1983), Wallis (1997), Pusey et al. (1997)
Kibale	11.1	15.4	–	7.2	–	Wrangham et al. (1996)
Mahale	10.6	14.6	0.28[c]	6	2.2	Nishida et al. (1990)
Captive facility						
Yerkes	8.9	–	–	–	–	Young and Yerkes (1943)
Taronga	6.5	9.8	–	4.1	–	Courtenay (1987), Littleton (2005)
Holloman	–	10.8	–	–	–	Smith et al. (1975)
CIRMF	–	11.2	–	4.2	–	Tutin (1994)
Sanwa		10		4.2		Cited in Sugiyama (1984)

Assuming that all mature females regularly give birth until 40 years of age

Sample size for survival to primiparity for Bossou was 163 (2009) and 122 (2001) female-years from 21 and 18 females, respectively, and >92 for Mahale (both sexes)

[a] IBI, mean interbirth interval after successful weaning of previous offspring

[b] LRS, lifetime reproductive success was calculated as: $[\{(40-\text{"Age at first birth"})/\text{"IBI"}\}+1]\times$ "Survival to primiparity"

[c] Survival rate beyond 8 years old was calculated based on that of the previous age class

the smallest known today. Over the years, females grew older and the average inter-birth interval increased. Among the four remaining oldest females, Jire last gave birth in 2009, Fana in 1997, and both Velu and Yo in 1991. Yo has even stopped showing evident signs of cycling altogether. Yet, we still lack a clear understanding of menopause in chimpanzees and why some females cease giving birth in spite of continuing to exhibit sexual swellings.

Chimpanzee society is typically characterized by male philopatry and female dispersal. However, the demographic features of the Bossou chimpanzees indicate that both sexes potentially emigrate (Sugiyama 2004). Although male migration has rarely been reported in chimpanzees, male temporary peaceful immigration has been documented in bonobos (*Pan paniscus*) (Hohman 2001).

Perhaps if intercommunity competition is low or even absent, cooperation among males and male bonding may not be essential for defending the community's territory. In such an environment, males as well as females may disperse and immigrate to other communities. In the Taï forest in Côte d'Ivoire, more than half the offspring were sired by males from adjacent communities (Gagneux et al. 1997), although this result is still being debated (Constable et al. 2001). Male dispersal may therefore ultimately ensue if males are initially able to sire offspring without actually immigration into a neighbouring community, as demonstrated in the case of Vui's father.

Acknowledgments We would like to thank all the researchers and students, as well as the local assistants, who have worked at Bossou and who have provided us with demographic and life history data. We are particularly grateful to Professor T. Matsuzawa and Dr. T. Humle for their editorial help with this chapter. Finally, we would like to thank the Ministère de l'Enseignement Supérieur et de la Recherche Scientifique, in particular the Direction Nationale de la Recherche Scientifique et Technologique (DNRST) and the Institut de Recherche Environnementale de Bossou (IREB), for granting us over the years the permission to carry out research at Bossou.

Chapter 4
The "Prehistory" Before 1976: Looking Back on Three Decades of Research on Bossou Chimpanzees

Gen Yamakoshi

4.1 Introduction

In 1976, Dr. Y. Sugiyama initiated the ongoing long-term research project for chimpanzees (*Pan troglodytes verus*) at Bossou in the Republic of Guinea. The project has been running continuously for more than 30 years and has yielded unique contributions to modern primatology, including discoveries about compact and cohesive social structures (Sugiyama and Koman 1979a), infrequent immigration and early primiparous ages (Sugiyama 2004; see Chap. 3), the first and detailed behavioral descriptions of nut-cracking behavior under naturalistic (Sugiyama and Koman 1979b) and experimental (Matsuzawa 1994; see Chap. 7) settings, and "endemic" behavioral patterns such as pestle pounding and algae scooping (Whiten et al. 1999; Yamakoshi 2001; see Chaps. 10, 11).

The most exceptional or unique quality of Bossou chimpanzees may be their coexistence with local humans. The chimpanzee is one of the totem animals sacred to the founding Manon clan of the Bossou village, and village residents worship and conserve chimpanzees as a reincarnation of their ancestral spirits (Kortlandt 1986; Yamakoshi 2006b). Any hunting or other harming of chimpanzees was strictly prohibited by local custom (Sugiyama 1978) long before modern conservation activity became influential in the area (Lamotte et al. 2003). Even today, forest patches in the vicinity of the village of Bossou, the chimpanzees' main habitat (Fig. 4.1), are not included in any type of national protected area, with the exception of the habitat incorporated into the "core area" of the UNESCO Monts Nimba Biosphere Reserve in 1991 (Wilson 1992).

Because of its reputation for chimpanzee conservation, several scientific expeditions visited the village of Bossou to conduct research before the project began. Therefore, Bossou chimpanzees were not "discovered" in 1976 when the ongoing project commenced. They were already enmeshed in complex relationships with the

G. Yamakoshi (✉)
Graduate School of Asian and African Area Studies, Kyoto University,
46 Shimoadachi-cho, Yoshida, Sakyo-ku, Kyoto 606-8501, Japan
e-mail: yamakoshi@jambo.africa.kyoto-u.ac.jp

T. Matsuzawa et al. (eds.), *The Chimpanzees of Bossou and Nimba*,
DOI 10.1007/978-4-431-53921-6_5, © Springer 2011

Fig. 4.1 The sacred forest of Gban and Bossou village, Guinea. Two closely related species, chimpanzees and humans, have neighboring habitats in this village (photograph by Gen Yamakoshi)

subsistence and worldview of the local people and with the scientists who had visited, observed, and consequently inscribed their influences to the chimpanzee community.

This chapter evaluates both the achievements of the past 30 years of research and the available comparative data recorded during pre-1976 scientific expeditions. In doing so, it may be possible to expand 30 years of documentation to 50 or 70 years. Although beyond the scope of this chapter, a comparison of scientific data and local knowledge would be of interest (see Yamakoshi 2006b for a preliminary attempt).

4.2 The Earliest Documents About Bossou Chimpanzees

Ostensibly, the first document to reliably mention Bossou chimpanzees was written by Dr. Maxime Lamotte (1942), a French zoologist who repeatedly visited the Nimba Mountains beginning in 1942 (Lamotte et al. 2003). The first brief description of the existence of a chimpanzee population in the forest of Bossou is as follows:

> Les Chimpanzés sont, dans certains villages, protégés par indigènes; à Bossou, par exemple, ils occupent une colline sacrée où s'est maintenu un îlot de belle forêt primitive. (Lamotte 1942: 155)

A group of French researchers led by Lamotte conducted naturalistic studies of the area in diverse fields including botany, zoology, geography, and geology; these studies contributed basic data for the establishment of "la Réserve Naturelle des Monts Nimba" (see Lamotte and Roy 2003). However, these researchers did not attempt any systematic behavioral observations of the Bossou chimpanzees.

Ethnologist B. Holas visited the Nimba area in 1949 and 1951 to research religious rites and related material culture, mostly among the Kono people, who had traditionally shared the Nimba Mountains territory with the Manon and the Dan (Holas 1952a, 1954). During his research in 1951, Holas visited Bossou and interviewed some villagers. He described the chimpanzees and the religious context surrounding their conservation as follows:

> Dans le canton voisin (des Manô de Bossou) cependant, la protection de cet animal n'a point besoin d'être assurée par des mesures administratives, parce qu'elle l'est déjà grâce aux croyances religieuses en vigueur. (Holas 1952a: 39–40)

It is interesting that both authors explicitly emphasized the local people's enthusiasm for conserving the chimpanzees, although in each case the description was only one sentence long.

In addition to Lamotte and Holas, French botanist J.-G. Adam, a highly respected author of an excellent series of thorough botanical descriptions of the flora of the Nimba Mountains (e.g., Adam 1971–1983), visited Bossou in 1943 and even climbed the sacred hill of Gban (Adam, personal communication, as cited in Kortlandt 1986: 91).

Dutch ethnologist A. Kortlandt first visited the village in 1960 with cultural anthropologist J. Suret-Canale. Kortlandt climbed the hill of Gban and observed chimpanzees' nests, food remains, and feces (Kortlandt 1986: 92–94). He revisited Bossou in 1965 and described his experience with chimpanzees as follows:

> The chimpanzees were still present, did not seem to be shy and could easily be observed. (Kortlandt 1986: 94)

This may be the first explicit description of direct observation of Bossou chimpanzees, although unfortunately it lacks detailed behavioral or demographic information. Even during these initial encounters with scientists, the chimpanzees appeared to be habituated to some extent to the presence of humans.

4.3 "The 6th and 7th Netherlands Chimpanzee Expeditions"

After conducting extensive surveys on wild chimpanzees in Guinea, Kortlandt organized two scientific expeditions to Guinea in the late 1960s that yielded the first substantial scientific data on the Bossou chimpanzees. His approach was unique because he used ethological methods to observe unhabituated wild chimpanzees and used the results to reconstruct hominization (or "dehumanization") processes in hominids (Kortlandt 1972). He applied classic ethological methods of field experiments "in accordance with the von Frisch-Lorenz-Tinbergen tradition" (from Kortlandt's unpublished letter "Chimpanzees in the wild: Videotapes now available"), observing the subjects under conditions that were as natural as possible, providing stimuli without changing other conditions, and observing the subjects' reactions.

Kortlandt's research group used a camouflaged hide to observe subjects without disturbing them (e.g., Kortlandt 1967). The best known stimulus in the series of experiments was an electric-powered stuffed leopard that could shake its head; its use was designed to induce the subjects' antipredatory responses. This research is highly respected, because instead of relying only on binoculars, field books, and pencils, the group used the most advanced technology available – high-quality 16-mm film (in addition to the electric-powered stuffed leopard). As a result, the group's data can be compared with data available since.

The first expedition Kortlandt organized to the Republic of Guinea (the sixth expedition) took place from 1966 to 1967 (Kortlandt 1968). The team was composed of three researchers: J. van Orchoven, R. Pfeijffers, and J.C.J. van Zon. They conducted field research at two study sites, Kanka-Sili (near Kindia in western Guinea) and Bossou, to observe both savanna- and forest-dwelling chimpanzee populations. Their goal was to compare and simulate the possible habitats of early humans. Unfortunately, the details of the experimental settings and quantitative data from the expeditions were never published.

This expedition lasted 4 months (Kortlandt, personal communication). The exact dates of the team's visit to Bossou were not explicitly stated, but a transcription of a lecture by van Zon in Stockholm in September 1967 indicates that the visit took place in March 1967:

> Film III, showing the responses by forest-dwelling chimpanzees of the same subspecies in eastern Guinea, filmed by J. van Orshoven and J.C.J. van Zon in March 1967.

The team first conducted field experiments at Kanka-Sili (according to the same lecture, they filmed at this location in December 1966). The first demonstration of the dummy leopard to Kanka-Sili chimpanzees was in the morning on December 22, 1966 (Kortlandt 1968: 14). The team seems to have invested more time at Kanka-Sili than at Bossou, where it appears they remained only several days. According to the film (Kortlandt et al. 1981), the team assembled their experimental setup somewhere on the sacred hill of Gban, just behind the village. They cleared the ground, arranged the stimulus (the stuffed leopard) with a certain amount of bait (probably mostly bananas), and constructed a film hide from which they recorded the scene using a 16-mm camera. The team successfully conducted the leopard experiment three times over three consecutive days (Kortlandt 1968: 15). For unknown reasons, the electrical components of the leopard did not function; however, according to the experimenter, the "frozen" leopard still managed to serve as a dummy predator (van Zon, from the transcription of the 1967 Stockholm lecture).

The expedition to Bossou yielded the first estimates of the size of the local chimpanzee population. Based on a personal communication from van Zon, Kortlandt (1986: 96) referred to the population size as 17 but noted the possibility of an underestimation because only individuals appearing in the leopard experiment film had been counted. In contrast, at the 1967 Stockholm lecture, van Zon referred to the group size as "more than 25 chimpanzees." It is not possible to determine the accuracy of either estimate, but a population of 17 appears to be a conservative estimate.

Given the success of the experiments, the chimpanzees appear to have been well habituated to the presence of researchers. Important information, such as the number of bananas provided as bait, was not recorded, but van Zon's anecdote below demonstrates that the Bossou chimpanzees were highly tolerant of humans:

> On the last day of our stay in the blind in the rain-forest of Guinea we decided to try to make contact with the apes. van Orshoven left the blind. Immediately the apes started to scream,

the youngsters fled to their mothers and all went deeper into the forest, glancing through the vegetation … Slowly the apes came nearer, after 15 minutes they were on the same distance as before … and after 30 minutes van Orshoven went on the chimpanzee-path, sat down and ate some bananas at a distance of perhaps 10 meters or so from a chimpanzee-male, also eating bananas.… there were no screams and everybody stayed where he was during more than half an hour. (van Zon, from the transcription of the 1967 Stockholm lecture; see also van Zon and van Orshoven 1967: 166)

Based on their 1967 observations, the expedition team concluded that the behavioral patterns related to an antipredatory response were "underdeveloped" in Bossou chimpanzees compared with Kanka-Sili chimpanzees. For example, they observed effective and "developed" types of fighting behaviors such as "over-arm clubbing" only at Kanka-Sili, but not at Bossou (Kortlandt 1972: 82). In addition, Kanka-Sili chimpanzees were judged to be more bipedal than their Bossou counterparts (van Zon and van Orshoven 1967: 162). In any case, these observations of antipredatory behaviors were the first documentation of object manipulation in the Bossou chimpanzees, although it is not clear whether the objects can be classified as tools: the manipulated sticks did not directly touch the enemy and only served a threatening function, if any. It is interesting that the team observed stick modification before use; a chimpanzee removed the stick's twigs and leaves, although the team did not observe this particular stick being used after modification (van Zon and van Orshoven 1967: 162).

The seventh expedition was dispatched to visit the same two sites (Kanka-Sili and Bossou) for 6 months in 1968 and 1969 (Albrecht and Dunnett 1971: 121). This time, the expedition included four researchers: H. Albrecht, S.C. Dunnett, P. Fera, and J. van Orshoven.

Once again, the precise timing and duration of the team's stay at Bossou were not recorded. The aridity of the filmed landscape suggests that the visit took place during the dry season (November–March). Because the expedition stayed at Kanka-Sili at least from November 20 to January 4 (Albrecht and Dunnett 1971: 133), the visit to Bossou likely occurred from January to March 1969. The experimental setup at Bossou appears to have been arranged at the top of the sacred hill of Gban (ibid: 11). The experimental setup was nearly identical to that of the previous mission, including bananas, a stimulus, and a hide. In addition to the stuffed leopard, a new stimulus was introduced: a live 4-year-old male chimpanzee. This chimpanzee, named Koos, was shown to the Bossou chimpanzees while being kept in a small cage. Koos was originally captured somewhere in Guinea (from neither Kanka-Sili nor Bossou; ibid: 104) and was probably taken to Europe; he was brought back "from Europe" (ibid: 14) for the purpose of this experiment.

The researchers worked hard to identify individuals. At Kanka-Sili they confirmed at least 45 individuals and named at least 20 individuals (ibid: 133). At Bossou they counted 18 individuals that they observed "fairly regularly": one mature male, two subadult males, five mature females with six infants, three subadult females, and one juvenile (ibid: 15). These were the first recorded details on the age and sex composition of the chimpanzee community. Whether the identified individuals were systematically named was not recorded, but two names explicitly appeared in the text: an alpha male, Hans, and an adult female, Xanthippe, who carried two infants.

The degree of habituation appears to have gone unchanged over the 2-year interval. "The chimpanzees at Bossou seemed to have exceptionally little fear of man ..." (ibid: 14). Compared with the Kanka-Sili chimpanzees, the Bossou chimpanzees appeared to be less attracted to provided food (i.e., bananas), more interested in artificial objects (e.g., matchboxes), and more interested in the caged conspecific Koos, suggesting a low degree of fear of artificial and experimental settings.

The research team recorded some interesting interactions between the Bossou chimpanzees and Koos, the caged juvenile male. Even before the researchers exposed Koos at the experimental site, some chimpanzees approached the researchers' camp in the village, apparently tracking Koos' scent. The wild and tame conspecifics then exchanged vocalizations. However, during the experiment, when Koos was "*formally* exposed to the chimpanzees at the respective observation fields of Kanka-Sili and Bossou, they showed disappointingly little interest" (ibid: 104). At Bossou, Koos was even released from his cage at the experimental site on the hill of Gban. He received "moderate" threats but was never attacked (ibid: 105).

The main literature published about the seventh expedition (Albrecht and Dunnett 1971) described for the first time many interesting naturalistic behaviors; this was the goal of the authors:

> The detailed analysis of these experiments is being performed by other workers. This paper is concerned primarily with naturalistic behaviour observed during the study period. (ibid: 9)

The report described naturalistic behaviors including nesting, grooming, charging display, copulation, sexual play, infant carrying by mother, mother's interference in young's rough play, indifference to the presence of a squirrel, and so on. The authors reported that the Bossou chimpanzees exhibited fewer aggressive behaviors than the Kanka-Sili chimpanzees, presumably because the Bossou group included only a single mature male (ibid: 32). A "massive build" female in the Bossou community impressed the authors with her "formalized display," which had previously been empirically attributed only to larger males (ibid: 34). The female Xantippe had two infants that differed in age by "at most three years, probably less." She occasionally carried the two infants together and was observed once allowing the older infant to suckle (ibid: 43).

The authors made particular note of one incident of stick-using behavior by young Bossou individuals:

> This consisted of pushing a twig into a knot-hole in a tree, rummaging in the hole, drawing out the stick, and then sucking it. (ibid: 45)

Unfortunately, the target and aim of this behavior were not determined, giving the impression that "this activity seemed more playful than dietary" (ibid: 45). As in the sixth expedition, the team observed several cases of stick use to threaten the dummy leopard. In one incident, an adolescent used a stick to touch the leopard's head (ibid: 111), but a photograph of the incident reveals that this may have been more inspection than weapon use (ibid: 95).

4.4   Implications for the Ongoing Study (1976 to Present)

4.4.1   Indigenous Conservation and Habituation

Of all the data collected during past studies of the Bossou chimpanzee community, the most consistent and impressive finding is the peaceful relationship between the chimpanzees and the local people. This relationship was highlighted in the very first comments made by Lamotte (1942) and Holas (1952a). Even today, the local humans are proud of their chimpanzees and are very positive about their ongoing relationship. Bossou residents explained that they worship chimpanzees because they consider them to be reincarnations of their own ancestors (Kortlandt 1986). They still refer to this belief when asked (e.g., Camara 1996; an article in a local newspaper). The peaceful outlook of the Bossou people toward chimpanzees is rooted in their history, mythology, and totemic religious beliefs (Holas 1952a; Kortlandt 1986; Yamakoshi 2006b).

These beliefs appear practiced to some extent and may have functioned to keep the relationship between humans and chimpanzees peaceful, judging from the degree of the chimpanzees' habituation to humans. Kortlandt's description of his 1965 visit suggested that some form of direct observation was possible even during what might have been the first encounter between scientist and chimpanzees at Bossou. Kortlandt was left with the impression that the chimpanzees were "not shy" (Kortlandt 1986). This impression was confirmed during the two University of Amsterdam expeditions in the late 1960s, which also demonstrated the Bossou chimpanzees' familiarity with artificial settings. The degree of habituation was similar in 1976 and 1977, when Sugiyama conducted his first substantial research at Bossou without provisioning (Sugiyama and Koman 1979a). It took him "only a month" from the start of his research to completely identify all 21 chimpanzees (Sugiyama 1981b: 56, 58).

This timing is unusual compared to most first encounters with chimpanzee populations. At other sites, chimpanzees generally first behaved cautiously, aggressively, and at best simply ignored observers (e.g., Goodall 1971; Johns 1996; Tutin and Fernandez 1991). An exception is the Goualougo Triangle chimpanzees, whose behavior toward researchers was described as "curious" (Morgan and Sanz 2003); these chimpanzees were likely naïve to humans and had not learned to avoid them.

In contrast, Bossou chimpanzees are far from naïve in terms of their historical experiences with humans. They live very near the village of Bossou, have been exposed to various human activities, and interact with humans on a daily basis. The chimpanzees frequently cross roads during foraging (Sakura 1994; Hockings et al. 2006; see Chap. 23), raid field crops (Yamakoshi 1999, Hockings et al. 2007, see Chap. 22), and occasionally accidentally injure human children on forest paths (Yamakoshi 2006b; Hockings et al. 2010a). Bossou locals, however, treat the chimpanzees very gently and respectfully. They avoid direct contact with the chimpanzees, because of the belief that the animals have dangerous supernatural powers (see Richards 2000). Women and children in particular are often advised to run away as quickly as possible if they see a chimpanzee on the road. If the animals try to raid crops, the humans make a moderate effort to chase them away, but any damage is explained as a kind of offering to "reincarnated ancestors" (Kortlandt 1986). In general, the Bossou people's descriptions of their peaceful coexistence with the local chimpanzees are consistent with observed chimpanzee behaviors (see also Chap. 10). Differences between chimpanzee populations are exemplified by Bossou chimpanzees' habituation to a humanized environment; Kanka-Sili chimpanzees reacted very differently to the artificial conditions and objects presented by the seventh expedition team. Recent researchers even observed Bossou chimpanzees deactivate snares (Ohashi 2006a).

4.4.2 Behavioral Continuity

Before the current research project (1976–present), the only documented behavioral observations were from "the 6th and 7th Netherlands Chimpanzee Expeditions" conducted in 1967 and 1969. These observations are very similar to observations documented after 1976.

Findings from the seventh expedition (1969) suggested that Bossou chimpanzees were less aggressive than Kanka-Sili chimpanzees, possibly because the Bossou community included only one fully grown adult male. Observations recorded since 1976 continue to suggest that this population is relatively less aggressive (e.g. BBC documentary "Wildlife on One. Chimpanzees: Toolmakers of Bossou"). In contrast, the impressive robustness and strength of the population's adult females, first observed in 1969, together with the later discovery of cohesiveness between females (Sugiyama 1988) and the virtual lack of infanticide among Bossou chimpanzees, suggest an overall similarity with the behavior of bonobos rather than other East African chimpanzee subspecies (Yamakoshi 2004b). The short interbirth interval implied in the 1969 observations of Xantippe (the female who carried two infants born approximately 3 years apart) supports subsequent long-term demographic data about this community (Sugiyama 1994a; see Chap. 3).

With regard to tool use, recent observers have noted most of the manipulative behaviors observed during the leopard experiments conducted in the sixth and seventh expeditions. Based on their leopard experiments, the team members of the sixth and seventh expeditions concluded that the forest-dwelling Bossou chimpanzees had less well developed fighting techniques than their savanna-dwelling conspecifics. This conclusion was based in part on the lack of observed over-arm clubbing at Bossou, but Sugiyama and Koman (1979b) observed six cases of over-arm throwing of 69 cases of aimed throwing. The frequency of aimed throwing has decreased recently because habituation has improved (cf. Tutin and Fernandez 1991). The young chimpanzee's use of a stick to inspect a tree hollow (observed in 1969) is similar to recent observations of tool use for obtaining food from a tree hollow (Sugiyama and Koman 1979b; Ohashi 2006a; Yamamoto et al. 2008; see Chap. 12), although unfortunately the target of this behavior was not detected in 1969.

4.4.3   Population Dynamics and Future Perspective

The unique population dynamics of the Bossou community have been intensively researched. This community has one of the smallest population sizes. It was relatively stable (17–23 individuals) between 1976 and 2003 (Sugiyama 2004). However, a recent epidemic caused a catastrophic decrease in size, as reported in Matsuzawa et al. (2004) (see also Chap. 32). Similarly, "the Netherlands Chimpanzee Expeditions" estimated the population at 17–25 individuals in 1967 and 18 individuals in 1969. These numbers may reflect the upper limit of the environmental carrying capacity of the Bossou chimpanzee community (Sugiyama 2004).

In addition to findings published in text form, the data sets of "the Netherlands Chimpanzee Expeditions" may contain large quantities of demographic information about the Bossou community in the 1960s. Recently, Koops and I accessed the original films recorded during the sixth and seventh expeditions, which are maintained in the National Museum of Natural History (Naturalis) (Leiden, The Netherlands; Fig. 4.2). After digitizing all the film materials (16-mm film and photographic slides), we found that they are of sufficiently good quality (particularly

Fig. 4.2 Slides and 16-mm film taken during the 6th and 7th Netherlands Chimpanzee Expeditions, preserved at Naturalis, Leiden, The Netherlands (photograph by Gen Yamakoshi)

Fig. 4.3 Photographic slide of the chimpanzees of Bossou in 1969. At least 16 individuals were confirmed in this scene (with permission from Naturalis (the National Museum of Natural History), Leiden, the Netherlands)

those from the seventh expedition) to enable identification of individuals, and in some cases it is even possible to recognize individuals still living in the Bossou community through identification of physical attributes such as facial scars. Using these data, we plan to obtain more accurate counts of the population size during the expeditions (Fig. 4.3) and to reconstruct each individual's life history by reevaluating the ages estimated by Sugiyama (see Sugiyama 1991) in 1976 (Yamakoshi and Koops 2008).

Acknowledgments This work was financially supported by KAKENHI [Grant-in-Aid for Young Scientists (B)] to the author (15710182), KAKENHI [Grant-in-Aid for Young Scientists (A)] to the author (18681036), JSPS-HOPE to T. Matsuzawa, KAKENHI [Grant-in-Aid for Scientific Research (A)] to S. Kobayashi (16252004), KAKENHI [Grant-in-Aid for Scientific Research (A)] to M. Ichikawa (17251002), Research Grant from Mitsui & Co., Ltd. Environment Fund (Soil and forests: 2007–) to S. Kobayashi, and JSPS-GCOE (E04: In Search of Sustainable Humanosphere in Asia and Africa). I thank la Direction Nationale de la Recherche Scientifique et Technologique of Republic of Guinea and l'Institut de Recherche Environnementale de Bossou, Republic of Guinea for their permissions for my research. I am grateful to Drs. Y. Sugiyama, T. Matsuzawa, all the colleagues of Bossou chimpanzee research projects, and the people in the Embassy of Japan in Guinea for their cooperation during my fieldwork. I also thank A.F. Molenkamp and people in Naturalis, Leiden, people in Grafimedia, Leiden University, K. Koops, and especially the late Dr. A. Kortlandt for their extraordinary open-mindedness and cooperation for digitizing old film materials.

Chapter 5
The Chimpanzees of West Africa: From "Man-Like Beast" to "Our Endangered Cousin"

Asami Kabasawa

5.1 Early Encounters

Chimpanzees (*Pan troglodytes*) are one of the most well known and popular animals in developed countries. They are exhibited in zoos and trained to appear on television and in movies. One can easily find pictures and abundant information about them on the Internet. Furthermore, they have been used for medical research because of their physiological similarity to humans, and as models to understand human behavior. Chimpanzees are only native to Africa; so, how did they end up all over the world?

The changing attitudes of the Western world toward chimpanzees have been documented by philosophers, anthropologists, and primatologists (Corbey and Theunissen 1995; Morris and Morris 1966; Peterson and Goodall 1993; Reynolds 1967). As the perception has changed from "man-like beast" to "an endangered species" and "human's closest relative," debates on ethical issues related to how chimpanzees were treated have increased in the Western world and other developed countries.

The Western world learned of the existence of "man-like beasts" from early explorers. Many accounts were given – some accurate, some more fantastical – about a mysterious creature living in the tropical forests; in many cases, the classifications were unclear (Yerkes and Yerkes 1929). Until the early twentieth century, people had difficulty distinguishing between chimpanzees, gorillas, and orangutans, and classification of the apes was a confused and complex matter. As a result of this confusion, it is very difficult to reconstruct the Western world's historical encounters with chimpanzees. Connections made between historical accounts and recent primatological observations have only now revealed that the creatures recorded by Eurafrican traders, priests, and others in the sixteenth and seventeenth centuries in Sierra Leone in West Africa were chimpanzees (Sept and Brooks 1994).

A. Kabasawa (✉)
Graduate School of Asian and African Area Studies, Kyoto University,
46 Shimoadachi-cho, Yoshida, Sakyo-ku, Kyoto 606-8501, Japan
e-mail: asamikabasawa@gmail.com

T. Matsuzawa et al. (eds.), *The Chimpanzees of Bossou and Nimba*,
DOI 10.1007/978-4-431-53921-6_6, © Springer 2011

The Western public had occasional opportunities to observe chimpanzees more closely in the mid-seventeenth century. It is likely that the first live chimpanzee (or bonobo) was brought to Europe from Angola in 1640 and presented to the Prince of Orange (Corbey 2005). In 1641, Nicolaas Tulp, a Dutch anatomist, described this animal. Although the word "Orang-outang" was used in the title of the account (Tulp 1641; cited by Yerkes and Yerkes 1929), the animal he described was probably a chimpanzee or a bonobo (Corbey 2005). In 1699, Edward Tyson, a physician in London, reported the first dissection of a chimpanzee, also from Angola (Tyson 1699; cited by Yerkes and Yerkes 1929). In 1738, a chimpanzee from "Guinea" was put on show in public for the first time (Anonymous 1738), and that was also the first time the word "chimpanzee" was used (Reynolds 1967). Exhibiting chimpanzees and orangutans began in Europe around the eighteenth century, but few of the animals survived very long because they were susceptible to human diseases and their handlers lacked knowledge and experience in keeping apes in captivity (Maple 1979).

5.2   Chimpanzee Trade from West Africa

The earliest record of chimpanzee trading from Africa was dated about 300 years ago (Corbey 2005). By the end of the nineteenth century, a French geographer, E. Reclus, recorded "Freetown is the chief West African market for wild animals, and here the agents of the European menageries come to purchase snakes, carnivore, gorillas and chimpanzees" (Reclus 1892: 210). At one time, chimpanzees were brought back as exotic souvenirs for aristocrats or as a rare creature to be put on show for the public, but the scale of chimpanzee trading escalated in the last century. The demand for live chimpanzees increased when it became apparent that they were suitable models for medical research. The trade initially started to supply zoos and the pet and entertainment industries, and it later expanded to provide experimental subjects for the medical industry.

5.2.1   Importation for Zoos, Pet Industry, and Entertainment

Collections of wild animals, or menageries, were widespread among the European aristocracy in the sixteenth century. In the nineteenth century, these collections became zoos created to serve entire nations (Baratay and Hardoun-Fugier 2002), so that the public and not just the aristocracy had the opportunity to see exotic animals. The first ape, a chimpanzee, was displayed at the London Zoo in Regent's Park in 1835, but it survived only 6 months (Blunt 1976). Chimpanzees and apes were admired by zoo visitors. A show called the "Primate tea party," in which young chimpanzees were trained and dressed up in clothes, became a popular and widespread attraction in many zoos, and was continued until the mid-1900s (Morris and Morris 1966; Rothfels 2002). As experience and knowledge in keeping and training chimpanzees grew, they were also used as entertainers in the circus and show business. Chimpanzees have starred in numerous well-known films such as the "Tarzan" series that began in 1932, "Bedtime for Bonzo" in 1951 (co-starring Ronald Reagan), "Project X" in 1987, and "MXP: Most Xtreme Primate" in 2003. Chimpanzees are also increasingly kept as pets (Brent 2004). Although chimpanzees were kept as pets previously, the practice has become easier as a result of the increasing availability of baby chimpanzees and the growing number of people who can afford them. Baby chimpanzees are viewed as status symbols among the wealthy (Harrison 1971); in fact, celebrities such as Elvis Presley and Michael Jackson had chimpanzees as pets.

5.2.2   Models for Scientific Research

Beginning in the early 1900s, chimpanzees and other primates were increasingly used as experimental subjects in biomedical, psychological, and space research. Primate use in biomedical research began in the early 1900s. The first biomedical research center using baboons and chimpanzees was the Sukhumi Primate Center built in the former Soviet Union in 1927 (Fridman and Nadler 2002). In the USA, the Yale Laboratories of Primate Biology (later renamed the Yerkes Regional Primate Research Center) was established in 1930. Its primate colony included 33 chimpanzees, 16 of which originated from the Pasteur Institute in Guinea (Fridman and Nadler 2002).

Around 1950, the chimpanzee trade entered a new era. The demand for live chimpanzees increased as the market shifted from zoos, the circus, and private pet owners to institutes and governments. In the 1950s, the US Air Force started a chimpanzee research and breeding program at Holloman Air Force Base in Alamogordo, New Mexico. The primates were used to test the effects of space flight on humans. In 1966, there were 150 chimpanzees in the facility's colony, ranging in age from 2 to 14 years old (Fineg et al. 1967). Although it is not mentioned where these chimpanzees originally came from, whether commercial importers and vendors (McRitchei 1967), chimpanzees were freely imported from Africa at that time, and it is likely that many of the animals at Holloman Air Force Base were wild caught.

The demand for chimpanzees to be used in biomedical research increased in the USA in the early 1960s and peaked in the middle of that decade. No record of the number of chimpanzees imported before 1965 is available; however, a total of 1,379 chimpanzees were imported for research between 1965 and 1969. They were mostly used in studies of infectious diseases such as hepatitis (Harrison 1971).

5.2.3   Western Chimpanzees Around the World

Four subspecies of the common chimpanzee are recognized: the western chimpanzee, *P. t. verus*; Nigeria chimpanzee, *P. t. ellioti* (formerly *vellerosus*); central chimpanzee, *P. t. troglodytes*; and eastern chimpanzee, *P. t. schweinfurthii* (IUCN 2009).

All four subspecies are listed as endangered by The World Conservation Union (IUCN) (IUCN 2009). The patterns of the historical chimpanzee trade and recent genetic analyses indicate that most of the captive chimpanzees in the USA, Japan, and Europe originated from West Africa. The current estimate of the western chimpanzee population in the wild is 38,000, with the greatest number located in the Republic of Guinea (Kormos et al. 2003a; see Chap. 40).

It is difficult to know the exact number of chimpanzees in captivity in the USA, as many are kept as pets or are in private collections. Ross et al. (2008) reported that about 2,300 chimpanzees live in North America. In the USA, it is estimated that about 1,300 individuals are in laboratories, more than 400 are in sanctuaries or roadside parks (Brent 2004), and 283 individuals are in zoological parks (Chimpanzee Species Survival Plan 2006). It is also impossible to know the origin or subspecies of the captive chimpanzees in the USA; however, a phylogenetic analysis of 218 feral chimpanzees, the source of the research populations, revealed that 95% of them were *P. t. verus* (Ely et al. 2005).

In Europe, the first chimpanzee studbook, compiled in 2006, identified 215 of 780 chimpanzees as *P. t. verus* (Carlsen and de Jongh 2006). The information from 3,315 specimens (all of the known chimpanzees in Europe including the 995 individuals) was analyzed, and 424 chimpanzees were identified as *P. t. verus* (Carlsen and de Jongh 2006). The remaining subspecies are not yet identified, but the genetic analysis of these primates is continuing and the number of *P. t. verus* may increase. In the past, chimpanzees were imported from various ports in Africa, and the population in Europe may be a mix of the four subspecies (Carlsen and de Jongh 2006); however, shipments of chimpanzees from West Africa, particularly Sierra Leone, were the most consistent (Peterson and Goodall 1993), and the majority of feral chimpanzees, the original population for European zoos, are likely to have originated from West Africa (Carlsen and de Jongh 2006).

In Japan, 61.7% of the 249 captive chimpanzees tested were genetically identified as *P. t. verus*, and 123 of 142 feral chimpanzees were *P. t. verus* (Shinoda et al. 2003). The origin of these chimpanzees is not known, but it is likely that they came from Sierra Leone and neighboring countries, as Japan imported several chimpanzees from Sierra Leone.

5.2.4   Exporting Chimpanzees

The records of chimpanzee exports are scarce and incomplete, and it is difficult to determine how many chimpanzees have been exported from West Africa to date. An examination of the trade records of exporting and importing countries indicates that the scale of the trade in the past century and its impact on the wild population in this region were immense.

In Guinea, the Pasteur Institute in Kindia was established in 1924 and is estimated to have exported 700 chimpanzees in 43 years (Harrison 1971). After the Guinean government took over the institute in 1959, chimpanzee exports were reduced (Harrison 1971), but chimpanzees captured in Guinea in the mid-1970s were smuggled to Sierra Leone for export to foreign countries (Sugiyama 1985). According to a report made to the IUCN, there were five concessionaries operating in Sierra Leone, each with a yearly quota of 96 chimpanzees (Harrison 1971). In his book *Jungle for Sale*, Henry Trefflich, a German animal importer in New York, stated that he bought and sold 3,980 chimpanzees between 1928 and the mid-1960s, although he did not reveal where and how they were captured. He set up the Trefflich Collection Center in Freetown, Sierra Leone, and it is likely that many of his chimpanzees were from this region. Trefflich provided chimpanzees for U.S. Space Agency research; among them was Ham, the first chimpanzee that went into space in 1961 (Trefflich and Anthony 1967).

Franz Sitter, a German businessman with a number of connections, was another major dealer. He lived in Sierra Leone from the 1950s until the early 1990s and exported an estimated 1,000–1,500 chimpanzees to the USA alone (Peterson and Goodall 1993). He was also in business with European countries and Japan.

Sierra Leone was a major exporter of live chimpanzees from the late 1950s to the early 1980s, according to the exportation records. Teleki (1980) compiled the wildlife export report for Sierra Leone. The records were only available for 10 years in two discontinuous periods, but they show that between 1959 and 1963, 516 chimpanzees were exported (103 chimpanzees per year), and 1,582 chimpanzees were exported between 1973 and 1979 (218 chimpanzees per year). Between 1973 and 1979, 44% of the chimpanzees were exported to the USA. Japan, the second largest chimpanzee importer, received 16% of the exports. These records showed that more than 2,000 chimpanzees were shipped from Sierra Leone, but the real numbers are larger, as no figures are available for the 10-year period between 1963 and 1973. Extrapolation of export figures during that period suggests that more than 4,000 chimpanzees were shipped from Sierra Leone in the 20 years between 1959 and 1979 (Teleki 1980).

5.2.5 Impact on the Wild Population and the Banning of Exportation

Young chimpanzees were captured because they were easier to handle. However, the capture of wild young and infant chimpanzees commonly meant killing the mothers and other group members (Sugiyama and Soumah 1988; Teleki 1980). Capturing young chimpanzees had a serious impact on the wild population, because for each chimpanzee exported, other chimpanzees were killed during the process of capture and transportation (Kortlandt 1966; cited by Harrison 1971; Sugiyama and Soumah 1988; Teleki 1980). After surveying the methods used to capture, handle, and transport chimpanzees in Sierra Leone, Teleki (1980) estimated that five to ten chimpanzees were killed to export a single one. This estimate means that if more than 4,000 chimpanzees were exported from Sierra Leone, then, even by a conservative estimate, some 20,000 chimpanzees were removed from the wild population within a 20-year period (Teleki 1980).

5.2.6 CITES: Controlling International Trade

The decreasing wild chimpanzee population has been of concern to primatologists since the late 1960s (Harrison 1971; Reynolds 1967). In 1975, when the Convention on International Trade in Endangered Species of Wild Fauna and Flora (CITES) was enacted, chimpanzees were classified as Appendix B (not necessarily threatened with extinction), and then changed to Appendix A (endangered species) in 1977. The USA, a major importer of live chimpanzees at the time, was among the first nations to join CITES. Japan joined in 1980, but it did not actually stop importing chimpanzees from Africa. For example, Japan imported 30 chimpanzees from Sierra Leone after joining CITES and signing a bilateral agreement. Although Sierra Leone had not yet signed CITES, chimpanzees were protected animals, and presidential bans on exporting chimpanzees were issued in 1978 and 1981 (Unti 2006). Japan was believed to have offered foreign aid to Sierra Leone in exchange for chimpanzees, thereby circumventing CITES (Teleki, personal communication; Peterson and Goodall 1993; Sugiyama 1985). Those imports were legal, as were others that were rushed through just before Japan ratified CITES in 1980. It was unfortunate that the principle behind CITES, that is, stopping the capture of animals to protect wild populations, was largely ignored by the Japanese government.

Sierra Leone became a signatory of CITES in 1995; however, the country's wildlife protection legislation was ineffective for decades. In July 2007, the country finally enacted a new law prohibiting the capture, killing, and possession of chimpanzees, and offenders face the penalty of a fine of up to $1,000 or jail (Species Survival Network 2007).

5.2.7 After CITES

After many countries joined CITES, new ways were sought to supply live chimpanzees for biomedical research. One suggestion was to breed chimpanzees that were already in captivity in nonhabitat countries; another was to create new breeding colonies in established laboratories within the habitat countries.

Chimpanzees have been widely used in biomedical research and have been particularly useful for the study of infectious diseases such as hepatitis. When the acquired immunodeficiency syndrome (AIDS) epidemic surfaced in the 1980s, the USA launched a research program to fight the disease. To supply animals for research, the USA started breeding chimpanzees in captivity in 1986 (Committee on Long-Term Care of Chimpanzees, Institute for Laboratory Animal Research, Commission on Life Sciences, National Research Council 1997). The breeding program was successful, but after discovering that human immunodeficiency virus (HIV)-infected chimpanzees rarely developed AIDS (Committee on Long-Term Care of Chimpanzees, Institute for Laboratory Animal Research, Commission on Life Sciences, National Research Council 1997), a problem with surplus chimpanzees emerged. There were increasing ethical concerns about the treatment of chimpanzees, and euthanizing the surplus individuals was not an option. Better captive conditions were required. With the high cost of keeping chimpanzees, laboratories could not afford to care for them when they were not being used in research (Committee on Long-Term Care of Chimpanzees, Institute for Laboratory Animal Research, Commission on Life Sciences, National Research Council 1997), and three laboratories with large colonies of chimpanzees have closed down since 1995 (Conlee and Boysen 2005).

Animal sanctuaries and rescue centers in the USA made space for ex-laboratory chimpanzees, but these nonprofit organizations, run on donations with limited financial means, could not accommodate the high number of surplus animals. In 2005, Chimp Haven, the first federally operated chimpanzee sanctuary, was created in the USA. More than 100 chimpanzees are currently housed at the facility (Chimp Haven 2010). Japan and Europe also have a problem with surplus chimpanzees. Japan established its first chimpanzee sanctuary, Chimpanzee Sanctuary Uto, in 2007. The facility currently holds 78 chimpanzees that were subjects in hepatitis experiments (Chimpanzee Sanctuary Uto 2010). In Europe, as most medical experiments on chimpanzees became illegal and the last laboratory with chimpanzees has closed down, chimpanzees retired from laboratories have been transferred to zoos and a sanctuary (Carlsen and de Jongh 2006). However, zoos in Europe are now moving away from chimpanzees in favor of displaying other great apes such as bonobos. Some institutions are trying to transfer chimpanzees to substandard zoos or "re-export" them to sanctuaries or other institutions in Africa (Carlsen and de Jongh 2006). The number of chimpanzees in laboratories in Japan and the USA is also decreasing. This trend shows great progress for advocates of chimpanzees' rights; however, biomedical researchers have raised the concern that there may not be enough chimpanzees for future experiments (VandeBerg and Zola 2005).

The establishment of laboratories in habitat countries within West Africa has had mixed success. Vilab II was set up in Liberia by the New York Blood Center in 1974. The facility carried out hepatitis research on 86 chimpanzees until the end of the 1970s (Van den Ende et al. 1980). The laboratory attempted to rehabilitate and reintroduce chimpanzees into the wild (Hannah and McGrew 1991). In the early 1980s, Immuno, a company based in Austria, planned to set up a biomedical laboratory in Sierra Leone with a large chimpanzee colony for hepatitis research. Primate protectionist Shirley McGreal showed that this plan was an illegal exploitation of the wild chimpanzee population as well as a way to circumvent CITES (McGreal 1983). When McGreal revealed this information, the company sued her and others for libel; however, Immuno lost the case and the colony was never established (Cherfas 1989; Peterson and Goodall 1993).

The concept of establishing laboratories in habitat countries is not new. The Pasteur Institute at Kindia in Guinea, established in 1923, was created to obtain and export chimpanzees to the Pasteur Institute in Paris. Later, the institute carried out experiments on primates and exported chimpanzees to laboratories and zoos in foreign countries (Sugiyama and Soumah 1988) as well as to Paris (Fridman and Nadler 2002). The primates at the Institute in Kindia were used by various groups of scientists for experiments on tuberculosis, polio, typhoid fever, and tropical

diseases (Fridman and Nadler 2002). From 1926 to 1927 one Russian scientist, I.I. Ivanov, even carried out experiments on hybridization between chimpanzees and humans at this institute in Kindia (Rossianov 2002). In the 1960s, the institute ceased to exist as a research institute (Fridman and Nadler 2002).

5.3 Scars from the Past: The Chimpanzee Pet Problem and Sanctuaries

Even after the species was recognized as endangered and international trade was banned, live chimpanzees continued to be captured to supply the local pet trade in habitat countries (Fig. 5.1). The history of chimpanzee exportation from this region has left serious scars that have not yet healed. Data collected at the Tacugama Chimpanzee Sanctuary in Sierra Leone showed that foreigners played a significant role in the pet chimpanzee trade after the civil war (Kabasawa 2009). It is not surprising that the locals continue to sell chimpanzees to "white" people, as this generates a cash income.

Chimpanzees are legally protected in West African countries (Kormos et al. 2003a), and pet chimpanzees are confiscated under the law (see Chap. 40). Furthermore, when young pet chimpanzees become too large and difficult to handle, their owners abandon them. This practice leaves a large population of unwanted chimpanzees, and euthanizing members of an endangered species is not a feasible option (Carter 2003; Harcourt 1987). Sanctuaries have been established to care for

Fig. 5.1 A young pet chimpanzee in Bo, Sierra Leone (photograph by Asami Kabasawa)

the unwanted, confiscated, and abandoned chimpanzees. There are currently 13 African chimpanzee sanctuaries caring for more than 700 chimpanzees. In 2006, four sanctuaries in West Africa held more than 200 chimpanzees, as reported in the PASA 2006 Management Workshop Report (Cress and Rosen 2006). The data are presented in Table 5.1.

The Tacugama Chimpanzee Sanctuary in Sierra Leone currently houses the greatest number of chimpanzees in West Africa (Figs. 5.2 and 5.3). Since its establishment in 1995, the facility is not only receiving and caring for orphaned chimpanzees, but it has been working diligently on conserving of wild populations

Table 5.1 Sanctuaries in West Africa

Name of sanctuary	Country	Habitat country	Year established	No. of chimpanzees (in 2005)
Chimpanzee Rehabilitation Project	Gambia	No	1974	79
Drill Rehab and Breeding Center (*Pandrillus*)	Nigeria	Yes	1991	28
Tacugama Chimpanzee Sanctuary	Sierra Leone	Yes	1995	84
Centre de Conservation pour Chimpanzés	Guinea	Yes	1996	45
–	–	–	Total	236

The numbers of chimpanzees held at the sanctuaries are based on the PASA 2006 Management Workshop Report (Cress and Rosen 2006)

Fig. 5.2 Feeding chimpanzees at the Tacugama Chimpanzee Sanctuary, Sierra Leone (photograph by Asami Kabasawa)

Fig. 5.3 Chimpanzees in a large forested enclosure at the Tacugama Chimpanzee Sanctuary, Sierra Leone (photograph by Asami Kabasawa)

through an education and sensitization program. The sanctuary has also established a strong positive relationship with local authorities and communities. This concern was proven during the civil war and after the incident in which the escape of a group of chimpanzees from the sanctuary resulted in the death of one local citizen and the injury of another. Instead of reacting negatively toward the sanctuary, the local authorities and communities collaborated with the sanctuary to locate the missing chimpanzees and provided moral support (Kabasawa 2008).

The role of sanctuaries has been debated, and the arguments reflect how people in developed countries view chimpanzees. Those who support sanctuaries claim that the facilities provide a humane and ethical solution for unwanted or confiscated chimpanzees and support the laws. They also argue that the sanctuaries promote conservation of wild populations by educating the local people (Carter 2003) and, in some cases, by releasing rehabilitated primates into the wild (Goossens et al. 2005). On the other hand, sanctuaries are criticized for their cost, and opponents argue that the long-term commitment required to rehabilitate confiscated animals takes time and resources away from other conservation efforts, such as preserving wildlife and their habitats (Soave 1982; Yeager and Silver 1999).

Both supporters and critics of sanctuaries generally agree that chimpanzees in the wild need to be protected, but they disagree on whether the welfare of each animal or protection of the population as a whole is more important. In other words, decisions are being based on whether people assign a higher value to the individual or the population. The views of animal welfare and population conservation advocates may seem similar on the surface, but their philosophies have a very different focus (Callicott 1983). Animal welfare proponents believe that each chimpanzee is worth protection under any circumstances, whereas population conservationists believe that the population as a whole is more important than individual chimpanzees, and they support channeling all available resources into protecting the wild population, even if it means that some individuals are neglected.

The sanctuaries in Africa and the chimpanzees they house face many challenges (Karesh 1995; Teleki 2001). Sanctuaries are struggling to generate enough funds to support the animals that are already in their care, and the number of chimpanzees brought to sanctuaries is still increasing (Mills et al. 2005). For the welfare of individual animals and conservation of the species, reintroducing rehabilitated individuals into the wild seems an ideal solution; however, this option is difficult to achieve (Beck et al. 2007). Successful programs must have suitable release sites, identify subspecies, test release candidates for infectious diseases, and be able to monitor animals after their release (Goossens et al. 2005; Tutin et al. 2001). These requirements and conditions are difficult to meet, which limits the number of chimpanzees that can be released back into the wild (Carter 2003). However, one hopeful report on reintroduction in West Africa is that of the Centre de Conservation pour Chimpanzés (CCC) in Guinea, which released ten wild-born and two sanctuary-born, ex-captive chimpanzees in the Parc National du Haut Niger in June 2008. These released individuals have been monitored regularly ever since, and it is reported that they are avoiding people and able to feed independently without provisioning (Pan African Sanctuary Alliance 2008; Humle et al., in press).

5.4 Chimpanzee Research: From Intellectual Inquisitiveness to Protecting the Species

Chimpanzees are one of the most studied animals. Their remarkable physical, anatomical, and physiological similarity to humans was noted from their first encounter with Western explorers, and recent genetic studies have revealed the evolutionary kinship between humans and chimpanzees (The Chimpanzee Sequencing and Analysis Consortium 2005). In the past 100 years, chimpanzee behavior and ecology have been studied to provide insight into human behavior. The first person to study a group of chimpanzees in captivity was Wolfgang Köhler. In 1912, he studied seven chimpanzees kept at an anthropoid station of the Prussian Academy of Science (Köhler 1925). Chimpanzees were first studied in the wild by Henry Nissen in 1930 near Kindia in Guinea (Nissen 1932), and long-term field studies were begun in 1960 by Jane Goodall in Tanzania, and in 1966 by a group of Japanese researchers led by Toshisada Nishida (Nishida 1990a). There are two long-term field research sites in West Africa: Taï in Côte d'Ivoire (Boesch and Boesch-Achermann 2000) and Bossou in Guinea. The study of chimpanzee behavior in the wild and in captivity has revealed that they use tools and have their own culture (McGrew 1992; Whiten et al. 1999), which includes wars (Goodall 1986) and politics (de Waal 1982). Their intelligence is demonstrated by their ability to learn language and complete cognitive tasks (Fouts 1997; Gardner et al. 1989; Matsuzawa et al. 2006; Rumbaugh 1977).

Reports from field research sites reveal that wild chimpanzees are severely threatened by habitat destruction and disease (see Chaps. 32, 39, and 40). Field researchers work to protect the wild populations in their study area, and the sites function as "in situ" conservation projects (see Chaps. 37 and 38). Studies on chimpanzees can help protect wild populations and improve their life in captivity. Understanding how they live in and use their natural habitat provides crucial information for planning effective conservation strategies; this knowledge is also used to construct the most natural conditions possible in a captive environment. The similarities between humans and chimpanzees help create public interest in this species and raise awareness of issues in population conservation and animal welfare.

5.5 Chimpanzees and Humans: Their Contribution and Our Responsibility

We now know that chimpanzees are closely related to humans in many respects and that those in the wild are facing the threat of extinction. The concepts of conserving chimpanzees in the wild and protecting the welfare of those in captivity are well known and supported in the Western world today, and many people would frown upon killing or mistreating chimpanzees. Reasons for supporting protection of these primates may vary depending how people view the chimpanzee: they may value the apes for their importance or uniqueness to the ecosystem and biodiversity, their close physical and genetic relationship to humans, or it could be because chimpanzees are sentient, social, and intelligent animals similar to humans (Cavalieri and Singer 1993).

Chimpanzees have made many contributions to our society. They have helped advance medicine, space research, the understanding of human behavior, and our place in the course of evolution. They have played a role in educating and inspiring the public about the African forest and have amused us as entertainers and exotic pets. Chimpanzees have been used to benefit mankind and have often taken our place when it was considered unethical to use humans, for example, infecting them with HIV. The role of chimpanzees has changed over time, depending on what humans needed from them and the prevailing ethical standards. Today, chimpanzees in the wild and in captivity are facing serious problems ranging from decreasing habitat and populations in the wild to lack of appropriate housing in captivity. It is now up to humans to act responsibly to find solutions and repay chimpanzees for their contribution to our society.

Acknowledgments Part of this study was supported by Japan Society for the Promotion of Science (JSPS) core-to-core program HOPE, the Matsushita International Foundation, and JSPS program, "Initiatives for Attractive Education in Graduate Schools: Training of Researchers for Applied Area Studies by on Site Education and Research." In Sierra Leone. I would like to thank Bala Amarasekaran, the staff at the Tacugama Chimpanzee Sanctuary, K. Bangura, M. Mansaray, the Wildlife Conservation Branch, and the Forestry Division under the National Commission for the Environment (the Government of Sierra Leone), and the Conservation Society of Sierra Leone. I also thank Dr. G. Teleki and Ms. S. McGreal (International Primate Protection League) for their hospitality and sharing valuable information on chimpanzee trade in the past. I am also grateful to Drs. T. Matsuzawa and G. Yamakoshi for their guidance.

Part III
Culture: Tool Manufacture and Use

Chapter 6
The Tool Repertoire of Bossou Chimpanzees

Tatyana Humle

6.1 Tool Use and Culture in Chimpanzees

Chimpanzees make and use a diverse and rich kit of tools and, with the exception of humans, they are the only living primates to at least habitually use and make tools and rely on tool use during their daily activities. Tool-use behavior in chimpanzees (*Pan troglodytes*) has been observed at all field sites where chimpanzees have been studied (Whiten et al. 1999, 2001). The use of tools is the most accessible form of culture among chimpanzees. Such elementary technology denotes the knowledgeable use of one or more physical objects as a means to achieve an end, and is termed material culture if standardized in a collective way that is characteristic of a group of individuals of a same species (McGrew 2004). Each community of chimpanzees has a unique cultural repertoire of tool-use behaviors within the feeding, social, and hygiene domains that differs from that of other communities (McGrew 1992; Whiten et al. 1999, 2001; Nakamura and Nishida 2006; see Chaps. 7–14). We see below how the tool kit of the Bossou community is quite remarkable in its diversity, complexity, and uniqueness.

6.2 Diversity of Tool Use at Bossou

6.2.1 Domains and Levels of Complexity

Tool use in chimpanzees may serve several purposes, including extracting, probing, reaching, expelling, wiping, cleaning, displaying, and pounding. In this sense, Bossou chimpanzees (*P. t. verus*) display a large repertoire of 24 different tool-use

T. Humle (✉)
School of Anthropology and Conservation, The Marlowe Building,
University of Kent, Canterbury CT2 7NR, UK
e-mail: t.humle@kent.ac.uk

T. Matsuzawa et al. (eds.), *The Chimpanzees of Bossou and Nimba*,
DOI 10.1007/978-4-431-53921-6_7, © Springer 2011

behaviors (Table 6.1). Among these, 14 (58%) are customary, meaning that they occur in all or most able-bodied members of at least one age-class and sex, or habitual, meaning they occur in at least two or more contemporary individuals (cf. Whiten et al. 1999). The majority of tool-use behaviors (71%) at Bossou concern subsistence activities (including water drinking); 13% relate to defense and 4% to communication, exploration, reaching, and comfort, respectively (see Table 6.1). Half of subsistence tool-use behaviors so far recorded at Bossou serve to enable access to insects or their products. Among customary or habitual behaviors recorded at Bossou, 65% concern subsistence (44% of which concern insects or their products and 22% access to fluids), whereas 21% concern defense and 7% exploration and communication, respectively. The four most commonly observed tool-use behaviors as a percentage of the sum of total observation time recorded among Bossou chimpanzees are nut-cracking (0.36%), ant-dipping (0.13%), pestle-pounding (0.07%), and algae-scooping (0.03%). The latter two tool-use skills are unique to the Bossou community and have so far never been recorded elsewhere (see Chaps. 10 and 11 for further details on these behaviors).

Chimpanzees at Bossou display use of five types of tool composites (see Table 6.2) (Sugiyama 1997). A tool composite here refers to "two or more tools having different functions that are used sequentially and in association to achieve a single goal" (Sugiyama 1997: 23). Among these, however, only one is customary: the use of hammer and anvil stones to crack open oil-palm nuts (*Elaeis guineensis*). Bossou chimpanzees are indeed mostly renowned for their use of a pair of stones as a hammer and an anvil to crack open oil-palm nuts (Sugiyama and Koman 1979b). Among all the tool-use behaviors observed in the wild, nut-cracking is probably the most sophisticated performed by any nonhuman animal as it requires the coordinated complementary use of both hands and the combination of three external objects – the nut and the hammer and anvil stones (see Chaps. 16 and 18). In addition, Matsuzawa (1991) observed three chimpanzees using a third stone as a wedge to stabilize an anvil stone. Matsuzawa termed this a meta-tool, that is, a tool that is used to improve the function of another tool (Matsuzawa 1991; see Chaps. 16 and 18).

6.2.2   Sex and Age-Class Distribution

Based on data gathered between 2003 and 2006, there was no sex difference among Bossou adults (>11 years old) in the percentage of observation time spent nut-cracking (independent samples t test: $t_8 = 0.034$, $P = 0.974$), ant-dipping (independent samples t test: $t_{10} = 1.045$, $P = 0.321$), or pestle-pounding (independent samples t test: $t_{10} = 0.509$, $P = 0.622$) (Fig. 6.1). However, adult males tended to algae-scoop significantly more than adult females (independent samples t test: $t_{10} = -2.417$, $P = 0.036$) (Fig. 6.1).

In addition, although no apparent age-class difference in percent time spent emerged for algae-scooping, the pattern was similar for ant-dipping, pestle-pounding,

Table 6.1 Tool-use behaviors observed at Bossou since 1976

Behavior	Date	Customary	Function	References
Nut-crack	–	Yes	Use anvil and hammer stone to crack oil-palm nuts	Sugiyama and Koman (1979b), Matsuzawa (1994)
Ant-dip	–	Yes	Use of stalk or stick to dip for driver ants	Sugiyama et al. (1988), Sugiyama (1995b)
Investigatory probe	–	Yes	Use of stalk of vegetation to probe or sniff	Ohashi (2006b)
Aimed throw	–	Yes	Throwing a branch or stone at specific target (e.g., human)	Sugiyama (1997)
Branch drag	–	Yes	Dragging a leafy branch in display, observed in males	Sugiyama (1997)
Flailing stick or club	–	Yes	Flailing a stick or branch and strike an object or conspecific	Sugiyama (1997)
Leaf clip	–	Yes	Strip leaf blade from petiole of one to several leaves with mouth	Sugiyama (1981a)
Honey dip	–	Yes	Use probing stick to collect honey from sweat bees or honeybees	Ohashi 2006b
Leaf-sponge	–	Yes	Use chewed-up leaves for drinking water from tree hole	Tonooka et al. (1994), Tonooka (2001)
Branch haul stick[a]	1977	No	Use long stick to haul another branch closer	Sugiyama and Koman (1979b)
Gum-collect[a]	1980	No	Use probing stick to collect gum	Sugiyama (1997)
Digging tool[a]	1980	No	Use of a stout stick to open the nest of driver ants	Sugiyama et al. (1988)
Push-pull[a]	1988	No	Use of a twig to push and retrieve leaf-sponge from water hole	Sugiyama (1995a)
Leaf spoon	1988	No	Use leaves to scoop water from tree hole	Sugiyama (1995a)
Pestle pound[a]	1990	Yes	Palm frond used to pestle palm heart	Sugiyama (1994c), Yamakoshi and Sugiyama (1995)
Anvil-prop[a]	1991	No	Meta-tool used in nut-cracking, used to wedge anvil	Matsuzawa (1991)
Leaf-fold[a]	1994	Yes	Use folded leaf to collect water from tree hole	Tonooka et al. (1994), Tonooka (2001)
Palm sap collecting	1994	No	Use of chewed fibrous material as sponge to gather palm sap	Sugiyama (1994c)

(continued)

Table 6.1 (continued)

Behavior	Date	Customary	Function	References
Algae scoop[a]	1995	Yes	Stalk or stick used to fish algae from pond surface	Matsuzawa et al. (1996)
Leaf cushion	1997	No	Leaves used to sit on wet ground	Hirata et al. (1998b)
Termite fish	1997	No	Stalk used to fish for termites	Humle (1999)
Larvae prize	2000	Yes	Stick used to reach bee larvae in dead branch	Matsuzawa, pers. obs.
Expel	2000	Yes	Stout stick used to expel animal prey from tree hole	Humle (2003b)
Ant-fish	2003	No	Use of stalk or stick to fish for arboreal ants	Yamamoto et al. (2008)

Date, the year the behavior was first observed

[a]The behavior not reported elsewhere, so far exclusively observed among Bossou chimpanzees; behavioral pattern in italics, anecdotal

Table 6.2 Tool composites observed at Bossou

Target	Tool composite	Customary	References
Driver ants	Digging stick/dipping wand	No[a]	Sugiyama et al. (1988)
Water in hollow	Leaf tool/push-pull stick	No[a]	Sugiyama (1995b), Tonooka et al. (1994)
Palm heart and sap	Pestle-pounding/fiber sponge	No[a]	Sugiyama (1994c)
Nut kernel	Stone hammer/anvil	Yes	Sugiyama and Koman (1979b)
Nut kernel	Stone hammer/anvil/wedge	No	Matsuzawa (1991)

Source: Adapted from Sugiyama (1997)
[a]Anecdotal

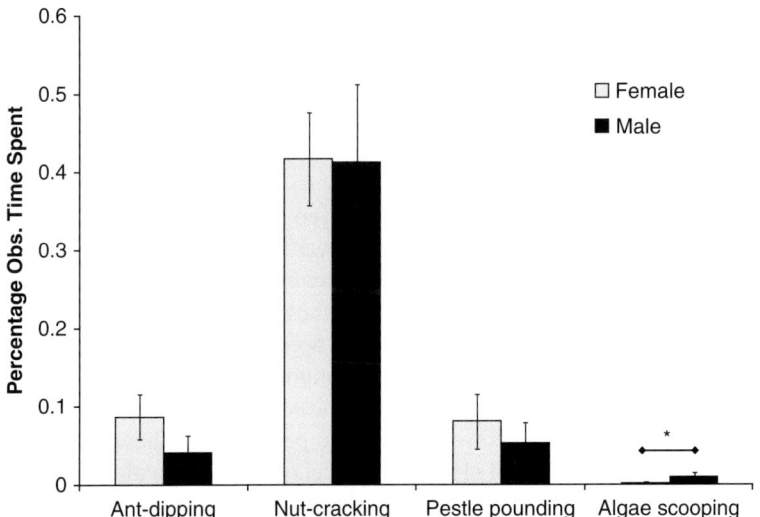

Fig. 6.1 Percentage of observation time spent performing ant-dipping, nut-cracking, pestle-pounding, and algae-scooping by adult female and male chimpanzees at Bossou (*$P < 0.05$). *Obs.* observation

and nut-cracking in that young animals between 5 and 11 years old spent significantly more time than young less than 5 years old performing these behaviors (see Fig. 6.2). Moreover, with the exception of nut-cracking, young between 5 and 11 years old also tended to spend more time than adults (those >11 years old) performing tool use (Fig. 6.2). This pattern reflects the high motivation of juveniles and adolescents to practice and perfect their tool-use abilities (see Chaps. 18 and 21).

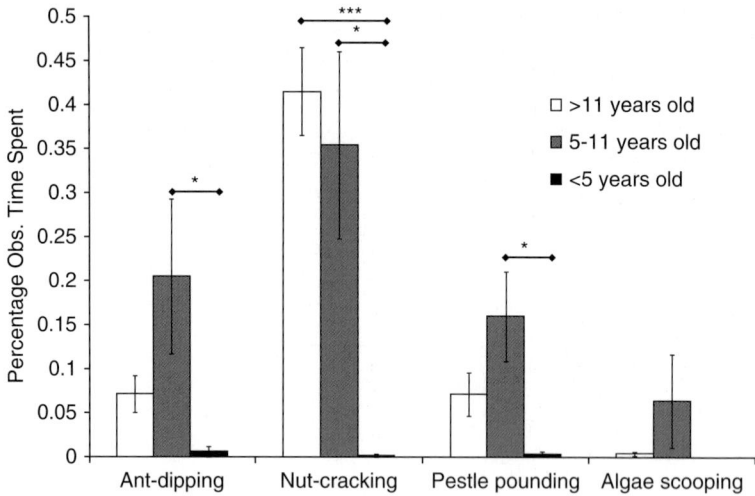

Fig. 6.2 Percentage of observation time spent performing ant-dipping, nut-cracking, pestle-pounding, and algae-scooping by age-class at Bossou (ANOVA Tukey post hoc test: *$P<0.05$; ***$P<0.001$)

6.2.3 Innovations and Cultural Evolution

Eight of the tool-use behaviors (35%) displayed by the Bossou chimpanzees are unique to this community and have thus far never been reported elsewhere. Among these, two – algae-scooping and pestle-pounding – are customary behaviors (Fig. 6.3; see Chaps. 10 and 11 for details on these two tool-use behaviors).

Although the chimpanzees of Bossou have been studied intensively since 1976, many tool-use behaviors performed by this community have been discovered only recently (see Table 6.1). The possibility that these behaviors represent recent innovations cannot be excluded; however, it is also possible that these behaviors were not observed earlier because of (1) the poor level of habituation of the chimpanzees before 1990, (2) the rarity of some of these behaviors, and (3) their seasonal occurrence. Indeed, both algae-scooping and pestle-pounding are highly seasonal behaviors that primarily occur during rainy season months; and before 1995 research at Bossou took place solely during dry season months.

Finally, nearly half the tool-use behaviors listed in Table 6.1 are anecdotal and were witnessed only once or twice, suggesting that these behaviors were most likely innovations that failed to diffuse to other members of the community. These behaviors are proof that chimpanzees can readily innovate and are equipped to exhibit cultural evolution through the ratchet effect, that is, cumulative modifications and incremental improvements thus resulting in increasingly elaborate technologies (Tomasello 1999). Bossou chimpanzees do indeed exhibit a great array of technical variants when water drinking (Tonooka et al. 1994; Tonooka 2001; also see Chap. 8) or dipping for army ants (Humle and Matsuzawa 2002;

Fig. 6.3 Jeje (adolescent male aged 8 years) algae-scoops (*Spirogyra* sp.) from the surface of a pond with the aid of a stalk of vegetation that he has just modified for this purpose (photograph by Tatyana Humle)

Yamakoshi and Myowa-Yamakoshi 2004; Humle 2006; also see Chap. 9). This heterogeneity in behavior may reflect ongoing cumulative cultural change. Sanz and Morgan (2007) at Goualougo, Republic of Congo, have reported that forest chimpanzees demonstrate novel twists on old patterns, for example, using tool sets to gather underground termites. Finally, field researchers would agree that modifications on prior technologies are common events in the wild. Only time will tell how these behavioral variants change in frequency within communities or may even be further modified across successive generations. Only now at Bossou are we able to start appreciating the possibility of witnessing such events, as our catalog of behaviors and knowledge of each individual's repertoire is more exhaustive and comprehensive than it was 30 years ago.

6.3   Tool Use and Handedness at Bossou

Hand use during tool-use performance at Bossou has been recorded across five tool-use behaviors (Matsuzawa 1994; Sugiyama et al. 1993a; Biro et al. 2006; Humle and Matsuzawa 2008; Sousa et al. 2009). These behaviors include pestle-pounding, ant-dipping, algae-scooping, nut-cracking, and leaf folding for drinking water (see Chap. 8 for more details on hand use in the latter two behaviors). Although Bossou chimpanzees demonstrate clear ambilaterality in hand use when performing simple unimanual non-tool-use food acquisition tasks, such as reaching, eating, or plucking fruit (Sugiyama et al. 2003), they exhibit a high degree of lateralization when performing tool use (Matsuzawa 1994; Biro et al. 2006; Humle and Matsuzawa 2008). Both Sousa et al. (2009) and Humle and Matsuzawa (2008) have revealed that young tend to exhibit less strength in hand use than adults (>11 years old) when performing tool use. Table 6.3 therefore only summarizes data on laterality in hand use across the five tool uses for individual adults. Nut-cracking, the most cognitively complex of the five behaviors and the only one requiring complementary coordinated action of both hands and the spatial and temporal combination of three external objects, yielded the greatest strength in hand use with all adults exclusively employing the same hand to manipulate the hammer when cracking nuts. Table 6.3 also indicates that ambi-laterality in hand use during tool manipulation is infrequent among adults across the other four tool-use measures of hand use.

Humle and Matsuzawa (2008) also importantly showed that there was no significant sex differences in lateral bias, although adult females tended to be slightly more strongly lateralized than adult males when pestle-pounding. This study also revealed that shared motor or grip patterns in tool-use skills failed to reveal any specialization in hand use at the individual level. Indeed only 4 among 14 adults (29%), for whom data were gathered across at least three measures of hand use during tool use, showed a consistent lateral bias in hand use; all four were biased to the right. Thus far, handedness data among Bossou chimpanzees indicate that only the most hazardous tool use, that is, ant-dipping, and a non-tool-use haptic task, that is, the extraction by hand of crushed oil-palm heart performed during pestle-pounding, are significantly laterally biased and both to the right (Humle and Matsuzawa 2008). Nevertheless, overall, the data suggest that Bossou chimpanzees demonstrate a significant population-level right-hand bias when manipulating tools, with 63% of adult individual task-specific data points significantly biased to the right and only 30% to the left (also see Humle and Matsuzawa 2008; see Table 6.3).

Table 6.3 Summary of laterality at Bossou across five measures of hand use in tool-use manipulation: leaf folding for drinking water, ant-dipping, nut-cracking, algae-scooping, and pestle-pounding, by individual adult (>11 years old) chimpanzees for which sufficient data were gathered for binomial testing

| Tool use | Number of individuals | | | | | |
	n	Exclusively right	Significantly right	Ambidextrous	Significantly left	Exclusively left
Leaf folding	12	2	5	3	1	1
Ant-dipping	11	2	7	0	2	1
Nut-cracking	14	9	0	0	0	5
Algae-scooping	6	2	2	0	0	2
Pestle-pounding	12	1	5	1	4	1
Totals		16	19	4	7	10

Binomial tests, two-tailed with $P<0.05$, were applied for individual measure to test significant departures from equal use of both hands
Based on data from Biro et al. (2006), Sousa et al. (2009), Humle and Matsuzawa (2009)

6.4 Chimpanzee Cultures: Comparison with Other Sites

Based on the list of tool-use behaviors produced in Whiten et al. (2001), which presents candidate cultural variants, including material culture, across seven long-term study sites, Bossou chimpanzees exhibit one of the richest tool kits. Compared with the other six communities, Bossou presents (1) a higher proportion of subsistence tool-use behaviors, (2) the greatest set of unique customary complex tool-use skills, and (3) no tool-use behaviors for purposes of hygiene (Fig. 6.4).

Patterns of sex differences among adult chimpanzees at Bossou do not seem to follow the emerging pattern observed at other sites. Indeed, several studies in chimpanzees suggest that adult females spend significantly more time in tool-use behaviors than adult males (e.g., termite fishing at Gombe: McGrew 1979; ant fishing at Mahale, Tanzania: Hiraiwa-Hasegawa 1989; nut-cracking at Taï, Côte d'Ivoire: Boesch and Boesch-Achermann 2000). However, as already described, Bossou females and males more than 11 years old do not show any sex differences in time spent nut-cracking, pestle-pounding, or ant-dipping, although males algae-scooped significantly more often than females (see Fig. 6.1).

Finally, Bossou is one of the few sites where chimpanzees crack nuts with an anvil and hammer. Nut-cracking had until recently only been reported among populations of the West African subspecies of chimpanzees (*P. t. verus*), situated west of the N'Zo-Sassandra River (Boesch et al. 1994). A recent account from the Ebo Forest in Cameroon, 1,700 km east of the N'Zo-Sassandra River, has revealed that

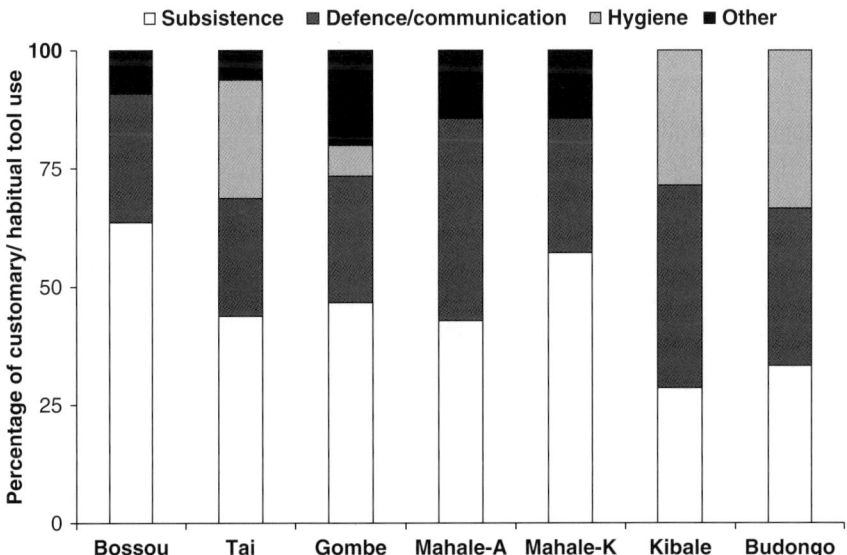

Fig. 6.4 Percentage of customary or habitual tool-use behaviors concerning subsistence, defense and communication, hygiene, and other domains across seven chimpanzee communities studied long term across Africa. (Adapted from data presented in Whiten et al. 2001)

a community of *P. t. ellioti* (formerly *vellerosus*) also cracks *Coula edulis* nuts using a stone hammer and the branch of a tree as an anvil, as well as rocky outcrops (Morgan and Abwe 2006; Abwe, personal communication). So far, nut-cracking has therefore only been reported in 14 populations of the West African subspecies of chimpanzee including both *P. t. verus* and *P. t. elliotii* (see Table 6.4). Nut-cracking has never been reported in the other two chimpanzee subspecies (i.e., *P. t. schweinfurthii* and *P. t. troglodytes*), although oil-palm nuts or other nut-bearing tree species, stones, wooden clubs, and tree roots are available at many sites across East and Central Africa (McGrew et al. 1997).

6.5   Conclusion and Summary

Bossou, similar to any wild chimpanzee community in Africa, presents its own unique tool-use repertoire, including two customary tool-use behaviors so far never observed elsewhere. Bossou chimpanzees employ all materials that are known to be used by chimpanzees in the wild to either generate a tool or use unmodified: such materials include organic materials such as leaves, leaf midribs, twigs, sticks, stalks, bark, petioles, stems, shoots, and boughs, or nonorganic matter such as stones. Bossou chimpanzees readily modify vegetation raw materials to satisfy the requirements of the task at hand (Humle 2003a). The tool-use repertoire of Bossou chimpanzees is heavily oriented toward subsistence behaviors, including behaviors enabling access to fluids such as water in a tree hole or *Raphia* palm wine from its collection vesicle.

As demonstrated by Table 6.1, even after 30 years of observation, we continue to make new observations of tool use at Bossou. Many of these novel tool uses are anecdotal and have only been observed once. They therefore represent innovations that other community members have either potentially not yet invented on their own or never acquired socially. These behaviors constitute an important cumulative data set that may allow us in the future to understand conditions promoting the emergence of innovations and why certain behaviors are not socially learned by other members of a community. Finally, the Bossou chimpanzees have been paramount in teaching us about various social, ecological, physical, and cognitive features of tool use in both extant and extinct primates, including ourselves.

Acknowledgments I thank the Ministère de l'Enseignement Supérieur et de la Recherche Scientifique, in particular the Direction Nationale de la Recherche Scientifique et Technologique (DNRST) and the Institut de Recherche Environnementale de Bossou (IREB), for granting us over the years the permission to carry out research at Bossou. I am particularly grateful to Yukimaru Sugiyama and Tetsuro Matsuzawa, the founders of the research on chimpanzees at Bossou and Nimba, for their continual support and advice, and all colleagues and students from the Bossou research team for increasing over the years our understanding of tool use among the Bossou chimpanzees. Finally, I am forever grateful to the local villagers and all our local assistants at Bossou for all their hard work and their invaluable contributions and collaboration.

Table 6.4 Nut species selected by chimpanzees for cracking at different study sites where nut-cracking has been confirmed

Country	Species	Site	Panda oleosa	Coula edulis	Parinari excelsa/glabra	Saccoglottis gabonensis	Detarium senegalensis	Elaeis guineensis
Guinea	Pan troglodyte verus	Diecké	Yes	Yes	?	?	?	⊙
		Bossou	No	No	⊙	No	No	Yes
		Seringbara	No	No	⊙	No	⊙	⊙
Côte d'Ivoire		Taï	Yes	Yes	Yes	Yes	Yes	⊙
		Mt Betro	Yes	?	⊙	?	⊙	⊙
		Scio (Mt. Zoa)	Yes	?	⊙	?	?	?
		Monogaga	Yes	?	⊙	?	?	?
		Banco	?	Yes	?	?	?	?
		Yealé	⊙	Yes	⊙	No	⊙	Yes
Liberia		Cape Palmas	?	Yes	⊙	?	?	⊙
		Sapo	Yes	Yes	Yes	Yes	?	⊙
		Mt Kanton	?	Yes	?	?	?	?
Sierra Leone		Tiwai	No	⊙	⊙	?	Yes	No
Cameroon	P. t. ellioti (=vellerosus)	Ebo	⊙	Yes	?	No	?	⊙

Yes, species is cracked; No, species is not cracked but is not available in the chimpanzees' habitat; ⊙, species not cracked although available; ?, availability unknown

References: Diecké: Matsuzawa et al. (1999); Humle and Matsuzawa (2001); Bossou: Sugiyama and Koman (1979b); Sugiyama (1981a); Sakura and Matsuzawa (1991); Fushimi et al. (1991); Yealé: Boesch et al. (1994); Joulian (1994); Matsuzawa and Yamakoshi (1996); Humle and Matsuzawa (2001); Taï: Boesch and Boesch (1981, 1983, 1984a, b); sites in Côte d'Ivoire: Boesch et al. (1994); Joulian (1994); Cape Palmas: Savage and Wyman (1844); Sapo: Anderson et al. (1983); Gamys, personal communication; Mt Kanton: Kortlandt and Holzhaus (1987); Tiwai: Whitesides (1985); Ebo: Morgan and Abwe (2006); Morgan, personal communication

Chapter 7
Stone Tools for Nut-Cracking

Tetsuro Matsuzawa

7.1 Introduction

Chimpanzees at Bossou are known to have a rudimentary form of lithic technology. They use a pair of stones as hammer and anvil to crack open the hard shell of oil-palm nuts (*Elaeis guineensis*) to consume the edible kernel within (Fig. 7.1). There are many oil-palm trees in the foothills surrounding the village of Bossou. Nut-cracking can be observed throughout the year. Field experiments have clarified many facets of this complex tool-use behavior (Matsuzawa 1994; Biro et al. 2003; Carvalho et al. 2009, see also Chaps. 15–18). This chapter focuses on stone-tool use and summarizes its importance in terms of comparative cognition. The topics covered include laterality, critical learning period, observational learning, possession, culture, planning, meta-tool use, and emergence of lithic technology.

7.2 Laterality

Bossou chimpanzees have their own unique repertoire of tool-use and manufacture (see Chap. 6 for summary). Bossou researchers have, over the years, investigated stone-tool use in great detail (Biro et al. 2003; Carvalho et al. 2009; Fushimi et al. 1991; Hayashi et al. 2006; Hayashi and Matsuzawa 2003; Humle and Matsuzawa 2001; Matsuzawa 1994, 1999; Matsuzawa et al. 2001; Sakura and Matsuzawa 1991; Sugiyama 1991).

Stone-tool use is a very interesting behavior in terms of comparative cognitive science. From the first observations of this behavior, researchers readily recognized chimpanzees' impressively perfect hand preference for hammering.

---

*Electronic supplementary material   The online version of this chapter (doi: 10.1007/978-4-431-53921-6_8) contains supplementary material, which is available to authorized users.

T. Matsuzawa (✉)
Primate Research Institute, Kyoto University, 41-2 Kanrin, Inuyama, Aichi 484-8506, Japan
e-mail: matsuzaw@pri.kyoto-u.ac.jp

Fig. 7.1 Jeje, a 12-year-old male, uses a pair of stones to crack open oil-palm nuts. He is left-handed for hammering. Each chimpanzee develops 100% hand preference for hammering during stone-tool use (photograph by Tetsuro Matsuzawa)

Expert nut-cracking chimpanzees at Bossou show a 100% hand preference, in clear contrast with other behaviors involving object manipulation such as reaching for fruit (Sugiyama 1991). My colleagues and I have been recording the use of the hammering hand longitudinally since 1988 (Matsuzawa 1994; Biro et al. 2003; see Table 7.1).

When chimpanzees start nut cracking at around 3–4 years of age, they tend to be ambidextrous. They use either their right or their left hand for hammering. It is very rare for them to use both hands. Use of both hands is only observed among very young chimpanzees when manipulating large and oblong hammer stones.

There is no explicit heredity in hand preference. All possible combinations of hand preference between mothers and offspring exist, that is, L–L, L–R, R–L, and R–R. There is, however, an interesting strong tendency for a shared hand preference for hammering among siblings (Matsuzawa 1999). Further evidence will be necessary to understand the determinants of hand preferences among Bossou chimpanzees in the context of stone-tool use.

Table 7.1 also shows two clear cases of individuals switching their hand preference. The first case is Fana, an adult female, who was left-handed until 1995 and who then shifted to using her right hand. This switch in preference occurred because she dislocated her left shoulder, possibly as a consequence of a severe fall. However, she rapidly adapted and shifted to using her right hand when hammering; her efficiency was actually not impaired. The second case is Joya, a juvenile female, who was right-handed in 2009 when she successfully started cracking nuts.

Table 7.1 Longitudinal record of stone-tool use by chimpanzees at Bossou, Guinea

Name	Sex	Age	Mother	Year observed																				
				88	90	91	92	93	94	95	96	97	98	99	00	02	03	04	05	06	07	08	09	10
Tua	M	Adult	Unknown	?	L	L	L	L	L	L	L	L	L	L	L	L	L	L	L	L	L	L	L	L
Kai†	F	(Adult)	Unknown	?	R	R	R	R	R	R	R	R	R	R	R	R	R	–	–	–	–	–	L	L
Kie*	F	34	Kai	?	R	R	–	–	–	–	–	–	–	–	–	–	–	–	–	–	–	–	–	–
Kakuru*	F	24	Kie	?	A	R	–	–	–	–	–	–	–	–	–	–	–	–	–	–	–	–	–	–
Nina*	F	(Adult)	Unknown	?	X	X	X	X	X	X	X	X	X	X	X	X	X	–	–	–	–	–	–	–
Na*	M	25	Nina	?	R	R	R	R	R	R	–	–	R	R	R	–	–	–	–	–	–	–	–	–
Nto*	F	17	Nina	–	–	–	–	–	X	X	X	–	–	–	–	–	–	–	–	–	–	–	–	–
Fana	F	Adult	Unknown	?	L	L	L	L	L	L	R	R	R	R	R	R	R	R	R	R	R	R	R	R
Foaf	M	30	Fana	?	R	R	R	R	R	R	R	R	R	R	R	R	R	R	R	R	R	R	R	R
Fotaiu*	F	19	Fana	–	–	–	X	X	X	AR	R	R	R	–	–	R	X	–	–	–	–	–	–	–
Fokaiye*	M	10	Fotaiu	–	–	–	–	–	–	–	–	–	X	X	X	R	R	R	R	R	R	R	R	a
Fanle	F	13	Fana	–	–	–	–	–	–	–	–	–	–	–	–	–	–	–	–	–	X	X	a	L
Flanle	M	3.5	Fanle	–	–	–	–	–	–	–	–	–	–	–	–	–	–	–	L	L	L	L	a	L
Jire	F	Adult	Unknown	?	L	L	L	L	L	L	L	L	L	L	L	L	L	L	L	L	L	L	L	L
Ja*	F	27	Jire	?	R	R	R	R	–	–	–	–	–	–	–	–	–	–	–	–	–	–	–	–
Jokro†	F	(3)	Jire	–	X	X	X	–	–	–	–	–	–	–	–	–	–	–	–	–	–	–	–	–
Juru*	F	17	Jire	–	–	–	–	–	X	X	X	X	X	r	r	r	–	–	–	–	–	–	–	–
Jeje	M	13	Jire	–	–	–	–	–	–	–	–	–	X	X	X	X	X	X	A	L	L	L	L	L
Jimato†	M	(1.5)	Jire	–	–	–	–	–	–	–	–	–	r	–	r	X	X	–	–	–	–	–	–	–
Joya	F	5.5	Jire	–	–	–	–	–	–	–	–	–	–	–	–	–	–	–	X	X	X	X	R	L
Jodoamon	F	0.5	Jire	–	–	–	–	–	–	–	–	–	–	–	–	–	–	–	–	–	–	–	–	X
Velu	F	Adult	Unknown	?	R	R	R	R	R	R	R	R	R	R	R	R	R	R	R	R	R	R	R	R
Vube*	F	28	Velu	?	L	–	–	–	L	L	L	L	L	L	L	–	–	–	–	–	–	–	–	–
Vui*	M	24	Velu	?	X	X	L	L	L	L	L	L	L	L	L	L	L	–	–	–	–	–	–	X
Vuavua*	F	19	Velu	–	–	–	–	–	–	–	AL	–	–	–	–	–	L	L	–	–	–	–	–	–
Veve†	F	(2.5)	Vuavua	–	–	–	X	X	–	–	–	–	–	–	X	X	X	X	–	–	–	–	–	–

(continued)

Table 7.1 (continued)

Name	Sex	Age	Mother	Year observed																				
				88	90	91	92	93	94	95	96	97	98	99	00	02	03	04	05	06	07	08	09	10
Yo	F	Adult	Unknown	?	L	L	L	L	L	L	L	L	L	L	L	L	L	L	L	L	L	L	L	L
Yunro*	F	26	Yo	?	X	X	X	l	–	–	–	–	–	–	–	–	–	–	–	–	–	–	–	–
Yera†	M	(0.5)	Yo	–	X	–	–	–	–	–	–	–	–	–	–	–	–	–	–	–	–	–	–	–
Yolo	M	19	Yo	–	–	–	X	X	X	X	L	L	L	L	L	L	L	L	L	L	L	L	L	L
Pama	F	Adult	Unknown	?	X	X	X	X	X	X	X	X	X	X	X	X	X	X	X	X	X	X	X	X
Pru*	M	30	Pama	R	R	R	R	R	R	R	R	R	R	R	R	–	–	–	–	–	–	–	–	–
Pili*	F	23	Pama	?	X	R	–	–	–	–	–	–	X	X	X	–	–	–	–	–	–	–	–	–
Pokuru*	M	14	Pili	–	–	–	–	X	X	X	X	X	X	X	X	–	–	–	–	–	–	–	–	–
Poni†	M	(9)	Pama	–	–	–	–	–	X	–	R	R	R	R	R	R	R	–	–	–	–	–	–	–
Peley	M	12	Pama	–	–	–	–	–	–	–	–	–	–	X	X	AL	L	L	L	L	L	L	L	L

Individuals are sorted according to matrilines. Oil-palm nut-cracking ability and hand used to hold hammer stone recorded for all individuals since 1988. The data were recorded by the collective efforts of the KUPRI-International team, especially Matsuzawa, Inoue-Nakamura, Dora Biro, Misato Hayashi, and Susana Carvalho with the help of local assistants

Notes: *, individual disappeared; †, individual confirmed dead; L, always uses left hand for hammer; R, always uses right hand for hammer; A, ambidextrous use of hammer; l, uses left hand to pound nut without a hammer; r, uses right hand to pound nut without a hammer; a, ambidextrous in pounding without hammer; X, no successful hammer use; ?, data unavailable due to lack of observation; –, data unavailable as subject had not yet been born, had disappeared, or died. Ages, where known, are shown as of mid-2010; figures in brackets represent age at which individual died

She then shifted to using her left hand in 2010, after her right hand was trapped in a wire snare in the summer of 2009. The wire remained stuck around some fingers of her right hand for several weeks. Because of this physical handicap, she was unable to use her right hand for hammering. She therefore started using her left hand. She finally removed the snare a few months later, but lost the tip of her little finger. Even after the snare was removed, Joya continued to use her left hand for hammering. These observations indicate that able nut-crackers can potentially readily switch their hand preference if necessary.

The afore-described two cases also clearly demonstrate that intermanual transfer of the hammering skill is almost perfect. The technique employed by each individual persisted even after the switch in hand use. Fana continued to prefer selecting small hammers. Her hammering technique before her injury was rather idiosyncratic, involving quick successions of hits of low amplitude. She maintained this technique even after switching hands. Although individuals may be able to switch their hand preference under special circumstances, it remains always skewed 100% either to the right or to the left. Such perfect lateralization in hand use may be because stone-tool use involves complex bimanual coordination, which also requires the remaining hand to pick up the nut, place it on the anvil, etc.

Hand preference in humans is strongly skewed to the right at the population level with approximately 90% of right-handers. This pattern in population-level handedness in humans is linked to our left hemisphere specialization, which controls language and contralateral manual dexterity. Among 36 Bossou chimpanzees I have observed during the past 25 years, 14 were right-handed and 10 left-handed. There is therefore no statistically significant task specialization in handedness for nut-cracking because only 58% of individuals are right-handed.

In sum, hand preference in stone-tool use is 100% skewed to one side. However, this preference can readily shift from one side to the other if necessary. In contrast to human laterality, there is no explicit hand bias for nut-cracking at the population level. Hand preference across different tool-use behaviors, however, may suggest a slight population right-hand bias (Humle et al. 2009; see also Chaps. 6 and 8).

7.3   Critical Period

When do chimpanzees start nut-cracking? It depends on the individuals. However, the answer is about 3–4 years old on average (see Table 7.1). There are no clear sex differences; however, our sample size is relatively small. Jeje, a male, took a long time to acquire the skill. He started cracking nuts at the age of 6. His older sister, Juru, also had difficulties in acquiring the skill, although their mother, Jire, is a very skillful nut-cracker. Her other offspring mastered the skill between 3 and 4 years of age. There are therefore large individual differences in the onset of the behavior.

There seems to be a critical learning period between the age of 3 and 7 years for acquiring the nut-cracking skill. At Bossou, we have had two adult females who never used stone tools: Nina and Pama. They consumed broken pieces of kernel cracked by other community members. When Nto cracked nuts, her mother, Nina, regularly used to steal her kernels. These two females were therefore motivated to feed on the kernels but failed to use a hammer to procure these themselves. Assuming female dispersal (Goodall 1986), we can postulate that these two females may have come from a neighboring community that lacks a tradition of stone-tool use, such as Seringbara in the Nimba Mountains (Biro et al. 2003; Humle and Matsuzawa 2001; see also Chap. 28). By the time they joined the Bossou community at puberty, they would have bypassed the critical learning period and therefore failed to acquire the skill. However, because these two females were already present in 1976 when research at Bossou began, we cannot confirm the validity of this hypothetical scenario.

There was a young female, named Yunro, who also failed to learn how to crack open nuts. At the age of 8 years in 1993, she still lacked the skill. She would place a nut on the anvil stone and hit it with the back of her left wrist or stamp it with her right foot (Matsuzawa 1994; Fig. 7.2). This kind of behavior is quite typical of

Fig. 7.2 Flanle, a 2.5-year-old male, is trying to crack open oil-palm nuts. He makes many attempts: he uses his right hand to hit the nut placed on the anvil stone (*top*), and he holds a hammer in his left hand to crack open the nut (*bottom*). He is very close to succeeding but not quite yet (photograph by Tetsuro Matsuzawa)

naive chimpanzees in captivity: they simply fail to use a hammer stone (Hayashi et al. 2004; Chap. 19). Why did Yunro fail to acquire the skill? As in the case of Joya, she was trapped in a wire snare when she was 4 years old. Because the wire caught her ankle, she would travel on the ground using both her forearms as crutches. This disadvantage at such a critical age might explain why she altogether failed to learn how to crack nuts.

7.4 Observational Learning

The study of the acquisition process of stone-tool use has helped us highlight and understand unique characteristics of observational learning in chimpanzees. We have coined this process "education by master-apprenticeship" (Matsuzawa et al. 2001; see Chap. 21). Education by master-apprenticeship hinges on three key aspects.

The foundation is the strong and enduring mother–infant bond. In this context, (1) mothers serve as a good model for their young, (2) the infants have an intrinsic motivation to copy their mother's behavior through intensive observation, and (3) mothers are highly tolerant of their infants. Young chimpanzees' copying behavior is at first never rewarded because at a young age they consistently fail to crack open nuts. Young chimpanzees persist regardless in their failed attempts until after successive series of observation and practice they finally succeed.

It is also important to note that observational learning is always unidirectional; young learn from studying older members of the community, never the reverse (Biro et al. 2003). In human society, observational learning can be bidirectional between members of the younger and older generations. However, this mode of transmission appears to be absent in chimpanzees.

7.5 Possession

Studies in the outdoor laboratory, that is, the field experiment site (see Chap. 16), have revealed that each chimpanzee has his or her own favorite hammer and anvil stone (Biro et al. 2003). When they move around in the outdoor laboratory, they will either carry their hammer/anvil set with them or repeatedly carry a handful of nuts to where they left their selected preferred set. When chimpanzees arrive at the outdoor laboratory, they will also tend to select the same stones over again, even though stones are randomly organized and are amply available.

The chimpanzees also have their favorite cracking place. When a group of chimpanzees arrives at the outdoor laboratory, each chimpanzee clearly has his or her own favorite location for cracking nuts. For example, Yo, an adult female, always stays on the right-hand side from the observer's perspective. Tua and Foaf, the two adult males, prefer to crack at the back. They seem to be cautious of human observers, especially those equipped with cameras.

Chimpanzees have three strategies available to them for obtaining stone tools: (1) displacing a subordinate individual and robbing them of their stones, (2) awaiting the departure of a dominant individual and then reusing their set, especially if these are preferred, and (3) tactically deceiving a stone-set owner, for example, by grooming them and obliging them to reciprocate and then stealing their stones (Matsuzawa 1999).

7.6   Culture

Each community of chimpanzees has its own unique set of traditions, which are cultural. "Culture" represents in this sense sets of behavior, knowledge, and values that are passed on from one generation to the next through nongenetic channels (Matsuzawa et al. 2001). The use of mobile anvil and hammer stones for cracking nuts is a cultural characteristic of Bossou chimpanzees.

Stone-tool use entails three key components: the target nut, the hammer, and the anvil. Bossou chimpanzees crack open oil-palm nuts using a mobile hammer and anvil stone. Yealé chimpanzees in the Nimba Mountains crack open oil-palm nuts and coula nuts (*Coula edulis*) but tend to prefer using a tree root or a stone outcrop as an anvil (Humle and Matsuzawa 2001, 2004; see Chap. 27). Diécké chimpanzees crack open both coula nuts and panda nuts (*Panda oleosa*) and also preferentially use rocky outcrops as anvils (see Chap. 30). Interestingly, environmental factors such as nut species and material availability cannot fully explain these differences. For example, Seringbara chimpanzees in the Nimba Mountains (see Chap. 28) do not crack open oil-palm nuts although trees and nuts, as well as suitable materials for tools, are available (Biro et al. 2003; Humle and Matsuzawa 2004; see Table 6.4 in Chap. 6). Nut-cracking behavior in chimpanzees therefore clearly demonstrates intercommunity cultural variation.

In chimpanzees, females are the most likely vehicles for intercommunity cultural diffusion, because females rather than males at puberty tend to disperse and immigrate into a new community. Females in that sense may then act as masters to new apprentices, especially their own offspring. Females can thus introduce a novel behavior, that is, an innovation, into her new community, such as cracking a nut species with mobile stones. Females are therefore the most likely candidates for cultural diffusion, carrying with them the knowledge of their natal community. We might therefore expect cultural zones comprising multiple adjacent communities that share elements of their tool-use repertoire.

7.7   Planning

In the case of termite fishing, Gombe chimpanzees in Tanzania manufacture a fishing tool in advance before reaching termite mounds (Goodall 1986). Tool or material transport is similarly observed in the context of nut-cracking. Bossou chimpanzees sometimes collect nuts in advance and then carry them to the cracking site. They may also pick up stones to carry them to a location with nuts.

In terms of planning, it is also important to note that adult chimpanzees often manipulate the anvil stone before starting to crack: rotating the anvil stone, placing it upside down, and so forth. It can be challenging to produce a flat surface. The oil-palm nut is round like a rugby ball, so it can also readily roll off the surface. Therefore, sensitivity to the flatness of the surface is essential. Let us suppose that a mature chimpanzee picks up two stones, a hammer and an anvil, and moves off to a place with nuts. She then sets the anvil stone on the ground and rotates it horizontally clockwise or anticlockwise. Bossou chimpanzees often employ such a strategy to obtain the flattest surface possible.

In fact, chimpanzees may utilize various techniques to keep the surface of the anvil flat: (1) rotation as just described, (2) turning the anvil stone upside down, (3) moving the anvil stone from one place to another to generate the flattest surface possible, and (4) placing the anvil stone on a smaller stone, which then serves as a wedge. The wedge stone stabilizes the anvil stone and keeps the surface flat. In all these cases, it must be noted that the chimpanzees adopt such strategies in advance before starting to crack. This observation means that they know that the surface should be kept flat and therefore plan accordingly.

7.8   Meta-Tool: Level Theory of Tool-Use

Similarly to grammatical rules in human language, we can conceive identical rules in human action. As we emit words in organized sequences, we also manipulate objects in our everyday life in a proper sequential way. An adequate sequence of action works, but if inadequate, it risks failing to produce the desired outcome: this is the underlying hypothesis behind "action grammar" (Matsuzawa 1997a). Action grammar decomposes behavioral sets into the four major components: agent, action, object, and location. In short, all behaviors can be described as who did it, how it was done, what for, and where. Action grammar implies that every behavior should follow some kind of rules comparable to grammatical rules of human language.

Action grammar predicts that the complex manipulation of multiple objects such as stone-tool use should follow their own grammatical rules. There are several different ways of describing action grammar (Hayashi 2007; Sanz and Morgan 2010), which differ according to which component of the behavior is considered. In my original "tree-structure analysis" (Matsuzawa 1996; see also Chap. 18), I had neglected the agent, action, and location; the focus was on the objects, that is, how are the objects related to one another during object manipulation.

Let us consider termite fishing. A chimpanzee uses a stick to get a termite. Initially, neither one of these objects is related to one another. However, once the agent, a chimpanzee, targets a stick toward a termite, then the stick is no longer a simple stick but becomes a tool to attain a goal. In this situation, tool-use can be defined as level 1, in which one item is related to another: a stick is related to a termite. Only one node connects the objects involved in this manipulation. Stone-tool use for nut-cracking is a good example of a level 2 tool-use. Two nodes connect the different objects involved. The first node links the nut to the anvil, that is, placing the nut on the anvil stone. The second node connects the hammer stone to the nut (that was placed on the anvil stone). Wedge-stone use represents in this sense level 3 because it involves an additional connection between the anvil stone and a smaller stone to keep the surface of the anvil stone flat.

Wild chimpanzees present a diverse array of tool-use behaviors. Each community possesses its own unique material culture. However, the majority of tool-use behaviors observed in chimpanzees can be classified as level 1, according to the proposed level theory of tool-use (Matsuzawa 1997a).

7.9 Emergence of Lithic Technology

During the first field experiments in the outdoor laboratory between December 1990 and February 1991, seven of the supplied stone tools, that is, hammers and anvils, eventually shattered into two or more pieces upon usage by chimpanzees. Four of the seven stones, initially used as anvils, were nevertheless reused as effective stone hammers. Such an example potentially denotes an elementary first step in stone-tool production (Matsuzawa 1994).

During the following years, we recorded many similar observations partly because many stone materials available in Bossou are composed of laterite, a soft mineral, which can readily fracture (Fig. 7.3). Therefore, an analysis of the fracturing process resulting from nut-cracking in chimpanzees may help us elucidate how lithic technology emerged among early hominids (see Chap. 15).

In short, what we observe in wild chimpanzees may be the first evolutionary step toward lithic technology. Nut-cracking serves to crack open the hard shell of nuts to obtain the edible kernel within; indirectly, this action results in the production of broken pieces of stone that eventually can be perceived as functional for other purposes; for example, a shattered anvil stone may then generate a useful hammer stone. Without intention, the result is stone-tool making. The cognitive ability for planning in combination with such kind of incidental occurrences may have led our ancestors to invent lithic technology. Further examination of stone-tool use is therefore likely to shed important further insights into how tools for survival emerged among our ancestors.

Acknowledgments The present study was financially supported by the following grants: MEXT #16002001, #20002001, JSPS-HOPE, gCOE-A06-D07. I wish to thank all my colleagues, students, and Guinean collaborators who have helped me understand stone-tool use in chimpanzees. I am most grateful to Takao Fushimi, Noriko Inoue-Nakamura, Dora Biro, Claudia Sousa, Misato Hayashi, and Susana Carvalho for all their efforts toward gathering longitudinal data on nut-cracking at Bossou.

Fig. 7.3 Jeje, a 10-year-old male, cracks open an oil-palm nut. However, because of his powerful strike, the anvil stone breaks in half. Stone fracturing in this way is commonly observed and represents the first evolutionary step in the emergence of lithic technology in early hominids (photograph by Tetsuro Matsuzawa)

Chapter 8
Use of Leaves for Drinking Water

Cláudia Sousa

8.1 Introduction

Chimpanzees (*Pan troglodytes*) are proficient tool-users, making and using a variety of tools in subsistence and nonsubsistence activities (McGrew 1992; Whiten et al. 1999). Each chimpanzee community has its own tool-use repertoire (McGrew 1992; Whiten et al. 1999, 2001). The chimpanzee community of Bossou is no exception, with a quite large repertoire of 24 different tool-use behaviors (see Chap. 6). Among these, two behaviors concern the use of leaves for drinking water.

Wild chimpanzees are known to use leaves for drinking rainwater from tree holes (Boesch and Boesch 1990; Ghiglieri 1984; Goodall 1964, 1968, 1986; McGrew 1977, 1992; Nishida 1990a; Quiatt and Kiwede 1994; Sugiyama 1989a, 1993, 1995a; Tonooka et al. 1994; Wrangham 1992). Although this behavioral pattern has been observed at many sites, the precise technique used varies considerably across populations. Usually the drinking behavior with leaf tools is described worldwide as the use of leaf sponges (Goodall 1968; McGrew 1977, 1992; Tonooka 2001; Tonooka et al 1994), a technique first described for Gombe chimpanzees. However, this is not the only technique used by wild chimpanzees. Wild chimpanzees are also known to use leaf spoons (Goodall 1968; McGrew 1977; Sugiyama 1995a) and a leaf-folding technique (Tonooka et al. 1994) for drinking water. We call the behavior leaf sponges when the leaves are crumpled in the mouth (Goodall 1968; McGrew 1977, 1992), leaf spoons when the leaves are used to scoop out the water, without crumpling them up (Goodall 1968; McGrew 1977; Sugiyama 1995a), and leaf-folding when the leaves are folded inside the mouth before use (Sousa et al. 2009), "at about 3-cm intervals" (Tonooka 2001: 326). The finished

C. Sousa (✉)
Department of Anthropology, Faculty of Social and Human Sciences,
New University of Lisbon, Avenida de Berna, 26-C, 1069-061 Lisbon, Portugal
and
Centre for Research in Anthropology (CRIA), Avenida das Forças Armadas,
Edifício ISCTE, 1049-029 Lisbon, Portugal
e-mail: csousa@fcsh.unl.pt

T. Matsuzawa et al. (eds.), *The Chimpanzees of Bossou and Nimba*,
DOI 10.1007/978-4-431-53921-6_9, © Springer 2011

tool is then dipped into a tree hole, retrieved, and put inside the mouth such that the water it carries can be drunk. The sequence is then repeated, using the same tool.

Although at Bossou all these three techniques are reported (Sugiyama 1995a; Tonooka 2001; Tonooka et al. 1994), the most frequent one is the leaf-folding technique for drinking water (Tonooka 2001), which is also exclusively observed among Bossou chimpanzees (see Chap. 6). To perform this tool-use behavior the chimpanzees also show high selectivity for the leaves, preferring the leaves of *Hypophrynium braunianum* as tools (Tonooka 2001).

Tool-use behaviors are assumed to be transmitted culturally between communities and across generations (Matsuzawa and Yamakoshi 1996), that is, they are learned from others who are more proficient than the novice (McGrew 2004). Nevertheless, few studies have explored the acquisition process underlying tool-use behaviors in chimpanzees. Those studies that exist have explored nut-cracking (Biro et al. 2003, 2006; Inoue-Nakamura and Matsuzawa 1997; Matsuzawa 1994, 1999), termite fishing (Lonsdorf 2005, 2006; Lonsdorf et al. 2004), ant-dipping (Humle 2006; Humle et al. 2009), and the use of leaves for drinking water (Biro et al. 2006; Sousa et al. 2009; Tonooka 2001).

Such studies may be scarce because it is difficult to conduct systematic direct observations on chimpanzees performing tool-use activities in the wild. Not all tool-use behaviors occur frequently enough to allow researchers to accumulate sufficient data for such in-depth analysis, and many may be seasonal or difficult to observe because of the nature of the chimpanzees' habitat. For these reasons, an "outdoor laboratory" (Matsuzawa 1994; see also Chaps. 16–18, 21) has been established in the home range of the chimpanzee community at Bossou, Guinea, increasing opportunities of individual group members to perform the behaviors, as well as the researchers' chances of observing them. As a result, we now have extensive annual records on various tool-using behaviors from every member of the Bossou community, including the use of leaves for drinking water.

In this chapter, I bring together the available information on the use of leaves for drinking water by the Bossou chimpanzees. I focus on the learning process underlying the acquisition of the skill, examining some developmental aspects involved.

8.2 Methods

8.2.1 Subjects and Study Site

Behavioral observations involved members of the chimpanzee community of Bossou (7°39′ N, 8°30′ W), situated in Guinea, West Africa, about 6 km northwest of the foot of the Nimba Mountains on the border with Côte d'Ivoire and Liberia (see Chaps. 27–29).

This community has been studied since 1976 (Sugiyama 1981a, 1984; Sugiyama and Koman 1979a, b; see Chap. 1 for further details) and was habituated to the presence of human observers without food provisioning. Since then the size of the community has remained relatively stable, varying between 16 and 23 individuals (see Chap. 3). The core area of the community's range encompasses 5–6 km², composed of primary and secondary forest that is surrounded on all sides by dry savannah which the chimpanzees rarely traverse (see Chap. 2).

Individual age ranged from 14 months to 53 years (estimated) old (Table 8.2), divided into four age classes (Goodall 1986; Sugiyama 1994a): adults (11 years old or more), adolescents (between 10 and 8 years old), juveniles (between 7 and 5 years old), and infants (4 years old or less). All individuals were recognized individually.

8.2.2   Data Collection

Observations were carried out in a small clearing in the forest – the "outdoor laboratory" (Matsuzawa 1994; see also Chaps. 16–18, 21) – located on the summit of a hill known as Gban, south of the village of Bossou. A hole with two openings (front and side) was made in the trunk of a tree (*Ricinodendron heudelotii*) (Fig. 8.1) and was refilled to the brim after each visit by chimpanzees to provide fresh water and to monitor the quantity of water consumed by the chimpanzees. The experimenters provided no leaves, so that chimpanzees had to use the vegetation available in the surrounding environment, as in a fully natural condition.

Over the years we recorded the use of leaves in every member of the community along with the identity of the hand used to hold the tool while performing the behav-

Fig. 8.1 The *Ricinodendron heudelotii* tree used in the present study, which contained a water-filled tree hole. A close-up of the two openings of the tree hole are shown: *F*, front opening; *S*, side opening

ior. Data were collected in the dry season, December to February, over a total of 20–30 h of observation each year. Leaf use for drinking water has been studied since 1994, although the data presented here correspond to four field seasons: January 2000, January 2003, January 2005, and January 2006. All chimpanzee visits to the outdoor laboratory were captured on videotape, using a Sony DCR-TRV20 or Sony DCR-TRV9 digital camera. Leaf tools were collected over one study period (January 2003), and systematic measurements (weight, size, and quantity of water carried) were obtained, along with information about tool-user identity, whenever possible.

Subsequent video analysis was conducted to collect further detailed information on the behavior. Only independent bouts were recorded, rather than every manual action. A new bout was said to have commenced if the subject changed hands or if the subject moved from one opening in the tree to the other to drink water.

8.3   Results: Technique Development

8.3.1   Stages of Learning and Use of Abandoned Tools

The use of leaves for drinking water involves two distinct phases, the manufacture of the leaf tool and its use. The use of the leaf tool is the first to emerge, at the age of 1.5 years (Sousa et al. 2009). The leaf tools used at this age are not the infants' own; rather, they use the ones discarded by older individuals after use (Table 8.1). Only at the age of 3.5 years do they start to use the leaf tools that they themselves manufacture, and they also continue to use those discarded by fellow group members for several more years (see Table 8.1).

The leftovers, either tools discarded by other individuals or remains that dropped during the manufacturing process, are picked up off the ground, or occasionally taken from the mother's hand, sucked on, and then placed into the water before being retrieved for drinking (tool reuse). Until the age of about 3.5 years, the infants exclusively use these leftovers. The percentage of total episodes of use of leftovers by

Table 8.1 Ability to use leaf tools (made and discarded by other individuals, or made by self) by chimpanzees in different age classes

Age class		Use	Use discarded tool	Make and use own tool
Adult	11–	Yes	No	Yes
Adolescent	8–10	Yes	No	Yes
Juvenile	5–7	Yes	Yes	Yes
Infant	3–4	Yes	Yes	Yes
	2–3	Yes	Yes	No
	1–2	Yes	Yes	No
	0–1	No	No	No

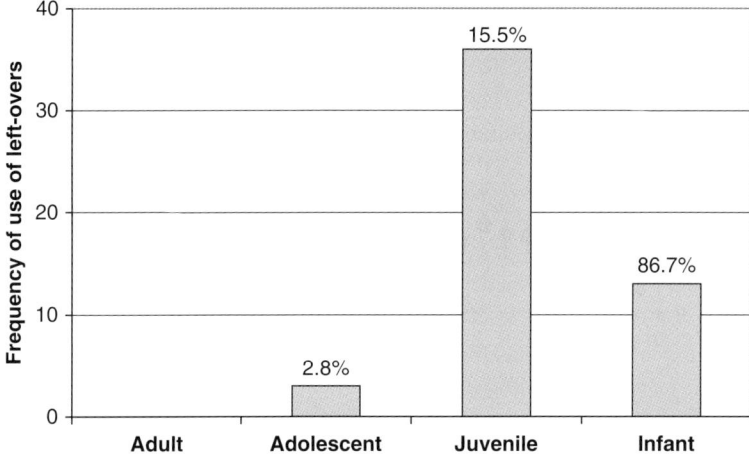

Fig. 8.2 Percentage of episodes in which the drinking tool was already used by other individuals or made from the beginning

infants is 86.7% (Fig. 8.2), consequent to two episodes of making and using a new tool by an infant already 3.5 years old. During this period, until reaching 3.5 years of age, infants also sometimes manufacture a small tool, but they never use it, dropping it immediately after making it, and picking then a leftover to use instead.

8.3.2   Development and Tool Efficiency

The leaf tools collected during January 2003 showed a significant correlation between tool weight and the quantity of water they could carry (Pearson's correlation; $n=31$, $r=0.83$, $P<0.001$), with larger tools holding more water (Fig. 8.3). These data also showed lower efficiency of tool-use in younger individuals when compared with the performance of adults. In the former, we find, for example, that leaf tools manufactured by juveniles carry a smaller volume of water per dip than those of adults (Biro et al. 2006).

8.3.3   Handedness, Adjusting, and Development

Our long-term records of leaf use for drinking water did not show strict individual-level laterality for the dipping action (Table 8.2, see Sousa et al. 2009 for more details). Of the 24 chimpanzees, 2 infants were never observed drinking water during the study period. Of the remaining 22 individuals, only 3 adults showed consistency in the use of a hand, although these results are based on very small sample sizes (five, two, and six episodes, respectively). The infant indicating the use of only one hand (Veve) only

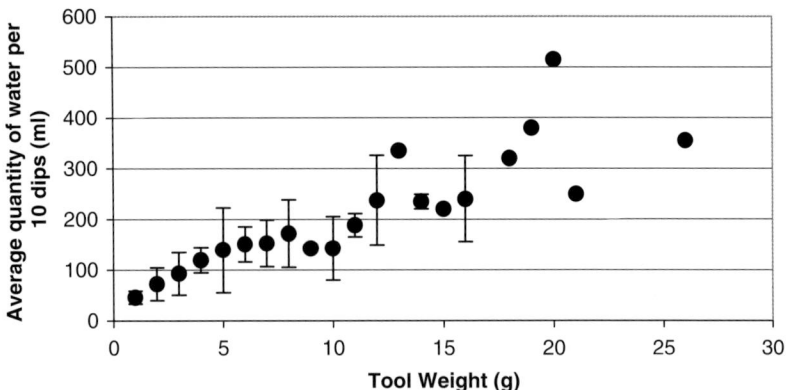

Fig. 8.3 Average quantity of water (ml) carried by each tool as function of tool weight. Data were collected in 2002. *Error bars* are standard errors of the mean

had one episode. The remaining 18 individuals were observed to use both the left and the right hand for the dipping action, although 5 of them showed a clear preference for one hand (Fig. 8.4).

A more detailed analysis of the results also showed that juveniles were the ones most likely to change hands during or between dipping bouts, followed by adolescents. Although adults also changed hands to dip, they did so much less frequently (Fig. 8.5). Regarding the use of the two openings of the tree hole (front or side), juveniles were also the ones who alternated most frequently between the holes (Fig. 8.5).

In only 11 (16.18% of all cases of hand change by juveniles) of the 26 episodes in which the juveniles changed holes was this followed by a change of hands. In adolescents, this happened in 5 of 12 episodes of hole changes (15.63% of all cases of hand change by adolescents) and in 1 of 2 episodes for adults (3.33% of all cases of hand change by adults). The only two episodes of hand change in infants occurred also after a hole change, although they were also observed changing holes twice without changing hands.

Figure 8.6 shows the distribution of hand use as a function of which opening of the tree hole individuals were using. Adults exhibit a tendency to use the right hand when dipping at the front opening of the tree hole and to use the left hand when dipping at the side opening. Adolescents and juveniles do not seem to show a preference for using either hand when dipping at the front opening but show a slight preference for the left hand when dipping at the side opening. For infants, data are not sufficient to draw conclusions.

8.3.4 Observing and Learning

Table 8.3 shows the total frequency of observational episodes by chimpanzees during the process of making and/or using drinking tools by other individuals

Table 8.2 Composition of the wild chimpanzee community at Bossou in each field season of the present study: record of handedness in dipping action during the use of leaves for drinking water

Name	Sex	Birth	Age (years:months)	Jan 2000		Jan 2003		Jan 2005		Jan 2006	
				Age class	Hand.	Age class	Hand.	Age class	Hand.	Age class	Hand.
Fana	♀	~1956	44	1	Ar	1	X	1	i	1	X
Fanle	♀	1997, Oct	3:3	4	A	3	A	3	r	2	Ar
Foaf	♂	1980, late	19	1	A	1	i	1	i	1	r
Fokaiye[a]	♂	2001, Jul	–	–	–	4	x	–	–	–	–
Fotaiu[a]	♀	1991, mid	9	2	Ar	1	r	–	–	–	–
Jeje	♂	1997, Dec	3:1	4	x	3	1	3	1	2	Al
Jimato[b]	♂	2002, Oct	–	–	–	4	x	–	–	–	–
Jire	♀	~1958	42	1	A	1	r	1	A	1	1
Joya	♀	2004, Sep	–	–	–	–	–	4	x	4	A
Juru[c]	♀	1993, Nov	6:2	3	r	–	–	–	–	–	–
Kai[b]	♀	~1950	50	1	i	1	X	–	–	–	–
Nina[b]	♀	~1954	47	1	i	1	r	–	–	–	–
Nto[c]	♀	1993, early	7	3	A	–	–	–	–	–	–
Pama	♀	~1967	33	1	X	1	A	1	i	1	A
Peley	♂	1998, Apr	2:9	4	x	3	A	3	A	3	A
Pili[c]	♀	1987, early	13	1	A	–	–	–	–	–	–
Pokuru[c]	♂	1996, Aug	3:5	4	A	–	–	–	–	–	–
Poni[b]	♂	1993, Feb	7:11	3	A	2	A	–	–	–	–
Tua	♂	~1957	46	1	A	1	A	1	i	1	X
Velu	♀	~1959	41	1	X	1	A	1	X	1	i
Veve[b]	♀	2001, May	–	–	–	4	i	–	–	–	–
Vuavua[a]	♀	1991, mid	9	2	A	1	A	–	–	–	–
Yo	♀	~1961	39	1	A	1	i	1	i	1	A
Yolo	♂	1991, mid	9	2	r	1	r	1	A	1	A

(continued)

Table 8.2 (continued)

Age: Age in January 2000

Age class: 1 = adult (11 years and above), 2 = adolescent (8–10 years), 3 = juvenile (4–7 years), 4 = infant (less than 4 years); age presented in years and months (years:months)

[a]Individual disappeared in 2004

[b]Individual confirmed dead

[c]Individual disappeared by 2001

Hand. (Handedness): L, always uses left hand; R, always uses right hand; A, ambidextrous; l, left biased ($P < 0.05$, statistically significant departures from equal use of both hands, two-tailed binomial test); r, right biased ($P < 0.05$, statistically significant departures from equal use of both hands, two-tailed binomial test); Al, ambidextrous with left-hand bias ($0.05 < P < 0.2$, statistically significant departures from equal use of both hands, two-tailed binomial test); Ar, ambidextrous with right-hand bias ($0.05 < P < 0.2$, statistically significant departures from equal use of both hands, two-tailed binomial test); X, no tool-use observed during period of observation (although individual had previously performed the behavior); x, no tool-use observed during period of observation or at any time previously; i, inconclusive because of the small number of observations

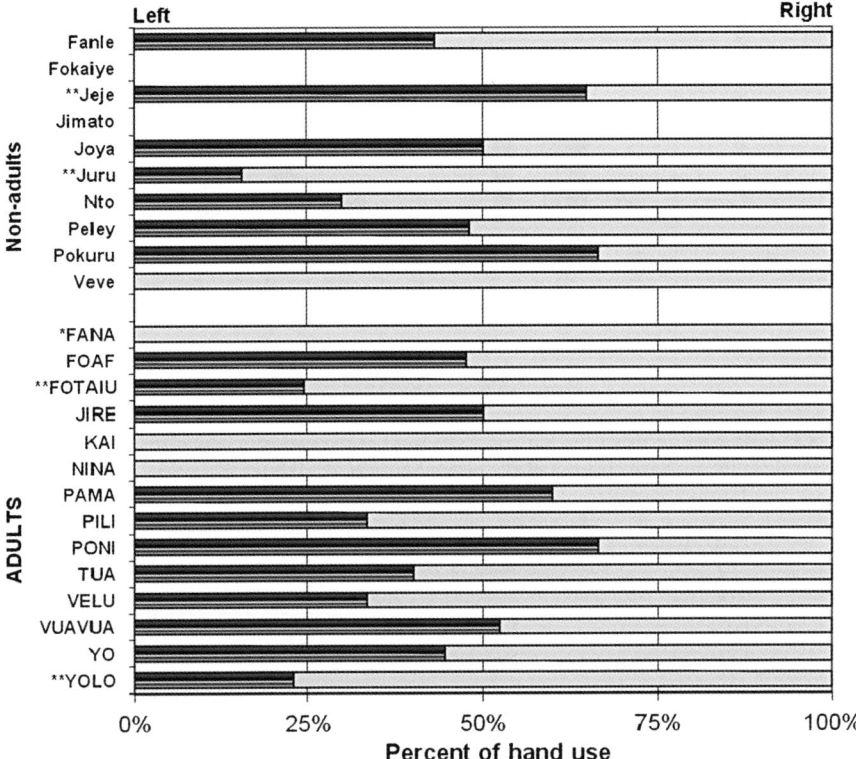

Fig. 8.4 Hand preference during the use of leaves for drinking water. The hand used for the dipping action was recorded for each bout. *All caps*, adults; *lowercase*, non-adults. Statistically significant departures from equal use of both hands: *$P < 0.05$, **$P < 0.01$, two-tailed binomial test (adapted from Sousa et al. 2009)

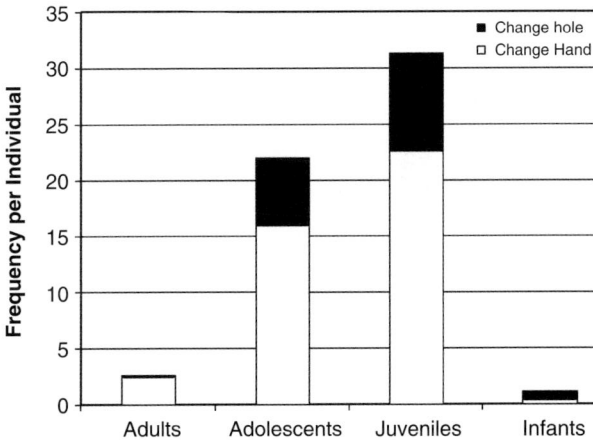

Fig. 8.5 Average number of times individuals in different age classes changed hands and changed hole openings during the use of leaves for drinking water at the outdoor laboratory (adapted from Sousa et al. 2009)

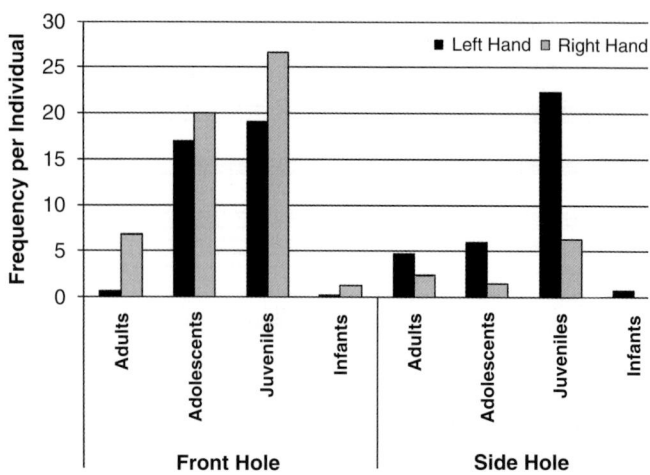

Fig. 8.6 Frequency of using the left and right hand at the two openings of the water hole by individuals in the four age classes (adapted from Sousa et al. 2009)

Table 8.3 Frequency of episodes of observations by chimpanzees

Observer	Performer/model				Total
	Infants	Juveniles	Adolescents	Adults	
Infants	0	3	7	26	36
Juveniles	1	1	5	11	18
Subadults	0	0	0	11	11
Adults	0	0	0	2	2
Total	1	4	12	50	

Fig. 8.7 A chimpanzee observing another chimpanzee drinking water with a tool made of leaves

(Fig. 8.7), collected during January 2000. Infants and juveniles were responsible for 54% and 27%, respectively, of observational episodes. The majority of the individuals observed performing the behavior were adults (75%). In only one episode was an individual observed to watch a younger individual: a juvenile observing an infant.

8.4   Discussion

This chapter provides an overview of the results obtained through the long-term study of a tool-using behavior by the chimpanzees of Bossou. Based on Matsuzawa's (1996) scheme (see also Chap. 21), a way of classifying tool-use according to its complexity, the use of leaves for drinking water is considered "level 1" tool-use. This classification corroborates the fact that young start leaf drinking at the age of 1.5 years, a younger age than a "level 2" tool-use, as less complex behavioral patterns are likely to appear earlier in development.

However, the question arises if we should consider leaf drinking alone, without the process of manufacturing the tool. The use of leaves for drinking water involves two distinct phases, the manufacture of the leaf tool and its use. If we take into account the manufacturing process, which must precede the drinking action, then this full behavior is only performed starting at the age of 3.5 years, when individuals begin to make and use their own tools. At 3.5 years of age, a "level 2" tool-use appears in development, that is, nut-cracking; this would suggest that the combination of tool-making and tool-using phases, which together facilitate the use of leaves for drinking water, considerably increases the cognitive demands of the task.

Although infants start to use leaves for drinking water at the age of 1.5 years, they do not make their own tools but use those made, used, and discarded by other individuals. Sometimes the infants also manufacture a small tool, but they never use it before the age of 3.5 years old, dropping it immediately after making it, and then picking up a leftover one to use.

From 3.5 years of age, they start to use their own tools for drinking water, but are still using the tools discarded by older individuals. The juveniles already know how to make and use the tool, but are still improving their technique, because they still use tools left by other individuals. Juveniles and adolescents also frequently shift hands when dipping for water with a leaf tool. In addition, juveniles often change from one opening of the tree hole to the other, and change hands more frequently after moving to the other opening.

The switching of hands, as well as changing openings of the tree hole during the performance of leaf-dipping actions, also represent important steps toward the learning of the behavior and suggest a trial-and-error process in acquisition by young individuals, the same way that the switching of tools during nut-cracking may represent young chimpanzees' attempts to learn about the properties and efficiency of different tools (Biro et al. 2006).

An analysis of the leaf tools also provides information on the development of the technique. Leaf tools manufactured by juveniles are smaller and thus carry a smaller volume of water per dip than those of adults (Biro et al. 2006), which also may explain why juveniles still use the discarded drinking tools of older individuals.

The use of leaves for drinking water is frequently performed in a social context, providing the youngster with plenty of opportunities to closely observe skilled tool-users, whether the mother or other adult members of the community, performing the behavior. Such observations may be an important part of the learning process.

Chimpanzees tend to observe conspecifics of the same age group or older, but not younger than themselves, the adults being the most popular targets for observation by individuals across all three age classes. Infants and juveniles are the ones observing older individuals more frequently, while adults are the ones observing less. The pattern of conspecific observation is similar to the one registered for nut-cracking by Bossou chimpanzees (Biro et al. 2006).

During the use of leaves for drinking water, chimpanzees exhibit a nonlateralized pattern, in contrast to nut-cracking activity (Biro et al. 2003), but much like ant-fishing (Marchant and McGrew 2007) (see also Chap. 6 for more details on hand use for other tool-use behaviors at Bossou). Marchant and McGrew (2007) showed a positive correlation between the frequency of hand changes and the incidence of major hand support, as ant-fishing is an arboreal activity. Although in our study leaf drinking is not completely arboreal, it is also not completely terrestrial, as the individual might be standing on the ground or hanging from lianas surrounding the tree containing water. Even if adult individuals do not change their hands frequently while dipping for water, as happens in ant-fishing (Marchant and McGrew 2007), their ambilaterality in hand use might be explained by the necessity for adopting different postural positions, depending on the locations of the hole to drink water, and hence using both hands in different situations. Our data on laterality during leaf drinking might thus suggest some adaptation in terms of handedness to postural position, as there was a tendency for individuals to prefer one hand over the other when dipping at one of the two openings of the tree hole.

Acknowledgments The following researchers contributed to the data collection described in this chapter: Tetsuro Matsuzawa, Rikako Tonooka, Dora Biro, Misato Hayashi, and Susana Carvalho. I am also grateful to Yukimaru Sugiyama who began the study of wild chimpanzees at Bossou and to local guides at Bossou who offered invaluable help in the field (Guano Goumy, Tino Camara, Paquilé Chérif, Pascal Goumy, Marcel Doré, Boniface Zogbila, Jiles Doré, and Henry Gbéregbé). The author thanks the Direction National de la Recherche Scientifique et Technologique, République de Guinée, for permission to conduct field work at Bossou. The research was supported by Grants-in-Aid for scientific research from the Ministry of Education, Science, Sports, and Culture of Japan (grants 07102010, 12002009, 10CE2005, and the 21COE program to Tetsuro Matsuzawa).

Chapter 9
Ant-Dipping: How Ants Have Shed Light on Culture

Tatyana Humle

9.1 Ant-Dipping and Culture

Probe-using behavior is one of the most prominent and diversified forms of tool-use among chimpanzees (*Pan troglodytes*) in their natural habitat. Based on data from long-term field sites, stick- or stalk-using for catching social insects on the ground and/or in trees is common to chimpanzees throughout their range with the exception of Budongo, Uganda (Whiten et al. 1999). However, the prevalence of each type of behavior differs by locality, implying cultural differences across chimpanzee communities (Whiten et al. 1999, 2001; Yamakoshi 2001). The ubiquity of stick- or stalk-using behaviors has been linked to the ready availability of diverse materials for tool making and the presence of potential target prey in all habitats in which chimpanzees live (McGrew and Collins 1985; Collins and McGrew 1987). Ant-dipping requires the manufacture and use of a stick or stalk of vegetation as a tool to gather army ants (*Dorylus* spp.). With the tool typically held between the index and middle finger, the chimpanzee performs a back-and-forth movement of the tool to stimulate the ants to attack the tool. Ants that climb the tool are then ingested. The reliance on a tool for ant-dipping has been proposed as a more efficient and less painful strategy for harvesting these biting ants (McGrew 1974).

Ant-dipping is a risky behavior because army ants are highly gregarious and mobile prey and can readily inflict painful bites to chimpanzees. These ants often migrate on the ground or move among low terrestrial herbaceous vegetation in great numbers, up to several million individuals, hunting for prey. They construct tunnel-nests underground that they use as a temporary bivouac. The entrance of the nest is often covered by a layer of fallen leaves and/or soil and can be readily penetrated manually.

Electronic supplementary material The online version of this chapter (doi: 10.1007/978-4-431-53921-6_10) contains supplementary material, which is available to authorized users.

T. Humle (✉)
School of Anthropology and Conservation, The Marlowe Building, University of Kent, Canterbury CT2 7NR, UK
e-mail: t.humle@kent.ac.uk

T. Matsuzawa et al. (eds.), *The Chimpanzees of Bossou and Nimba*,
DOI 10.1007/978-4-431-53921-6_10, © Springer 2011

The early descriptions of ant-dipping emerging from Gombe in Tanzania and Taï in Côte d'Ivoire soon revealed that the chimpanzees at these two sites employ tools of significantly different lengths when dipping (e.g., Boesch and Boesch 1990; Goodall 1986). Tools used by chimpanzees at Gombe [$n=13$; mean$=66$ cm (range, 15–113 cm)] (cf. McGrew 1974) are indeed significantly longer than those used at Taï [$n=35$; mean$=23.9$ cm (range, 11–50 cm)] (cf. Boesch and Boesch 1990). Differences in ant-dipping between Gombe and Taï are not only restricted to tool length but also concern the technique employed in consuming the ants off the tool. At Gombe, chimpanzees use one hand to hold the stick among the attacking ants and, once these have swarmed about halfway up the tool, the chimpanzee usually withdraws the stick and sweeps it through the closed fingers of its free hand, a technique known as *pull-through*. The mass of ants is then rapidly transferred to the mouth and chewed (McGrew 1974). Chimpanzees at Gombe on rare occasions take ants directly from the tool by *direct-mouthing*, that is, by directly pulling the tool sideways through the lips (McGrew 1974). At Taï, on the other hand, the chimpanzee holds the stick among the soldier ants with one hand until they have swarmed about 10 cm up the tool (Boesch 1996a). On withdrawal of the tool, the chimpanzee then typically twists the hand holding the tool and directly nibbles off the ants with the lips, thus always performing a frontal version of direct-mouthing (Yamakoshi and Myowa-Yamakoshi 2004). Because of these differences in ant-dipping technique and tool length between Gombe and Taï, ant-dipping was for a long time cited as one of the best examples of culture in chimpanzees (Boesch and Boesch 1990; McGrew 1992).

At Bossou, Sugiyama (1995b) reported that chimpanzees employ a direct-mouthing technique when dipping for ants, similar to that observed on rare occasions at Gombe. In contrast to the frontal version of direct-mouthing observed at Taï, when employing this technique, Bossou chimpanzees nearly exclusively pull the tool sideways through the lips to remove the ants. Bossou chimpanzees only more rarely perform a frontal version of this technique. Recent observations of ant-dipping from Bossou indicate that some members of the community also occasionally employ another technique, that is, the pull-through technique observed at Gombe (Humle and Matsuzawa 2002; Yamakoshi and Myowa-Yamakoshi 2004). The pull-through technique at Bossou was first noticed in 1997 in a juvenile individual named Fotaiu, aged 6 years at the time; however, we cannot reliably ascertain whether this technique was prevalent within the Bossou community before then.

Several hypotheses have been put forth in explaining the differences in tool length and technique between Bossou, Gombe, and Taï. Because Bossou chimpanzees exhibit both the direct-mouthing and the pull-through technique, this community offered the potential to explore variables that might influence tool length and technique employed by chimpanzees during ant-dipping. Sugiyama (1995b: 203) proposed that differences in ant-dipping techniques, tool length, dipping posture, and material selection may depend on variations in prey characteristics, most particularly the aggressiveness of the prey, *Dorylus* spp., across these different study sites. Hashimoto et al. (2000) further suggested that differences in the length of tools might reflect the difference in techniques used for catching ants. In this chapter, I present evidence that these two hypotheses explain much variation that is observed both within and between communities of chimpanzees.

Nevertheless, I will additionally show that in ant-dipping both microecology and cultural processes intermingle to produce intra- and inter-site variations and that socially biased learning significantly dictates the learning trajectory of young.

9.2 Variations in Ant-Dipping

9.2.1 The "What"?

Army ants are ubiquitous across all sites where chimpanzees have been studied. Sites may exhibit a range of one to six different species of army ants, and wild chimpanzees are known so far to consume a total of 12 species (Schöning et al. 2008). Each species of army ant consumed by chimpanzees can be classed into two different lifestyles (Schöning et al. 2008). The species with an epigaeic lifestyle hunt for animal prey on the ground and in the vegetation and produce conspicuous trails and earth nests. The other species have an intermediate lifestyle; these species hunt in the leaf litter but never in the vegetation. Although they also form conspicuous trails in open areas, their nests are not as readily detectable as those of epigaeic species.

9.2.2 The "Where"?

Army ant consumption has so far been observed at 5 study sites across Africa and reported present at 11 others, while absent at 5 long-term study sites (Schöning et al. 2008) (Fig. 9.1). Chimpanzees may target ants at their nest or at trails (whether foraging or migration trails). At sites where army ants are consumed with tools, dipping context may vary between sites (Fig. 9.2). Although dipping at nests has been confirmed at all sites where ant-dipping has been recorded, dipping at trails has more rarely been reported. Chimpanzees at some sites either never or only rarely dip at trails, or there has been no discovery of tool artifacts to indicate that they do dip at trails.

9.2.3 The "How"?

One striking characteristic of army ant consumption by chimpanzees is that it also exhibits a great deal of variation in how it may be performed. On the one hand, chimpanzees may target the eggs and brood via manual extraction, which involves the insertion of the hand and arm directly into the ants' nest. On the other hand, as discussed previously, chimpanzees may instead harvest the adults with the aid of a tool, that is, ant-dipping. Tool length can vary remarkably between sites, with mean

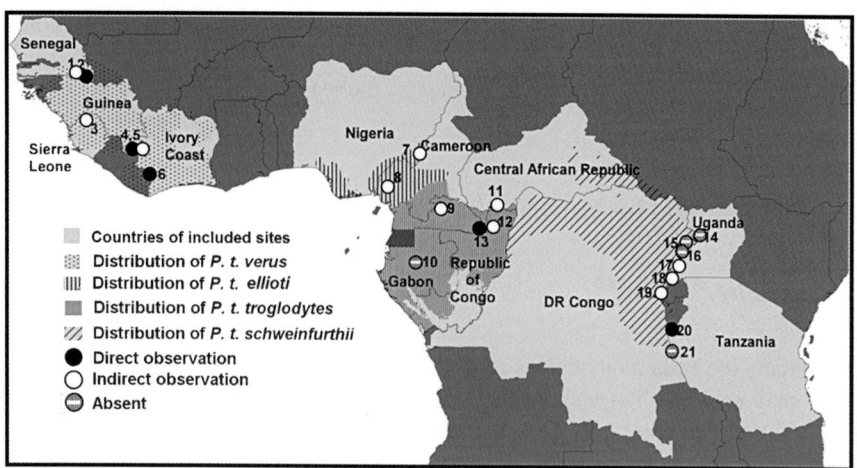

Fig. 9.1 Map of the sites across equatorial Africa where chimpanzees have been observed either directly or indirectly (based on the recovery of tool artifacts) ant-dipping and long-term study sites where ant-dipping has never been reported (adapted from Schöning et al. 2008). Sites: *1*, Assirik, Senegal; *2*, Fongoli, Senegal; *3*, Tenkere, Sierra Leone; *4*, Bossou, Guinea; *5*, Seringbara, Guinea; *6*, Taï, Ivory Coast; *7*, Gashaka, Nigeria; *8*, Ntale, Cameroon; *9*, Dja, Cameroon; *10*, Lopé, Gabon; *11*, Ngotto, Central African Republic; *12*, Ndakan, Republic of Congo; *13*, Goualougo, Republic of Congo; *14*, Budongo, Uganda; *15*, Semliki, Uganda; *16*, Kibale, Uganda; *17*, Kalinzu, Uganda; *18*, Bwindi, Uganda; *19*, Kahuzi-Biega, DR Congo; *20*, Gombe, Tanzania; *21*, Mahale, Tanzania

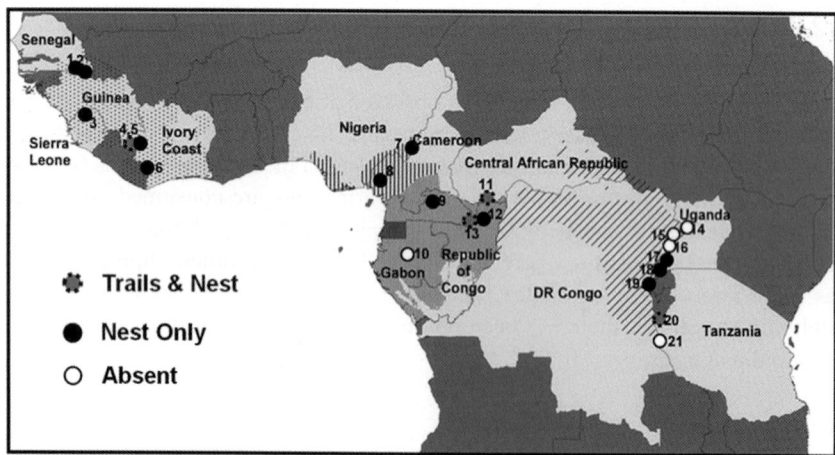

Fig. 9.2 Map of the sites across equatorial Africa where chimpanzees have so far been confirmed to dip for army ants either only at the nest or at both the nest and trails (foraging or migratory) (based on either direct observation or the recovery of tool artifacts). Sites: *1*, Assirik, Senegal; *2*, Fongoli, Senegal; *3*, Tenkere, Sierra Leone; *4*, Bossou, Guinea; *5*, Seringbara, Guinea; *6*, Taï, Ivory Coast; *7*, Gashaka, Nigeria; *8*, Ntale, Cameroon; *9*, Dja, Cameroon; *10*, Lopé, Gabon; *11*, Ngotto, Central African Republic; *12*, Ndakan, Republic of Congo; *13*, Goualougo, Republic of Congo; *14*, Budongo, Uganda; *15*, Semliki, Uganda; *16*, Kibale, Uganda; *17*, Kalinzu, Uganda; *18*, Bwindi, Uganda; *19*, Kahuzi-Biega, DR Congo; *20*, Gombe, Tanzania; *21*, Mahale, Tanzania

site tool lengths ranging from 23.9 to 84.6 cm (Schöning et al. 2008). In addition, there may be significant differences between sites in the relative frequency of the pull-through (McGrew 1974) versus direct-mouthing (including two variants described above, i.e., swiping sideways through teeth or lip and frontal plucking) techniques.

9.3 Microecological Influences on Ant-Dipping

Ant-dipping at Bossou has been studied in detail since 1997 (Humle and Matsuzawa 2002; Yamakoshi and Myowa-Yamakoshi 2004; Humle 2006). Chimpanzees at this site dip on five different species of army ants (three epigaeic species and two inter-mediate species) both at nests and at trails (Humle 2006). Because ants are present in high densities at nests with colonies containing up to nine million individuals, dipping at nests poses a greater risk to the chimpanzees than dipping at trails (Humle 2006). In addition, whether based on the behavior of the ants or determined via a series of human ant-dipping field experiments (Humle and Matsuzawa 2002) or morphological data (cf. Schöning et al. 2008), epigaeic species are also more gregarious and potentially more aggressive than intermediate species. When dip-ping at the nest, Bossou chimpanzees clearly adopt specific behavioral strategies to circumvent these risks, by either (1) positioning themselves more above ground when dipping at nests than at trails or (2) using longer tools, particularly when dealing with ants at nests or with the more aggressive epigaeic species (Humle and Matsuzawa 2002; Humle 2006). However, no significant difference emerged in tool length used between the two lifestyles at trails (Schöning et al. 2008) (Fig. 9.3). In addition, the pull-through technique was almost exclusively associated with tools more than 50 cm long, whereas tools 50 cm long or less were solely associated with direct-mouthing (Fig. 9.4). Both the pull-through and the direct-mouthing tech-niques were observed with the use of tools more than 50 cm long.

9.4 Between-Site Comparison

Schöning et al. (2008) explored the relationship between tool length and technique as a function of prey lifestyle and dipping condition, that is, nest versus trail, across 13 sites across eastern, central, and west Africa (4 where ant-dipping was directly observed and 9 where the behavior was only recorded indirectly). As found at Bossou, epigaeic species at nests were dipped with longer tools, typically associ-ated with the pull-through technique (e.g., Gombe, Tanzania), and intermediate species tended to be dipped with shorter tools, coupled with the direct-mouthing technique (e.g., Taï, Côte d'Ivoire). Nevertheless, several important variations remained that could not be accounted for by microecological variables alone. The most remarkable differences lie between Bossou and Taï, two long-term study sites

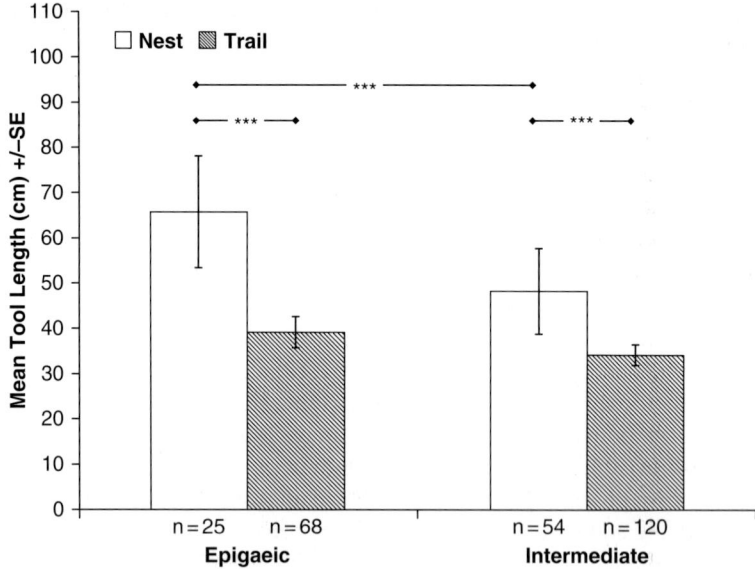

Fig. 9.3 Mean tool length as used by chimpanzees more than 11 years old at Bossou between 2003 and 2006 when ant-dipping on either epigaeic or intermediate army ant species (*Dorylus* spp.) at either nests or trails. ***$P < 0.001$

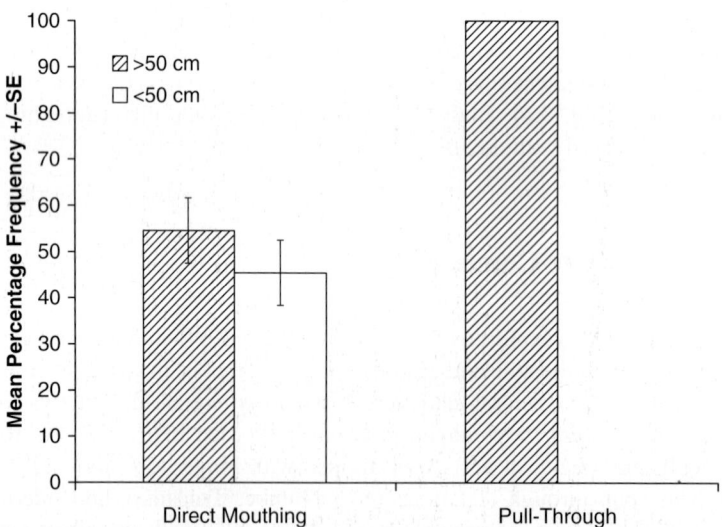

Fig. 9.4 Mean percentage frequency of either the direct-mouthing or pull-through technique employed by Bossou chimpanzees in consuming the ants from the tool when using a tool either 50 cm or less or more than 50 cm long

of chimpanzees where the same five species of *Dorylus* ants that chimpanzees target are available (Schöning et al. 2008). Taï chimpanzees do not dip on epigaeic species at the nest (see Fig. 9.2). In addition, although both Taï and Bossou chimpanzees dip on intermediate species, Taï chimpanzees use significantly shorter tools than Bossou in this context. Taï chimpanzees also do not dip on ants at trails, whereas Bossou chimpanzees do so customarily and preferentially. The only recorded instance of an individual dipping at a trail at Taï was performed by a young female chimpanzee (Boesch, personal communication). Finally, more than 70% of sessions at Taï are dedicated to brood and egg extraction on epigaeic species nests, whereas, at Bossou, brood and egg extraction, although also predominantly focused on epigaeic species, is typically observed in less than 35% of sessions.

Möbius et al. (2008) assessed whether environmental differences in the availability, density, and behavior of the two types of species consumed at Bossou and Taï could account for these differences between the two study sites. We assessed differences in the speed and yield of the ants through a series of human ant-dipping experiments (sensu Humle and Matsuzawa 2002), performed surveys to establish the availability and density of nests and trails of each army ant species at both sites, and tested for differences in accessibility of eggs and brood at nests. We found no significant differences in the availability and density of nests and trails of both intermediate and epigaeic species between the two sites. Although insufficient data for analysis were gathered for intermediate species, the results for epigaeic species showed no significant differences in yield or speed of the ants and brood and eggs accessibility that could satisfactorily explain observed variations in army ant consumption between Bossou and Taï. These results suggested to us that socially biased learning might indeed play an important role in explaining these behavioral variations in army ant consumption between the two sites.

9.5   Development of Ant-Dipping

In a study exploring the role of mothers in the acquisition of ant-dipping among the chimpanzees of Bossou, Humle et al. (2009) confirmed that for the first 5 years of a young chimpanzee's life, mothers were the prime model for their offspring, although accessibility and exposure to other social models increased post weaning (>5 years old). Mothers clearly influenced the learning opportunity of their young who were 5 years old or younger by dipping significantly more often at trails than at nests. Whether intentional or not, all mothers therefore provided their offspring with less hazardous conditions in which to observe and practice ant-dipping, even though dipping at nests would supply a greater yield per unit time than dipping at trails.

In addition, young with higher learning opportunity, as reflected by the mother's percent observation time spent ant-dipping, started observing ant-dipping sooner than young with less learning opportunity. High-opportunity young also first acquired ant-dipping earlier than low-opportunity young. However, we cannot with certainty show that early observation influenced the onset of the behavior. Based on our understanding of the role of observation in the acquisition of other tool-use behaviors during ontogeny in chimpanzees (Lonsdorf 2005; Inoue-Nakamura and Matsuzawa 1997), however, it is likely that observation of ant-dipping was vital in its acquisition, alongside the opportunity to practice the behavior under less risky conditions.

Learning opportunity also influenced dip success and proficiency in young. Young with greater learning opportunity performed fewer dipping errors (dips yielding no ants), especially during their formative years between 2 and 6 years old. Dip duration was used as a measure of proficiency, that is, the number of ants gathered per dip, as during a series of human ant-dipping experiments (Möbius et al. 2008) this measure correlated well with greater yield regardless of the ants' lifestyle or condition. Young between 5 and 10 years old with greater learning opportunity demonstrated longer dip durations compared to low-opportunity young. Finally, mother's proficiency and time spent ant-dipping correlated positively with that of their offspring more than 5 years old, indicating for the first time in chimpanzees a relationship in time spent performing tool-use and competence level between mothers and their progeny (Humle et al. 2009).

However, mothers and offspring did not match in tool length, although there was a trend for high learning opportunity young matching more their mother's tool length than other young. Only a single mother more than 13 years old ever exhibited the pull-through technique when using tools greater than 50 cm long, whereas all young more than 5 years old and adult males exhibited this technique. Young, therefore, did not acquire the pull-through technique by observation of their mother. Young between 5 and 10 years old experienced 61.8% (59/86) of their ant-dipping sessions in the presence of other ant-dipping members of the community than their mother, and, therefore, had ample opportunity to observe others employ the pull-through technique with tools more than 50 cm long (Humle et al. 2009). Therefore, it is possible that this technique is acquired by young through the observation of others.

Although the study of Humle et al. (2009) could not reveal precisely the social learning mechanisms at work, it highlights the importance of socially biased learning opportunities in the acquisition of behavior. It also reveals some behavioral matching related to behavioral competence between mother and offspring. Finally it additionally provides some preliminary support for van Schaik et al. (2003)'s hypothesis that individual differences in competence and time spent in tool-use behaviors among adults mirror the developmental experience of the individual. This hypothesis could also explain the absence of sex differences observed in adulthood (Humle et al. 2009; see Chap. 6).

9.6 Conclusion and Summary

Finally, these studies taken together reveal, as first suggested by Sugiyama (1995b), that chimpanzees flexibly adjust their tool length and technique in response to microecological conditions, as reflected by differences in prey density and/or belligerence, that is, biting risk, between lifestyles and conditions, that is, nest or trail (Humle 2006). Considering the remarkable cognitive abilities of chimpanzees in selecting suitable materials and/or manufacturing appropriate tools for various purposes (Boesch and Boesch 1990), our results therefore do not seem very surprising. However, it is apparent that some variations cannot be explained on that basis, and we therefore cannot rule out that the observed variations in army ant consumption between sites reflect cultural differences among chimpanzee communities.

Indeed, practice and greater exposure to ant-dipping positively influenced the learning trajectory and skill level reached by young, confirming the importance of the education by the master-apprenticeship process in tool-use acquisition by young (Matsuzawa et al. 2001; see Chaps. 18 and 21) and of the socially biased learning opportunities provided especially by the mother during ontogeny.

Clearly, ecological, developmental, and social influences and constraints may intermingle in shaping culture in chimpanzees (Humle et al. 2009). However, this is not a unique feature of chimpanzee culture. Many cultural anthropologist or paleoanthropologists would argue that many aspects of human culture are similarly shaped. The results emanating from studies of army ant consumption within and between study sites illustrate the complexity of these interactions and how various cognitive abilities of chimpanzees may liaise in producing unique cultural community profiles.

Although these interactions render the study of culture in field settings quite complex, they also open up many interesting avenues for further exploration, important in promoting our evolutionary understanding of the origins of culture in humans. The homogeneity highlighted here in army ant consumption by Taï chimpanzees, and the rare observations of the direct-mouthing technique and the prime reliance on the use of long wands and the pull-through technique at Gombe (although Bossou chimpanzees also use the direct-mouthing technique when employing long tools at ants' nests), suggest that group norms likely exist in chimpanzees in the wild, especially when alternative behavioral variants are available. Why some communities exhibit greater homogeneity than others still remains to be investigated. Why do Bossou chimpanzees appear to behave so heterogeneously? Socioecological differences might allow us to elucidate some of these variations.

Acknowledgments I wish to thank the Ministère de l'Enseignement Supérieur et de la Recherche Scientifique, in particular the Direction Nationale de la Recherche Scientifique et Technologique (DNRST) and the Institut de Recherche Environnementale de Bossou (IREB), for granting me the permission to carry out research at Bossou. I am particularly grateful to Tetsuro Matsuzawa, Charles Snowdon, and William McGrew for their advice and support, and to Caspar Schöning, Kathelijne Koops, Gaku Ohashi, Gen Yamakoshi, Yasmin Möbius, Christophe Boesch, and all the local assistants at Bossou for their invaluable contributions and collaboration. Finally, I would like to acknowledge the financial support of the Ministry of Education, Science, and Culture, Japan (nos. 07102010, 12002009, and 10CE2005 to T. Matsuzawa), a Leakey Foundation Grant, and an NIH Kirschstein-NRSA Postdoctoral Fellowship (no. MH068906-01) to T.H.

Chapter 10
Pestle-Pounding Behavior: The Key to the Coexistence of Humans and Chimpanzees

Gen Yamakoshi

10.1 Introduction

Cultural behaviors among wild chimpanzees (*Pan troglodytes*) have been studied extensively at several long-term research sites across tropical Africa (McGrew 1992; Whiten et al. 1999). Researchers continue to observe new behaviors (e.g., Pruetz and Bertolani 2007; Yamamoto et al. 2008). Some behaviors are observed widely across subspecies and throughout ecoregions: for example, ant-dipping behavior (Schöning et al. 2008). Other behaviors appear to have a limited regional distribution: use of large pounding clubs to open a beehive for honey has only been reported in the central African region (Sanz and Morgan 2007), and nut-cracking behavior appears to be confined to an area west of the Sassandra River (Boesch et al. 1994). The latter finding is currently in dispute, based on new findings in Cameroon, located about 2,000 km east of the Sassandra River (Morgan and Abwe 2006). These geographic distributions may be affected by multiple factors, such as the innovation, diffusion, and transmission of new behavioral patterns, the frequency and efficiency of performance and learning of the behaviors, and various ecological factors (Whiten et al. 2001).

Pestle-pounding behavior by chimpanzees (*P. t. verus*) at Bossou, Guinea (Sugiyama 1994c) is a unique tool-using behavior in terms of its local distribution. To date, this behavior has only been observed at a single research site. Although rare, the behavior is ecologically important and is therefore performed quite frequently (Yamakoshi 1998). The target is the pith of the oil palm (*Elaeis guineensis*), and tool materials used are also taken from the oil-palm tree (Yamakoshi and Sugiyama 1995). The tree is distributed widely in the West African forest environment (the so-called palm belt; see Hartley 1988: 6) arguably as a result of local farming practices (Andah 1993). Clarifying the ecological background of this behavior, and why it

*Electronic supplementary material   The online version of this chapter (doi: 10.1007/978-4-431-53921-6_11) contains supplementary material, which is available to authorized users.

G. Yamakoshi (✉)
Graduate School of Asian and African Area Studies, Kyoto University,
46 Shimoadachi-cho, Yoshida, Sakyo-ku, Kyoto 606-8501, Japan
e-mail: yamakoshi@jambo.africa.kyoto-u.ac.jp

developed only at Bossou, should help in an examination of the complex relationship between geography and chimpanzee culture.

10.2 Description of Pestle-Pounding Behavior

The entire behavior consists of two discrete processes: the first does not require tools, but the second does. The behavior typically begins when a chimpanzee enters the crown of an oil-palm tree. The chimpanzee then tries to push the radiating mature leaves down to expose the top of the trunk from which a new shoot is growing. Next, it grasps a bunch of young shoots and pulls it out with force. These shoots are difficult to remove, so the chimpanzee often fails even after several successive pulling trials and must rest before attempting it again. When successful, it turns the bundle of shoots over and bites the base with a crunching sound (Fig. 10.1). The chimpanzee then discards the shoots and may repeat the process. This circuit of procedural components does not require tools and will be referred to as petiole-feeding behavior, as opposed to the following process, which is termed pestle-pounding behavior (Humle and Matsuzawa 2004).

The next process typically begins after the chimpanzee has spent some time engaged in petiole feeding, after the top of the trunk is sufficiently exposed and some shoots have been pulled out, leaving a vertical cylinder-shaped hole. Some of the shoots bitten and abandoned during the petiole feeding fall to the ground, but often some are left on the horizontally radiating leaves. The chimpanzee picks up an abandoned shoot, inserts it into the vertical hole, and then pounds several times, deepening the hole (Fig. 10.2). After this pounding, the petiole is withdrawn and the chimpanzee licks its basal end, to which juice and white fibrous matter is attached. It then inserts its arm, often up to the shoulder, into the hole to extract white matter to eat.

Fig. 10.1 An adult female chimpanzee feeding on oil-palm petiole (photograph by Gen Yamakoshi)

Fig. 10.2   A juvenile chimpanzee holding a pestle tool (photograph by Gen Yamakoshi)

The petiole-feeding process is considered to be independent from the pestle-pounding process because nearly half of all petiole-feeding events end without the chimpanzee proceeding to the pestle-pounding process. However, the petiole-feeding process is indispensable to the pestle-pounding process, with the exception of cases in which a chimpanzee reuses another individual's abandoned workshop. On average, an entire sequence, including both petiole feeding and pestle pounding, takes 25.7 min, and includes 4.1 rounds of the pestle-pounding circuit (pounding–licking–extracting), with 10.0 pounding acts in each round (Yamakoshi and Sugiyama 1995).

The size of pestle tools varies according to the maturity of shoots, but the tool is generally around 1–3 m long (Yamakoshi and Sugiyama 1995). It is certainly one of the largest and heaviest tools used by chimpanzees.

Bossou chimpanzees use one or both hands to maneuver these large tools; preference varies by individual. Handedness or lateralization in hand use is less clear during pestle pounding than other tool-using behaviors (such as nut-cracking), probably because of the fatigue caused by manipulating such heavy tools (Humle and Matsuzawa 2009).

Pestle-pounding behavior is certainly a habitual behavior at Bossou. Of the 18 individuals present during a 1995 research period, 3 were infants, less than 3 years old, who were considered too young to perform the behavior. Of the remaining 15 individuals, I observed 13 engaged in pestle-pounding behavior. The two non-performers were a mother–infant pair (Yo and Yolo), who apparently never performed pestle pounding after my observation (Humle and Matsuzawa 2009), although both were skillful in ant-dipping and nut-cracking.

In contrast, all individuals older than 8 years engaged in the petiole-feeding process, including Yo, who never progressed to pestle pounding. The juvenile chimpanzees around the age of 4 had already started to pestle-pound, but no individual of this age was able to engage in petiole feeding. The task, particularly pulling out the tight shoots, appeared to be too difficult for these young individuals. Typically, these youngsters would wait for adults (usually their mother) to complete the entire petiole-feeding and pestle-pounding process and leave the crown. Because they must follow their mother, the youngsters had little time to spare, so they would quickly take over the area, pick up an abandoned tool, pound, and eat for as long as possible.

10.3   Ecology of Pestle Pounding

The oil palm is the single most important food species for Bossou chimpanzees (Yamakoshi 1998). The chimpanzees consume its ripe fruits both by swallowing whole seeds and also by wadging. They also eat the kernel, using two stones as a hammer and an anvil to crack the hard shell of the nut (Sugiyama and Koman 1979b; see Chaps. 7 and 18). During petiole feeding, they consume the bases of young shoots, and during pestle pounding, they consume the palm heart, which is deeper in the trunk. In addition to these major uses, they also sometimes exploit the palm's dead trunk, eating decayed fiber, and often finding the prized delicacy of fat beetle larvae of the red palm weevil (*Rhynchophorus ferrugineous*) (Fig. 10.3).

In 1995, the Bossou chimpanzees spent 9.6% of their annual feeding time engaged in petiole feeding and pestle pounding. They spent 6.3% of their feeding time engaged in nut-cracking and 1.1% in eating ripe fruit pulp. Therefore, these processes accounted for 17% of their total annual feeding time (Yamakoshi 1998). The oil palm is important to the ecology of Bossou chimpanzees, not only because it is consumed in such large quantities, but also because it is primarily used during food shortages. Most of the fruit foods in the Bossou forest are highly seasonal, producing fruit only from January through April. Thus, the Bossou chimpanzees face a lean period for several months every year, when almost no fruit is available; it is during this period that petiole feeding and pestle pounding are most likely to be observed (Yamakoshi 1998). Thus, the oil palm serves as a fallback food (Marshall and Wrangham 2007), providing qualitatively and quantitatively sufficient food for the chimpanzee community.

In terms of landscape management, the oil palm is a key species for human–chimpanzee coexistence at Bossou. When scientists first came to the village during the colonial era, they often described it as a site of harmonious coexistence between humans and chimpanzees (see Chap. 4). How was it possible for a group of wild chimpanzees to live in a rural, agricultural village landscape? It turned out that the two species (humans and chimpanzees) used a "traditional buffer zone system" to maintain a level of segregation from each other (Yamakoshi 1999).

The Bossou people (of the Manon ethnic group) are agricultural and live on cassava and upland rice, produced using a traditional shifting cultivation system. Thus, the village environment is a mosaic of the settlement area, active cultivated fields, fallow bush in various stages of regeneration, and patches of forest that often have religious importance. The Bossou chimpanzees regularly stay in these sacred groves, which are often situated at a hilltop or a riverside. This habitat is a kind of a refuge for the chimpanzees, because almost no human activity (or even presence) is allowed in the area. The groves are surrounded by fallow bush

Fig. 10.3 An adult male chimpanzee (*above*) inspecting a hole on a dead trunk of an oil-palm tree (photograph by Gen Yamakoshi)

at various stages, where agricultural activities have been temporarily suspended and villagers may use the area to some degree for gathering useful plants and animals. The chimpanzees have relative freedom to use the fallow bush, where they find various kinds of important foods, including oil palm (Yamakoshi 1999). The oil palm is superabundant in both active fields and fallow bush because farmers do not cut down these useful species during the slashing process of shifting cultivation (Fig. 10.4). The oil palm is resistant to fire (Swaine 1992) and grows well under conditions of intense cultivation in West Africa (Andah 1993). Thus, the fallow bush constitutes an important part of the chimpanzee habitat: they can forage with minimal human interference and find plenty of oil-palm trees, the most important fallback food during the annual lean period. Additionally, local people are very tolerant about the chimpanzees using oil palms in the fallow bush. This tolerance may be related to the fact that the tree is superabundant, or because the villagers consider oil-palm trees to be common property; everyone is free to

Fig. 10.4 An active cassava field in Bossou village. Oil-palm trees are left uncut (photograph by Gen Yamakoshi)

harvest fruits or leaves as construction materials. They may extend these rules to the chimpanzees (Yamakoshi 1999).

The geographic segregation achieved between humans and chimpanzees at Bossou appears to fit nicely with UNESCO's biosphere reserve model (Batisse 1982). According to this model, it is advisable to set buffer zones of less-intense human activities around a core area where most human activity is inhibited. Bossou village has a deliberately arranged local design, which may be considered a traditional buffer zone system. The system was, no doubt, based on the local human culture in palm-belt areas of West Africa (Hartley 1988). On the other hand, Bossou chimpanzees actively make use of the rich oil-palm habitat using their own tool-using "culture," by nut-cracking and pestle pounding. The peaceful coexistence between humans and chimpanzees is made possible by the dynamic interaction between the two cultures (Yamakoshi 1999).

As already discussed, oil palms certainly benefit the Bossou chimpanzees, but the reverse may not be true. Pestle pounding may damage the tree because important parts of the plant are eaten, including young shoots and leaves, and more importantly, the heart of the palm. In 2004, Humle observed 127 oil-palm trees at Bossou and the same number at Yealé, Côte d'Ivoire. Of the entire sample, 19 trees were subjected to petiole feeding and 22 were subjected to pestle pounding. Only four trees died during the study period, and none of these deaths was attributed to damage from chimpanzees. On average, new leaves emerged 3.0 and 3.3 months after disturbances from petiole feeding and pestle pounding, respectively (Humle and Matsuzawa 2004). Thus, petiole feeding and pestle pounding by Bossou chimpanzee do not appear to cause the immediate deaths of the target palm trees. However, any long-term effects remain unknown (Fig. 10.5).

Fig. 10.5 Newly sprouting young shoots some months after damage from petiole feeding (photograph by Gen Yamakoshi)

10.4   Possible Scenario of the Origin of Pestle Pounding

Pestle-pounding behavior by chimpanzees was first observed by Dr. Y. Sugiyama on January 7, 1990, at Bossou (Sugiyama 1994c). Soon after this discovery, more detailed accounts were published, including 16 more observations from 1990 to 1994 (Yamakoshi and Sugiyama 1995). Sugiyama's discovery of pestle pounding occurred after he had conducted 14 years of intensive initial research, which yielded various discoveries: nut-cracking (first observed on November 29, 1976; Sugiyama 1990), leaf-clipping (December 1976; Sugiyama 1981b: 172), use of leaves to drink water (May 17, 1977; Sugiyama and Koman 1979b), meat eating (January 18, 1980; Sugiyama 1981a), and ant-dipping (October 28, 1987; Sugiyama et al. 1988) (see Chap. 6). The first sighting of pestle-pounding behavior was followed by only one new discovery of major cultural behavior in the Bossou area: algae scooping (August 28, 1995; Matsuzawa et al. 1996).

The relatively lateness of the discovery of pestle-pounding behavior is strange, considering that the behavior is now the most frequent tool-using behavior observed at Bossou (Yamakoshi 1998); it is also quite conspicuous, with a great deal of noise and tool movement occurring at the top of the palm trees. In contrast, petiole feeding was observed from the very onset of Sugiyama's research (December 12, 1976; Sugiyama 1981b: 174–175). This circumstance may indicate that pestle-pounding behavior is a recent innovation, which commenced only a few years before 1990, and developed from the well-established petiole-feeding tradition (Yamakoshi and Sugiyama 1995). However, it is also possible that the initial research was insufficient to document the entire range of major behavioral patterns over all seasons. Sugiyama and colleagues (T. Matsuzawa and O. Sakura) spent a cumulative total of 34 in situ research-presence months across five different expeditions from 1976 through 1988, but these visits were biased seasonally; most took place from December to March when relatively little pestle pounding occurs, and no visits were made in the "high season" of June–August. That the behavior appeared to be widespread and habitual in 1994 may support the latter possibility: that both petiole feeding and pestle pounding have long been a part of the cultural repertoire of Bossou chimpanzees.

It is premature to try to trace the origin and establishment of this complex palm-feeding culture observed at Bossou. However, it seems plausible and logical that petiole feeding preceded the origin of pestle pounding. Researchers have confirmed that chimpanzees in several communities undergoing long-term research eat the petiole of the oil-palm frond; these areas include Taï, Côte d'Ivoire, Lopé, Gabon, and Gombe, Tanzania (see Humle and Matsuzawa 2004). Other studies from the western part of the Republic of Guinea and Sierra Leone have also confirmed petiole feeding by chimpanzees (de Bournonville 1967; Sept and Brooks 1994; Leciak et al. 2005;

Fig. 10.6 A female Bossou chimpanzee making a nest in an oil-palm tree (photograph by Gen Yamakoshi)

Leblan 2008), although the details of this behavior are often not available. Interestingly, the geographic areas in which this behavior has been documented are almost identical to the areas in which chimpanzees use the oil palm as a nest tree (de Bournonville 1967; Barnett and Prangley 1996; Gippoliti and Dell'Omo 1996; Leciak et al. 2005; Leblan 2008). Based on this regionally established "oil-palm cultural complex" of petiole feeding and nest building (Fig. 10.6), diffused around the western edge of the palm belt (Hartley 1988), it is quite likely that pestle pounding was established at Bossou after a single innovation.

Further studies will be needed to clarify the historical and ecological interactions between the palm cultures of humans and chimpanzees in West Africa, where a large number of chimpanzee populations live outside protected areas and coexist with local people (Butynski 2003). We have just begun to ask how this coexistence originated historically, and what kinds of social and ecological mechanisms are involved.

Acknowledgments This work was financially supported by KAKENHI (Grant-in-Aid for International Scientific Research Program) to Y. Sugiyama (0441066, 07041135), KAKENHI [Grant-in-Aid for Young Scientists (B)] to the author (15710182), KAKENHI [Grant-in-Aid for Young Scientists (A)] to the author (18681036), KAKENHI [Grant-in-Aid for Scientific Research (A)] to S. Kobayashi (16252004), KAKENHI [Grant-in-Aid for Scientific Research (A)] to M. Ichikawa (17251002), Research Grant from Mitsui & Co., Ltd. Environment Fund (Soil and forests: 2007–) to S. Kobayashi and JSPS-GCOE (E04: In Search of Sustainable Humanosphere in Asia and Africa). I thank la Direction Nationale de la Recherche Scientifique et Technologique of Republic of Guinea and l'Institut de Recherche Environnementale de Bossou, Republic of Guinea, for their permissions for my research. I am grateful to Drs. Y Sugiyama, T. Matsuzawa, and T. Humle, all my colleagues of the Bossou chimpanzee research project, and the people of the Embassy of Japan in Guinea for their cooperation during my fieldwork.

Chapter 11
Algae Scooping Remains a Puzzle

Tatyana Humle, Gen Yamakoshi, and Tetsuro Matsuzawa

11.1 Introduction

Along with pestle pounding and leaf folding for drinking (see Chap. 10), algae scooping is a tool-use signature marker of the Bossou chimpanzee community, as it is unique to this community and has never been observed at any other chimpanzee field site in Africa, although the species of algae targeted and consumed, *Spirogyra* sp., occurs elsewhere. For example, *Spirogyra* sp. occurs at Mahale, Tanzania (Nishida, personal communication), and a young adult female, migrant into the Mahale M group, was observed feeding on algae by hand without the use of a tool (Sakamati 1998).

Spirogyra sp. is a widespread free-floating species of filamentous algae belonging to the division of eukaryotic algae known as Chlorophyta, that is, the green algae. Also known as water-silk, mermaids' tresses, or pond scum, *Spirogyra* grows to such great numbers that it forms a thick scum at the surface of ponds, as well as streams and lakes (van den Hoek et al. 1995).

Algae scooping is customarily performed by all able-bodied members of the Bossou community. Algae scooping was observed for the first time at Bossou in 1995 during the rainy season when algae including *Spirogyra* sp. thrive at the surface of ponds, primarily located at the edges of rice paddy fields (Matsuzawa et al. 1996). Before 1995, research at Bossou primarily took place during dry season months (November to March) (see Chap. 4). This bias may explain why this tool-use behavior had previously never been recorded as these ponds are often dried up during the dry season. However, it is also possible that algae scooping represents a recent innovation that diffused via social learning to other members of the community.

T. Humle (✉)
School of Anthropology and Conservation, The Marlowe Building, University of Kent,
Canterbury CT2 7NR, UK
e-mail: t.humle@kent.ac.uk

G. Yamakoshi
Graduate School of Asian and African Area Studies, Kyoto University, 46 Shimoadachi-cho,
Yoshida, Sakyo-ku, Kyoto 606-8501, Japan

T. Matsuzawa
Primate Research Institute, Kyoto University, 41-2 Kanrin, Inuyama, Kyoto 484-8506, Japan

11.2   Description of Algae Scooping

Generally, during algae scooping, the chimpanzee selects a stalk or stick, which he/she breaks off using his/her teeth and modifies in length. Then, he/she half-cups one hand at the stem base and strips the leaves off the tool with a swift, upward motion of the hand. A tool more or less devoid of protruding leaves is thus obtained and used for scooping up the algae from the pond surface (Fig. 11.1). During the manufacture of an algae-scooping tool, three separate components of tool making can thus be discerned: (1) detaching from substrate with teeth or hands; (2) cutting to a specific length; and (3) removing leaves.

Fig. 11.1 Algae-scooping tools from Bossou, illustrating (**a**) the two most common plant species employed (*left*: *Tectaria* sp.; *right*: *Costus afer)* (photograph by Tetsuro Matsuzawa) and (**b**) the two tool types uncovered (first six tools starting from top: "Smooth:; tool situated below: "Hooked") (photograph by Tatyana Humle)

Fig. 11.2 Algae scooping performed by a Bossou chimpanzee: (**a**) the chimpanzee scoops the algae from the surface of the pond; (**b**) the chimpanzee ingests the recovered algae by wiping the length of the tool through its mouth and lips (photograph by Tatyana Humle)

The tool is then most frequently held between the thumb and the index finger (Fig. 11.2), although on occasion the chimpanzees will also grip the tool between the middle and index fingers. Both types of grip are also often observed during ant-dipping (see Chap. 9 for more details on ant-dipping). The stalk is then inserted, distal end first, into the water, and a gentle swiveling action of the wrist usually follows, scooping up the surface algae (Fig. 11.2). The stalk or stick is then brought up to the mouth. Two techniques may be used to remove the algae from the wand for consumption. Most often the proximal end of the tool is first held in the mouth and the algae are licked off over the length of the tool, similarly to direct-mouthing in ant-dipping (see Chap. 9). More rarely, the chimpanzee gathers the algae off the stick using his/her free hand and then licks the algae off his/her hand. On occasion, some chimpanzees bypass the use of a tool and collect the algae from the pond surface directly by hand. However, because the algae are very filamentous and slimy, these latter two techniques appear less efficient than the former. Algae scooping occurs typically either while sitting on the ground leaving the non-dominant hand idle or tripedally, as also observed occasionally during nut-cracking or more commonly during ant-dipping. Nevertheless, some chimpanzees have also been observed algae scooping or feeding manually on algae while standing in the water, thus suggesting that the use of a tool in this context is not necessarily linked to water avoidance. Tools may be discarded after several dips, and a new one is subsequently fashioned or an old one lying nearby is reused. After use, all the tools are left at the site; these are sometimes reused by newcomers.

11.3   Algae-Scooping Tools in Perspective

Algae-scooping tools were measured and raw material identified over five study periods: July–October 1997, July–September 1999, June–September 2000, June–September 2001, and July–September 2005. A total of 131 tools were thus systematically collected immediately after the behavior was observed and after departure of the chimpanzees. Occasionally, algae-scooping tools were found during daily tracking of the chimpanzees or after arriving at a site where chimpanzees had previously been scooping algae. All tools retrieved thus "indirectly" were found in small ponds or within 2 m from the edge of the water surface where *Spirogyra* algae could be found.

The sample of 125 taxonomically identified algae-scooping tools recovered from five ponds at Bossou (Fig. 11.3) comprised 12 different plant species from 10 families (Fig. 11.1, Table 11.1). Five tools could not be identified, and one was labeled according to its Manon vernacular name. *Tectaria* sp. and *Costus afer* represented, on their own, 67.2% of all recovered tools, suggesting a preference for these two species, although their availability was not evaluated and is high across all five algae-scooping sites. Almost 90% of tools were made of herbaceous material; the remaining 10.7% were made of woody material gathered from small trees, shrubs, and vines.

Previous observations of algae scooping have distinguished between two tool types, that is, "smooth" and "hooked" (Matsuzawa et al. 1996). "Hooked" tools are more pliable than "smooth" tools and can also be distinguished from the sturdier "smooth" tools by having small "hooks" projecting along their length, that is, remnants of petiole ends after the leaves had been stripped from the whole length of the tool (Fig. 11.1). Nearly 50% of recovered tools were hooked tools, which were predominantly made of herbaceous or terrestrial herbaceous vegetation (THV) materials. The other half of the tools were smooth tools composed of an admixture of woody as well as herbaceous materials (see Table 11.1).

The mean length of algae-scooping tools was 56.4 ± 15.0 cm ($n = 131$; range, 25–105 cm) and the mean width (measured at half length) was 6.9 ± 4.1 mm ($n = 110$; range, 3–32 mm). Although there was no significant different in tool width

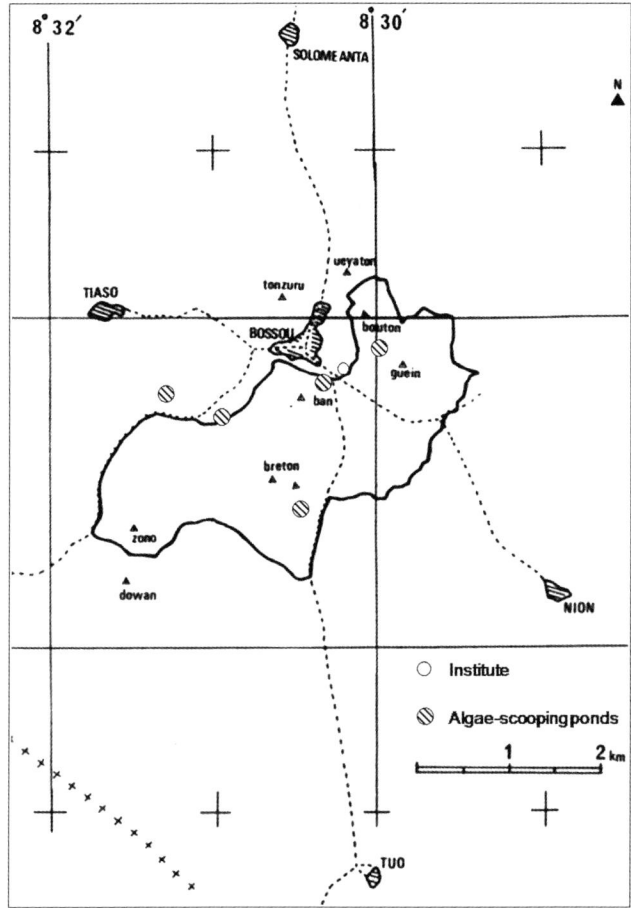

Fig. 11.3 Map of Bossou area with sites where algae scooping has been observed and evidence recorded since 1995

between hooked and smooth tools (hooked: $X \pm SE = 6.1 \pm 5.7$ mm; $n = 43$; smooth: $X \pm SE = 7.0 \pm 2.5$ mm; $n = 67$; independent samples t test: $t_{108} = -1.074$, $P = 0.285$), hooked tools were significantly shorter than smooth tools (hooked: $X \pm SE = 51.8 \pm 15.1$ cm; $n = 57$; smooth: $X \pm SE = 59.9 \pm 14.1$ mm; $n = 74$; independent samples t test: $t_{129} = -3.146$, $P = 0.002$). In addition, hooked tools had a smaller width-to-length ratio than the smooth tools (Mann–Whitney U test: $z = -3.193$, $P = 0.001$). Although not quantified, during the few direct observations of algae scooping, these two tool types appear to be used in different contexts. Indeed, the smooth tools appear to be used when algae is abundant at the pond's surface. Because these tools are longer and sturdier, the chimpanzee can potentially scoop further away from the pond's edge without falling into the water and gather a greater quantity at a time. The more flexible hooked tools are more susceptible to fracture under the weight of the algae when *Spirogyra* is plentiful. However, the

Table 11.1 Plant species used as algae-scooping tools

Species used	Family	Plant type	Tool type	Frequency	Percent
Tectaria sp.	Tectariaceae	Herb	"Hooked"	50	38.2
Costus afer	Zingiberaceae	Herb	"Smooth"	38	29.0
Eupatorium sp.	Compositeae	Herb	"Hooked"	10	7.6
Alchornia cordifolia	Euphorbiaceae	Shrub	"Smooth"	7	5.3
Polypodium aureum	Polypodiaceae	Herb	"Smooth"	7	5.3
Triumfetta sp.	Tiliaceae	Herb	"Smooth"	7	5.3
Aframomum cuspidatum	Zingiberaceae	Herb	"Hooked"	2	1.5
Acacia pennata	Mimosaceae	Vine	"Hooked"	1	0.8
Commelina bengalensis	Commelinaceae	Herb	"Hooked"	1	0.8
Dracaena adamii	Dracaenaceae	Tree	"Smooth"	1	0.8
Sida sp.	Malvaceae	Herb	"Hooked"	1	0.8
Siolongolo (m)		Herb	"Smooth"	1	0.8
Unknown		Tree	"Smooth"	5	3.8
Total				131	100.0

protruding hooks are very useful when finer scooping is required and when smoother, thicker tools may be less appropriate.

11.4 Future Perspectives

The hypothesized pattern of tool selection pending on algae abundance still requires further quantified analysis of video recordings of algae scooping, complemented by direct observation and tool analysis. Different tools should also be tested experimentally for their properties and efficiency. Individual variations in the different algae-feeding techniques described here also should be further explored. The patterns of intracommunity patterns of algae-feeding techniques may correlate with observational learning and proximity between specific individuals during algae-feeding sessions and thus purport a social learning mechanism in their transmission. However, because algae scooping is the least frequently observed customary tool-use behavior at Bossou (see Chap. 6), more empirical data are needed to test the foregoing. Tool selectivity in terms of species choice also needs to be further explored, as plant species availability beside ponds has yet to be evaluated to test whether the chimpanzees are preferentially selecting some species over others. Finally, the reasons for the sex difference in time spent algae scooping over total time observed (males>females) noted in Chap. 6 also require further investigation. An analysis of the nutritional and calorific contents of *Spirogyra* should also be performed to reveal the benefits gained by the chimpanzees.

Acknowledgments We wish to thank the Ministère de l'Enseignement Supérieur et de la Recherche Scientifique, in particular the Direction Nationale de la Recherche Scientifique et Technologique (DNRST) and the Institut de Recherche Environnementale de Bossou (IREB), for granting us over the years the permission to carry out research at Bossou.

Chapter 12
Ant Fishing in Trees: Invention and Modification of a New Tool-Use Behavior

Shinya Yamamoto, Gen Yamakoshi, Tatyana Humle, and Tetsuro Matsuzawa

12.1 Introduction

Chimpanzees (*Pan troglodytes*) use tools habitually. Behavioral differences between communities suggest significant cultural variation (Whiten et al. 1999). Such cultural variation is seen in tool use aimed at catching ants (McGrew 1992; Whiten et al. 1999). Ant dipping, a tool-use behavior for catching army ants (*Dorylus* spp.) on the ground, has been observed at other sites, including Gombe National Park, Tanzania (Goodall 1986) and Taï, Côte d'Ivoire (Boesch and Boesch 1990), but never at Mahale, Tanzania. In contrast, ant fishing, a tool-use behavior for catching carpenter ants (*Camponotus* spp.) in trees, has primarily been observed among the chimpanzees of Mahale (Nishida 1973). At Bossou, Guinea, ant dipping on the ground is customary (Sugiyama et al. 1988; Sugiyama 1995b; Humle and Matsuzawa 2002; Yamakoshi and Myowa-Yamakoshi 2004; see Chap. 9), but we never observed ant fishing before 2003 (Yamakoshi, unpublished data; Yamamoto et al. 2008). Because both army ants and carpenter ants are available across all sites, such variations appear unlikely to be the result of differences in local environmental conditions.

Several studies have investigated differences between field sites (e.g., Whiten et al. 1999; Schöning et al. 2008) and social transmission of tool-use behaviors (in

S. Yamamoto (✉)
Primate Research Institute, Kyoto University, 41-2 Kanrin, Inuyama, Aichi 484-8506, Japan
and
Great Ape Research Institute, Hayashibara Biochemical Laboratories, Inc., 952-2 Nu,
Tamano, Okayama 706-0316, Japan
e-mail: syamamoto@pri.kyoto-u.ac.jp

G. Yamakoshi
Graduate School of Asian and African Area Studies, Kyoto University,
46 Shimoadachi-cho, Yoshida, Sakyo-ku, Kyoto 606-8501, Japan

T. Humle
School of Anthropology and Conservation, The Marlowe Building,
University of Kent, Canterbury CT2 7NR, UK

T. Matsuzawa
Primate Research Institute, Kyoto University, 41-2 Kanrin, Inuyama, Aichi 484-8506, Japan

T. Matsuzawa et al. (eds.), *The Chimpanzees of Bossou and Nimba*,
DOI 10.1007/978-4-431-53921-6_13, © Springer 2011

captivity: Hirata and Celli 2003; Whiten et al. 2005; in the wild: Biro et al. 2003; Humle et al. 2009; Inoue-Nakamura and Matsuzawa 1997; Lonsdorf 2005, 2006; see also Chaps. 8, 9, 16–18, 21). However, we have little knowledge of how such "cultural" tool-use behaviors appear at each site and how these are modified over time. Here we report two cases of ant fishing at Bossou, a tool use never observed before at this site in spite of 27 years of observation. This chapter aims (1) to provide a detailed description of ant fishing by a Bossou chimpanzee and (2) to discuss the process of innovation and modification of a new tool-use behavior.

12.2 Description of Cases of Ant Fishing at Bossou

Two cases of a chimpanzee's ant-fishing behavior were observed in trees at Bossou, first by G.Y. in March 2003, and then by S.Y. in January 2005. The chimpanzee observed fishing ants in both instances was a juvenile male (named Jeje; 7 years 2 months old as of January 2005), and the target ant species was arboreal carpenter ants (*Camponotus brutus*). The behavior in both cases was recorded on videotape, supplemented by direct observation and ad libitum recording of the behavior. Here we describe in detail these two cases of ant fishing, and then compare the tools and ant species characteristics between ant fishing and ant dipping observed in 2003 and 2005 in the same individual.

12.2.1 Ant Fishing: Case 1

The first observation was as follows: on March 6, 2003, G.Y. found Jeje (5 years 4 months old at the time) fishing carpenter ants nesting in a hollow in a tree trunk (*Carapa procera*) 3 m above the ground (Fig. 12.1a). This tool-use session lasted 12 min 59 s. In this case, 14 bouts were observed. A bout was defined as a sequence of behavioral components that begins with the insertion of a tool into the entrance of the ant's nest and ends with either the ingestion of ants (successful bout) or the cancellation of the sequence (unsuccessful bout). Jeje successfully captured and consumed ants in 3 of the 14 bouts observed. In each successful bout, he held the wand with one hand, swept the wand directly with his lips, and consumed approximately three ants. He held the wand with his right hand in 13 of the 14 bouts. In 11 bouts, he held the wand between his thumb and the side of his index finger. In the other three bouts, he held the tool between his palm and fingers. Jeje was observed shaking some ants off the wand once and sweeping them off twice. He was bitten by ants three times. When bitten, the bite seemed painful, and Jeje dropped the wand each time.

12.2.2 Ant Fishing: Case 2

Two years later, on January 4, 2005, S.Y. found Jeje (7 years 2 months old at the time) fishing for carpenter ants nesting in a hollow in the trunk of a tree (*Pseudospondias*

Fig. 12.1 Jeje ant-fishing up in a tree. (**a**) Ant fishing was first observed when Jeje was 5 years 4 months old in 2003 (photograph by Gen Yamakoshi). He used a long rigid tool that is similar to tools used for ant-dipping on the ground. (**b**) Ant fishing with a short tool when Jeje was 7 years 2 months old in 2005 (photograph by Shinya Yamamoto) (reproduced from Yamamoto et al. 2008)

microcarpa) 8 m above ground (Fig. 12.1b). His ant fishing lasted approximately 7 min, and three bouts during 3 min 57 s were video-recorded. Jeje succeeded in one of the three bouts in eating approximately three ants. During his successful bout, Jeje moved the wand back and forth 12 times in the nest and then removed and ate the ants directly with his lips. S.Y. witnessed an additional successful bout before filming the behavior. During this observation, Jeje held the wand between his thumb and the side of his index finger of his left hand. In this case, he was never bitten by the ants.

12.3   Comparison of Tools and Evolution in Ant-Fishing Tools

We collected the tools for ant fishing in the trees used by Jeje after the observation in 2003 (Fig. 12.2) and in 2005 (Fig. 12.3). Jeje was also observed dipping for army ants by T.H. in 2003 and by T.M. in 2005. For obtaining comparable data, we also collected the wands for ant dipping on the ground used by the same subject at the corresponding period. All tools for both ant dipping and ant fishing were rigid and straight and stripped of leaves.

Figure 12.4 shows that the tools used by Jeje at age 7 for ant fishing in the tree were significantly shorter than those he used for ant fishing at age 5 (ant fishing at age 5: $n=5$, mean$\pm$SE$=33.7\pm2.3$ cm; ant fishing at age 7: $n=3$, mean$\pm$SE$=16.4\pm2.3$ cm; t test: $t_6=-4.82$, $P=0.0029$). At this age, he used wands of similar length for both ant fishing in the tree and ant dipping on the ground (ant fishing at age 5: $n=5$, mean$\pm$SE$=33.7\pm2.3$ cm; ant dipping at age 5: $n=9$, mean$\pm$SE$=42.8\pm3.3$ cm; t test: $t_{12}=-1.90$, $P=0.081$). In contrast, the wands used for ant fishing at age 7 were significantly shorter than those he had used for ant dipping during the same period (ant fishing at age 7: $n=3$, mean$\pm$SE$=16.4$ cm±2.3; ant dipping at age 7: $n=4$, mean$\pm$SE$=42.3\pm6.6$ cm: t test: $t_5=-3.20$, $P=0.024$). The wands used for ant fishing at age 7 were also significantly shorter compared to the community average wand length of 50 cm recorded for ant dipping [46.7 cm (Sugiyama 1995b) and 53.7 cm (Humle and Matsuzawa 2002)]. The detailed analysis of the video-recorded ant fishing at age 7 shows that Jeje had shortened at least one of the three wands during use.

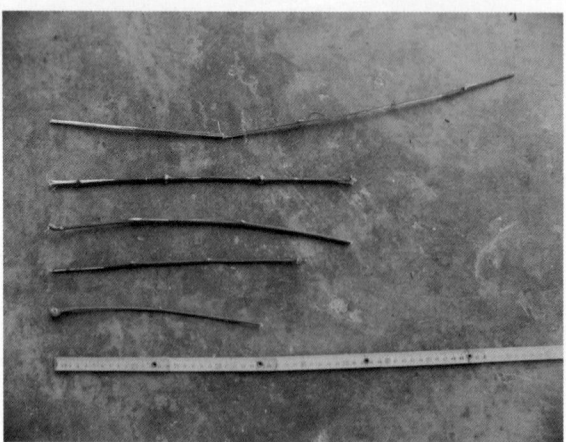

Fig. 12.2 Tools used by Jeje for ant fishing in trees in 2003 (photograph by Gen Yamakoshi)

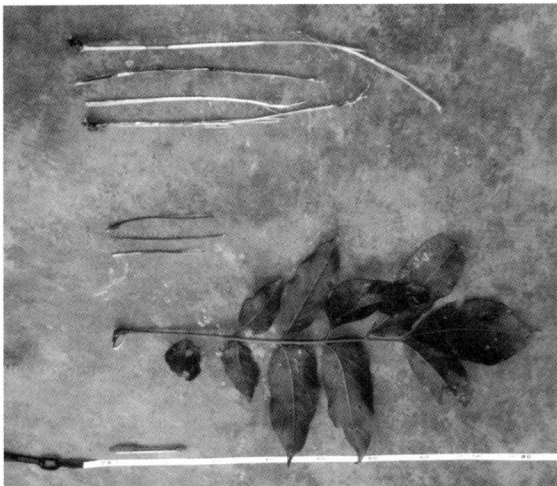

Fig. 12.3 Tools used by Jeje in 2005. The upper four wands (terrestrial herbaceous vegetation) were used for ant dipping on the ground; the middle three wands (leaf stem of *Pseudospondias microcarpa*) were used for ant fishing in trees. The *lower one* is a leaf of *Pseudospondias microcarpa*, which was available at the site where Jeje fished ants in trees (photograph by Shinya Yamamoto)

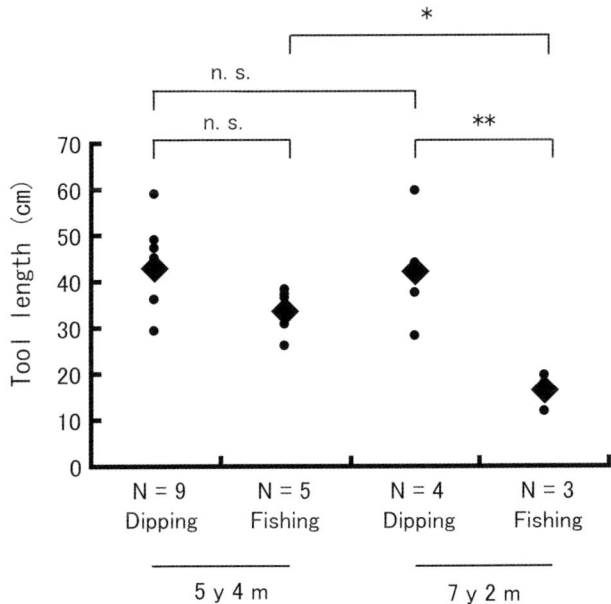

Fig. 12.4 Length of tools used by the chimpanzee (Jeje) for ant fishing in trees (*Fishing*) and ant dipping on the ground (*Dipping*), each at the age of 5 years 4 months (5 y 4 m) and at the age of 7 years 2 months (7 y 2 m). The *small dots* represent each datum point of the tool length; the *large diamonds* represent the mean length. *<0.05; **<0.01; n.s., not significant

12.4 Differences in Gregariousness Between Army Ants and Carpenter Ants

After the observations, we collected the ant species Jeje had fished or dipped for subsequent identification. Samples of ants were identified to the species level by several ant specialists, Dr. B. Taylor and Dr. C. Schöning (for army ants collected in 2003), Mr. B. Bolton (for carpenter ants collected in 2003), and Dr. S. Yamane (for all ants collected in 2005, based on previously identified specimens).

Jeje dipped for army ants (*Dorylus* spp.) on the ground and fished for carpenter ants (*Camponotus brutus*) in trees. Army ants construct underground bivouacs, which can reach up to 1 m in diameter and 0.5 m in depth. They often migrate on the ground or among low terrestrial herbaceous vegetation in great numbers (up to several million individuals) hunting for prey (Gotwald 1972) (see Chap. 9 for more details on army ants). In contrast, carpenter ants (*Camponotus brutus*) nest in the hollow of tree trunks. The nest found in 2005 was 5 cm long, 10 cm wide, and 33 cm deep. The soldier ants of *Dorylus* spp. were on average 15 mm in length compared to 18 mm for *Camponotus brutus*. Both species are very aggressive, and their strong mandibles can cause painful bites to chimpanzees.

In contrast to their comparable aggressiveness, gregariousness significantly differs between the two species. Army ants are so gregarious that they readily attack and cling together on the tool. According to the field experiment conducted by Humle and Matsuzawa (2002), 39–64 ants could be captured per dip as performed by a human experimenter. Sometimes more than 500 ants clung together on a wand. On the other hand, carpenter ants are much less gregarious. The detailed analysis of the videotaped records of ant-fishing in 2003 and 2005 revealed that no more than four ants climbed the length of a tool at a time, and that the ants never clung together.

12.5 Discussion and Implications

In Bossou, before these two cases, we had never observed ant fishing in trees, although we have been carrying out almost year-round research, following the chimpanzees practically daily from nest to nest for 27 years (Sugiyama and Koman 1979a, b; Sugiyama 2004; Matsuzawa 2006a). In addition, we never witnessed the immigration of a chimpanzee into the community during this period. Taken together, these points suggest that Jeje innovated this tool-use behavior.

When Jeje invented this tool-use behavior, he seemed to apply an "ant-dipping" tool and technique for capturing ants in trees; this happened after the subject was already successful at dipping army ants on the ground. The length, as well as the thickness and rigidness, of the wands used by Jeje at age 5 for ant fishing were similar to the wands he used for ant dipping at the same period.

At this age, chimpanzees are highly motivated in using tools and learning and practicing tool-use skills (Biro et al. 2003). Chimpanzees learn various tool-use behaviors when they are young. By the age of 6 years, although less efficient than adults, Bossou chimpanzees are able ant dippers (Humle 2006). As for nut cracking, Bossou chimpanzees ordinarily learn tool use between the ages of 3 and 5 years (Inoue-Nakamura and Matsuzawa 1997; Biro et al. 2003). Jeje was first observed ant dipping on the ground when he was 2 years 8 months old (Humle 2006), and his first observed nut cracking was when he was 7 years old (Biro et al. 2006). In 1995 G.Y. had observed two cases of insertion of wands into carpenters' nests by 8-year-old juvenile chimpanzees (named Pili and Vui, respectively); however, none of these attempts led to the consumption of the ants (Yamakoshi, unpublished data). Young chimpanzees are highly motivated in using tools, especially ubiquitous stick tools. The motivation to use tools may not only encourage young chimpanzees to learn culturally transmitted tool-use behaviors but may also lead them to innovate new tool-use behaviors through individual exploration.

After the invention of this new tool use, Jeje seemed to modify the length of tools. He made and used significantly shorter tools for ant fishing in 2005 than in the first case in 2003. Research was continuous at Bossou; researchers were present at the site for 20 of 22 months between the two observations, and field assistants also entered the forest throughout the term regardless of whether researchers were present or not. Therefore, any other ant-fishing event was either absent or rare. There is thus little possibility that Jeje already used short tools during another ant fishing event in 2003, the year when we first observed his ant fishing.

It is possible that he learned the characteristics of the target ants. The length of wands for ant dipping is significantly influenced by the characteristics of the target ants (Humle and Matsuzawa 2002). There is a significant difference in gregarious-ness between army ants and arboreal carpenter ants. To catch highly gregarious army ants, long and sturdy wands are most suitable, whereas to catch the less gre-garious carpenter ants, shorter and more readily maneuverable wands are more appropriate. Jeje used less suitable long wands at age 5 for ant fishing, when he innovated the behavior, than later at age of 7. Humle et al. (2009) showed that, when ant dipping, young chimpanzees 5 years old or less used shorter wands than when they were more than 5 years old in similar contexts. However, in the case of ant fishing described here, Jeje reduced his wand length between 5 and 7 years of age, indicating that he adjusted and adapted his tool length to this new arboreal context aimed at a less gregarious ant species.

His efficiency in obtaining carpenter ants improved between 5 and 7 years of age. At age 5, with the long wands, Jeje succeeded in eating carpenter ants in 3 of 14 bouts (21.4%) and was bitten by ants three times. At age 7, with the short wands, he succeeded in eating the ants in one of three bouts (33.3%), and he was never bitten by the ants.

This chapter presents the invention and the modification of a new tool-use behavior. At this time, we cannot tell whether this novel tool-use behavior will disappear or spread among other members of the community. So far we have not yet witnessed any other member of the community perform this behavior, which might be because of the rarity of the behavior and the social context. It is unlikely that Jeje performed ant fishing any other time except during these two cases. We can at least be confident that this behavior occurred only extremely infrequently. Moreover, during both observations, Jeje performed ant fishing solitarily. These two aspects may have limited the observation opportunity of this novel tool-use behavior by other community members and may have thus hindered its diffusion to others. From this point of view, we propose that other members may socially learn the behavior if Jeje continues to perform ant fishing and if they have an opportunity for socially biased learning through observation of his behavior. Future observations of ant fishing at Bossou will help us further understand the mechanism of cultural innovation and social propagation in wild chimpanzees.

Acknowledgments This work was financially supported by MEXT and JSPS KAKENHI (#16002001, #20002001 to T.M.; #22800034 to S.Y.) and JSPS-HOPE, JSPS-gCOE (A06, D07), and a fund from the Ministry of the Environment of Japan (F-061). We are grateful to the Direction Nationale de la Recherche Scientifique et Technologique (DNRST), and the Institut de Recherche Environnementale de Bossou (IREB) of the Republic of Guinea for granting us permission to carry out this research and approving. We would also like to thank the people of Bossou, all the guides who helped during this research period, and also Mr. B. Bolton, Dr. C. Schöning, Dr. B. Taylor, and Dr. S. Yamane for identifying the ant samples. Dr. Y. Sugiyama and many other colleagues kindly supported our fieldwork and gave us useful comments.

Chapter 13
Log Doll: Pretence in Wild Chimpanzees

Tetsuro Matsuzawa

13.1 When Jokro was Sick

The episode described below highlights the strong bond between mother and infant chimpanzees. A 2.5-year-old female infant, named Jokro, succumbed in 1992 to flu-like symptoms. Her mother continued to carry her dead infant even after the corpse was completely mummified. This event occurred between January and February during the dry season. I observed the mother–infant pair for 16 days before the death and for 27 days following Jokro's death.

The mother, Jire, was at the time approximately 35 years old. Jokro's older sister, Ja, was 7.5 years old. Their older brother, Jieza, would have been 13 years old then but had disappeared 2 years prior. He may have emigrated to a neighboring community in the Nimba Mountains.

Although Bossou is located only approximately 7.6° north of the equator, it is cool at night (minimum around 14°C), especially during the dry season (November to February). However the temperature may reach 30°C during the daytime. There is therefore a large temperature differential between night and day during this period. During this season, it is also dry and dusty, so that the chimpanzees sometimes catch colds.

Jokro appeared to be affected by a severe cold or flu. She almost ate nothing, and suckled only on occasion. She would often sit alone and look placidly at the sky above. Her sister, Ja, regularly approached Jokro to initiate play, but Jokro declined every time. Once, Jire extended her arm and touched Jokro's forehead, as if to check whether she had any fever.

Jokro gradually weakened and finally collapsed on the ground in front of me. Her sister Ja took Jokro's hand and tried to prop her up. Then, Jire came and picked the infant by the hand and placed her on her back. Jokro was still alive; however, she could no longer sit up by herself. Subsequently, Jire always delicately placed Jokro onto the ground, which was often covered with fallen leaves. Finally, Jokro died during the night of January 24, 1992.

T. Matsuzawa (✉)
Primate Research Institute, Kyoto University, 41-2 Kanrin, Inuyama, Aichi 484-8506, Japan
e-mail: matsuzaw@pri.kyoto-u.ac.jp

T. Matsuzawa et al. (eds.), *The Chimpanzees of Bossou and Nimba*,
DOI 10.1007/978-4-431-53921-6_14, © Springer 2011

13.2   Jokro After Death

On January 25, in the afternoon at 1400 hours, Jire visited the outdoor laboratory for nut-cracking where we usually remain all day. She carried her dead infant on her back. Jokro was no longer clinging to her mother's fur. Jire took Jokro's left hand and braced it between her neck and shoulder. Jire walked away tilting her neck slightly toward the left side, thus carrying the corpse of her dead infant.

Chimpanzees typically build nests in trees at night. Jire made a nest by bending and breaking the branches of a tree. She lay in a supine position and then held the dead infant to her chest. She groomed the infant's face for a while before falling asleep. During this period during the daytime, Jire would wave her hand over the rotting corpse to carefully chase away the flies. The corpse eventually dried up and became mummified.

Because the dead infant no longer suckled, Jire resumed her menstrual cycle and started swelling again. Her body was ready to conceive another baby. Adult males in the community started to court Jire, standing bipedally with an erect penis, and stamping the ground with their heel. Jire began to rest during the daytime with the alpha male of the community while putting aside her mummified infant. However, she would always return to her dead infant after resting. She would then pick up the body, place it on her back, and walk away.

Not all chimpanzee mothers might behave like Jire for several reasons. First, Jokro's death occurred during the dry season so that the rotting corpse failed to decompose and retained its shape, thus providing the mother with opportunity to keep hold of her dead infant for a long period of time. Second, Jokro was 2.5 years old. She therefore had not yet reached weaning age and was still dependent on her mother. Mothers who have infants of this age are typically reluctant to be parted from their offspring for any length of time and are very protective. There may be some additional reasons that accidentally coincided which caused Jire to keep caring for and carrying her dead infant for this long. Although physiologically she was ready to have another baby, she demonstrated a strong bond to her infant that persisted beyond her first menstrual cycle. Her perseverance in carrying the corpse potentially reflected her deep affection for her infant.

13.3   Pretence of Taking Care of a Sick Infant

While Jokro was really ill, I witnessed a very interesting incidence of "pretence." Jire was carrying her sick infant, and moved from one tree to the other. Ja, Jokro's older sister, was following her mother in the canopy. Ja then stopped in a huge *Aningeria* tree, took a dead branch of the tree, and produced a wooden club or rod about 50 cm long and 10 cm in diameter. Ja then placed the club on her shoulder and followed her mother. Ja shifted the club from her shoulder to her armpit to get a better grasp. She then stopped on a big horizontal branch to take a rest, balancing the club on the branch. She slapped it softly with one hand several times, just like when mothers softly slap the back of their infant during play. Ja then moved off through the trees carrying the club for at least 100 m until she was out of sight.

Ja seemed to actually manipulate the club or rod as if it were a log doll (Fig. 13.1). The native Manon people of the village of Bossou actually also have log dolls like these with hair braids appended on top, mimicking hair. Manon girls usually hold these rods on their backs and play with them like dolls (Fig. 13.2). It appeared that Ja was pretending to take care of her sick sister, using a log doll, just as she had witnessed her mother do.

Fig. 13.1 Ja, a 7.5-year-old female chimpanzee, from Bossou, carries a log doll by holding it with her hand and foot. This play behavior can be described as pretence (photograph by Tetsuro Matsuzawa)

Fig. 13.2 A girl from the village of Bossou carrying a wooden doll on her back (photograph by Tetsuro Matsuzawa)

All the details of the episodes of Jokro's death have been described in an article written up in *Pan Africa News* (Matsuzawa 1997b). Most of the episodes were video-recorded and photographed. Some of the episodes were put together to produce a film that aired in Japan in 1992. The foregoing summarized description, however, fails to describe some additional interesting episodes. For example, the alpha male of the community once used Jokro's corpse as a tool for a charging display. A 6.5-year-old male played with the corpse while the mother slept in a tree during the day. The juvenile male took the carcass and climbed up a tree and then dropped it from about 5 m high. Then, he rushed down to the ground to retrieve it, before rushing up again, dropping it, and retrieving it again repeatedly. This play behavior episode was reminiscent of an actual chase–play game often observed between juvenile and infant chimpanzees.

13.4   Two More Cases of Dead-Infant Carrying in 2003

Between November and December 2003, the Bossou chimpanzee community was hit by a flu-like epidemic. Five chimpanzees passed away: two old females (Kai and Nina), a 10-year-old young male (Poni), and two infants (Veve and Jimato). The two cases of the dead infants were very similar to Jokro's first case (Biro et al. 2010)

In the case of the death of Jimato, we witnessed other instances of play with the corpse. A 4-year-old male named Fokaye played with the carcass of Jimato while the mother, Jire, was nut-cracking. The juvenile male dragged the corpse by the hand while running around a tree trunk as if playing chase.

Vuavua, who performed hyrax toying at the age of 9, gave birth to her daughter, Veve, at the age of 10. Veve passed away 2.5 years later. Vuavua continued to carry Veve's corpse after having watched Jire carry her dead infant for a month. Vuavua's carrying persisted for 19 days, which is in clear contrast with her carrying of the hyrax's carcass, which lasted only 1 day. It is clear that Vuavua discriminated between her daughter's corpse and the hyrax's carcass. However, it is difficult to infer anything about death awareness from these simple observations.

13.5   Dead-Infant Carrying: What is Unique in Chimpanzees?

There are several reports of dead-infant carrying in nonhuman primates such as gorillas, chimpanzees, baboons, Japanese monkeys, and rhesus monkeys (for review, see Nakamichi et al. 1996). Among these described instances, dead-infant carrying behavior has been fully analyzed in Japanese monkeys. Sugiyama and his colleagues analyzed more than 6,000 births among the Takasakiyama troop during the course of 24 years. Among the births, about 23% died within the first year of life. Among the dead infants less than 1 year old, about 10% of them were carried by the mother. However in 91.1% of cases, the mothers abandoned their infant's corpse within a week and the majority (67.8%) within 3 days. Maximum carrying duration was 17 days. In total, they recorded 157 cases of dead-infant carrying, suggesting that dead-infant carrying is in fact not uncommon among nonhuman primates.

At Bossou, however, among the three infant deaths witnessed over the course of more than 30 years of research, the mothers carried all three (100%) for several weeks post death. At Bossou, without exception, the mothers carried their dead infants until their corpses were mummified (see Chap. 25 for additional details).

Another important difference between chimpanzees and Japanese macaques, as described in this chapter, is their ability for pretence. We described above the case of Ja, a juvenile female, carrying a log doll, and thus behaving in a similar way as her mother carrying her sick younger sibling. This kind of behavior has never been reported in Japanese monkeys. Pretence, based on imitation, might thus be unique to chimpanzees and absent in monkeys.

Acknowledgments   Thanks are due to Y. Sugiyama for his guidance and to J. Koman, T. Camara, and G. Goumy for their assistance in the field research. The video of Jokro's story (19 min) may be viewed for free at: http://www.pri.kyoto-u.ac.jp/chimp/Bossou/Jokro.html. The English narration was done by Dr. Dora Biro and the French narration was done by Dr. Laura Martinez.

Chapter 14
Animal Toying

Satoshi Hirata and Yuu Mizuno

14.1 Introduction

Researchers have observed chimpanzees preying on mammals in populations throughout Africa, suggesting that chimpanzees (*Pan troglodytes*) regularly eat meat (Uehara 1997). However, only a few episodes of predation on mammals have been observed at Bossou, and in the majority of cases, the prey species was a tree pangolin (*Manis tricuspis*) (Sugiyama 1987, 1989a). Systematic studies at several long-term research sites have revealed that chimpanzee predation differs, for example, in prey selectivity, hunting frequency, cooperative hunting, killing technique, and tendency to share meat (Boesch and Boesch 1989; Uehara 1997). The availability of potential prey may play a role in differing prey selectivity at different locations. However, Boesch and Boesch (1989) reported that chimpanzees at Taï have a highly specialized "prey image" that accounts for the difference in prey species. Therefore, local differences in predation and animal species targeted and recognized as prey may also be influenced by the social tradition of each community (McGrew 1983). Cases of chimpanzees capturing a mammal, but not eating it, contradict the ecological perspective that attribute the occurrence of predation to the availability of prey in an environment, and these cases may support the idea of prey image in chimpanzees. However, only a few reports have described chimpanzees' actions toward a mammal that they have captured and abandoned without eating. This chapter discusses three such cases among the chimpanzees of Bossou (*P. t. verus*) (Hirata et al. 2001).

S. Hirata (✉)
Great Ape Research Institute, Hayashibara Biochemical Laboratories, Inc., 952-2 Nu, Tamano, Okayama 706-0316, Japan
e-mail: hirata@gari.be.to

Y. Mizuno
Department of Children, Faculty of Children Studies, Chubu-Gakuin University,
30-1 Nakaoita, Kakamihara, Gifu 504-0837, Japan
e-mail: yumizuno@chubu-gu.ac.jp

T. Matsuzawa et al. (eds.), *The Chimpanzees of Bossou and Nimba*,
DOI 10.1007/978-4-431-53921-6_15, © Springer 2011

14.2   Case 1

On January 8, 1995, observers heard chimpanzees screaming in a patch of forest on
the southeastern side of the hill of Gban (see Chap. 2 for map indicating location).
A party of chimpanzees were observed in a tree, including Tua (alpha male), Foaf
(beta male), Na (adolescent male), Vui (adolescent male), Kai (adult female), Velu
(adult female), and Vuavua (3-year-old juvenile female). A small Western tree
hyrax (*Dendrohyrax dorsalis*) suddenly fell from the tree, but whether the chimpanzees' initial screams were caused by the hyrax was not clear. Shortly after the hyrax
fell, all the chimpanzees descended from the tree, and the two adolescent males, Na
and Vui, approached the hyrax. Vui slapped the ground near the hyrax with both
hands. Na performed the same behavior, then grabbed a nearby sapling, and bent it
in such a way that the tip of the sapling flailed the ground and hit the hyrax eight
times intermittently. Vui and Na remained near the hyrax and the other chimpanzees observed the scene from a distance for some time, but then started to leave the
area. Soon after, Na retreated from the hyrax, which was still alive, and joined the
other members.

14.3   Case 2

In the evening at around 5 P.M. on January 18, 2000, observers heard a chimpanzee scream from a bush near the top of Gban. A party of chimpanzees was
observed there, including two adult males (Tua and Foaf), five adult females
(Fana, Jire, Kai, Velu, and Yo), one adolescent male (Yolo), two adolescent
females (Vuavua and Fotaiu), one juvenile female (Juru), and two infants (Fanle
and Jeje). Shortly after the scream, Yolo emerged from the bush and climbed a
tree holding a live western tree hyrax in his hand. Juru followed him immediately and remained near Yolo, who swung the hyrax against the tree, beat it
against branches several times, and wandered about the tree with a playful
expression. The adults were not interested in this event and continued to feed on
nearby fruits and leaves. After about 1 min, Yolo dropped the hyrax into the
bush, apparently by accident, and then immediately descended, followed by
Juru. The ensuing sounds from the bush suggested that at least one chimpanzee
was hitting the hyrax. After some time, Vuavua emerged from the bush, hyrax in
hand, and climbed a nearby tree (Fig. 14.1). At this point, the hyrax appeared
dead. Vuavua hit the hyrax repeatedly with her hand and foot, swung it into the
air, and moved about in the tree, carrying the hyrax on her shoulder or in her
groin pocket. Fotaiu and Fanle approached Vuavua and stayed near her for some
time. Vuavua continued to carry the hyrax and made a nest in the tree around
5:50 P.M. She stayed in the nest with the hyrax for about 15 min, but then abandoned this nest and made another nest in another tree, about 30 m from the first
one. Vuavua exhibited several different behaviors toward the hyrax in her nests:
plucking its hair with her mouth (being careful not to bite the skin), slapping it,

Fig. 14.1 Vuavua carrying the hyrax (photograph by Satoshi Hirata)

pressing on it with her hands, poking it with her fingers, grooming it with her fingers and mouth, and raising it into the air with her hands and feet. During this time, Pokuru (3-year-old juvenile male) approached Vuavua and the hyrax and observed them. Vuavua slept with the hyrax in the nest. The next morning, she was observed with the hyrax in a tree near her nest; she was intermittently grooming it with her mouth and fingers (Fig. 14.2). After a while, Pokuru and his mother Pili approached Vuavua and the hyrax, observed them for a short period, and then went away. Vuavua also began to travel with the other members of the community; she carried the hyrax for some distance, but abandoned it about 300 m from the tree where she had nested. Upon examination, the hyrax had several minor lacerations, probably caused when the chimpanzees smashed it on the ground and against tree branches, but nothing indicated that any chimpanzee had tasted its flesh. Throughout this episode, all individuals that had exhibited interest in the hyrax or the holder were younger than 8 years old, with the exception of one young adult female (Pili).

Fig. 14.2 Vuavua grooming the hyrax (photograph by Satoshi Hirata)

14.4   Case 3

On February 8, 2004, Yolo (12-year-old male) captured a live western tree hyrax and carried it up a tree. He beat it against a tree branch several times, holding the hyrax by one of its hindlimbs. He bit off a tiny portion of the hindlimb at intervals. He then climbed down the tree carrying the hyrax, and beat it against the ground. He appeared to be excited; his mouth was wide open as he hit the hyrax with his fists. After a while, he ran off, dragging the weakened hyrax by its hindleg. He stopped at one point, left the hyrax on the ground, and beat it four times with a dry branch that he found nearby. Subsequently he began to run again, dragging the hyrax, but eventually left it on the ground. The hyrax was weak but still alive. No other chimpanzees observed this episode, and no one else touched it. During the episode, Yolo did not bite any vital body parts of the hyrax, such as the neck or head.

14.5   Conclusion

These three cases clearly indicate that Bossou chimpanzees do not regard the tree hyrax as prey, although researchers at Mahale Mountains have observed chimpanzees consuming another species of hyrax (*Heterohyrax brucei*) on two occasions (Nishida and Uehara 1983). Local differences in prey selectivity of chimpanzees might represent social traditions of communities, whereby "prey images" are established and transmitted from one generation to the next. The behaviors exhibited by the adolescent female in case 2 can be classified as iconic play, in the sense that the dead animal was handled like a doll (see Chap. 13). The Bossou chimpanzees in the above-described instances appeared to use the hyrax for play rather than prey.

Acknowledgments We thank the Direction Nationale de la Recherche Scientifique et Technologique (DNRST) and the Institut de Recherche Environnementale de Bossou (IREB) for the permission to carry out research at Bossou. Thanks are also due to all the collaborators during our stay at Bossou. This study was financed by a Grant from the Ministry of Education, Science, Sports, and Culture, Japan (#07041135, 07102010, 12301006, 16002001, and 20680015) and from JSPS-HOPE.

Part IV
Stone-Tool Use: Observation and Experiments

Chapter 15
Extensive Surveys of Chimpanzee Stone Tools: From the Telescope to the Magnifying Glass

Susana Carvalho

15.1 Stone-Tool Surveys in Tropical Forests: A Methodological Challenge

Surveys are procedures used to find and to evaluate sites that can unveil remnants of ancient material cultures in a certain area (e.g., Roskam 2001). A survey is the first stage to any long-term archeological project, that is, an essential preexcavation technique, but surveys can also be used as a primary method to study aspects of the past. Mainly developed by archeologists to detect ancient human signs, a survey can rely on either noninvasive methods, such as surface field walking or ground-based remote sensing, or more invasive techniques, including shovel testing, chemical mapping, coring, and evaluating trenches (Banning 2002). The surface survey, a detailed prospection through field walking (e.g., reconnaissance, spaced at equal distances, transect lines), is one of the most popular and essential techniques, indispensable when research is to begin in any unstudied environment. Moreover, only regional surveys can provide data on the use of landscapes, or on the lifestyles of populations that are dispersed through space (Banning 2002). Thus, surveys present multiple benefits.

Despite these considerable benefits, surveys have not been much extended to tropical rainforest areas, and there are several reasons for this. Tropical rainforest is usually considered a complex and atypical environment in which to carry out archeological work, whether surface surveys or excavations (Renfrew and Bahn 1998; Banning 2002; Mercader 2003; Mercader et al. 2002, 2007). Until recently, "mysterious" tropical rainforests were even considered to be such extreme habitats that traces of human cultures were not likely to be found (see Mercader 2003 for an update on archeological work performed "under the canopy"). It has been argued that the first hominins avoided these risky environments, and Bailey et al. (1989) hypothesized that these areas were colonized only with the advent of agriculture.

S. Carvalho (✉)
Leverhulme Centre for Human Evolutionary Studies, Department of Biological Anthropology,
University of Cambridge, Henry Wellcome Building, Fitzwilliam Street, Cambridge CB2 1QH, UK
e-mail: scr50@cam.ac.uk

T. Matsuzawa et al. (eds.), *The Chimpanzees of Bossou and Nimba*,
DOI 10.1007/978-4-431-53921-6_16, © Springer 2011

However, these arguments are debatable, based on evidence that come from the fossil record: *Australopithecus afarensis* presents a mosaic of anatomical features that includes adaptations to forested habitats (Johanson and Edgar 1996), and *Ardipithecus ramidus* was found in association with flora and fauna remains, also suggesting that the species inhabited forested areas (White et al. 1995). Unfortunately, empirical evidence from forests to either support or refute these arguments is thus far lacking (Mercader 2003).

Tropical soils are acidic and have significant bioturbation (Bown and Kraus 1993; Johnson 2002), and the dynamic processes (Tixier 2000) that motivate erosion, transport, and deposition of sediments seem to develop at faster rates when compared with nontropical environments. All these factors diminish the probability of detecting preserved organic remains. As a result, until very recently, tropical forests were ignored in archeological research.

In addition to the fundamental problems with conducting archaeological work in tropical forests, these environments are methodologically challenging. Methods for surveying, as most archeological techniques, were developed in response to the demands of field conditions in nonforested zones where key sites needed to be investigated on a larger scale (Renfrew and Bahn 1998). Consequently, the methods have not been adjusted to accommodate the special features of tropical rainforests, and the currently existing survey methods are not always suitable to tropical rainforest habitats. Therefore, forested habitats present novel challenges for archaeologists attempting to employ survey methods. Typical problems relate to dense vegetation, leading to no substrate (and sometimes no aerial) visibility, and limited ability to perform field walks in teams spaced at equal distances. In addition, many physical obstacles obscure looking for the remains of the past (Fig. 15.1). However, if survey methods are adjusted to respond to habitat challenges and constraints, reliable and informative surveys can be done in tropical rainforests (Carvalho et al. 2008; Biro et al. 2010).

If few researchers have chosen the rainforest as the place to do archeological work (Sept 1992; Mercader 2003; Mercader et al. 2007), even fewer have carried out archeological surveys at sites used by nonhuman primates with concentrations of stone tools currently in use to crack nuts (Wynn and McGrew 1989; Joulian 1996; Carvalho et al. 2007, 2008). Interestingly, primatologists pioneered work on the spatial analysis of the stone tools used by wild chimpanzees (*Pan troglodytes*), focusing on tool transportation (Boesch and Boesch 1984b; Sakura and Matsuzawa 1991), whereas archeologists have chosen to excavate abandoned nut-cracking sites (Mercader et al. 2002, 2007). Etho-archeology is a recent field of research [see Marchant and McGrew (2005) for a detailed description on interdisciplinary works] that requires the combination of direct and indirect methods for investigating individual behavior (Martin and Bateson 2007; Kühl et al. 2008). Archeological methods and equipment need to take into consideration two uncommon variables: (1) areas of activity to be recorded are still in use, that is, not abandoned; and (2) *tool-users* may enter the site during data collection.

Fig. 15.1 Surveying the forest of Diécké and searching for nut-bearing trees near watercourses (photograph by Henry Gbéregbé)

15.2 Why Survey Chimpanzee Stone Tools?

Surveying chimpanzee nut-cracking sites is a valuable research tool because *Pan* and *Homo* share ancestral traits, and it is reasonable to assume that tool-use was one of the traits of our last common ancestor (LCA) (Panger et al. 2002).

The oldest archeological sites (Semaw et al. 2003; Delagnes and Roche 2005) do not provide hominin fossils in direct association with the assemblages, and the oldest fossils are found mainly in open-air sites that allow for little preservation of the activity areas. Therefore, every new approach that brings new insights into exploitation strategies, use of the territory, raw material procurement, and reuse of sites and tools may open a window into the behavior of our LCA.

The main goal of this research was to investigate typological and technological variation in tool-use between different chimpanzee (*P. t. verus*) communities inhabiting different habitats, in two geographic areas (Bossou and Diécké, Guinea, West Africa), and to explore, with reference to early hominins, issues of diversity and regionalism and the emergence of culture. To find chimpanzee nut-cracking sites was fundamental, both at Bossou and at Diécké. Because part of the study relied upon indirect observation, recording and monitoring of chimpanzee nut-cracking tools depended on the results of these surveys.

In the absence of direct observation of behavior, early human artifacts have been described as reliable indicators of their activities and abilities (Isaac 1986). When surveying current chimpanzee stone tools and nut-cracking sites, it is possible to obtain data, directly and indirectly, on their *activities* and *abilities* (McGrew 1992, 2004). Additionally, by monitoring daily these activity areas, one can conduct fine-detailed spatial analyses and infer strategies for the exploitation of resources or have access to all the displacements of each tool (Carvalho et al. 2008).

15.3 Surveys in the Forest of Bossou and Diécké

15.3.1 Geology

Southern Guinea is, generally, part of the meridional portion of the West African Craton, defined by an underlayer of Precambrian rocks of the Kenema-Man domain and the Paleoproterozoic Birrimian system (Schlüter 2006). This region is characterized by granite-greenstone associations of the West African Archaean, which comprises both gneiss-migmatite basements and intrusive granites (Wright 1985). In the Bossou area, the predominant rocks, such as gneisses, schists, quartzitic schists, ferruginous quartzites, and biotite, are considered to be of Neoarchean age [2,500 million years ago (Ma)–2,800 Ma]. The Diécké area is generally of Paleoproterozoic age (1,600 Ma–2,500 Ma), with granodiorites and granites (see also Chap. 30). During the surveys, I commonly encountered igneous rocks such as dark and white granite (coarse grained), white granodiorite (coarse grained), diorite (medium- to coarse grained), dolerite (medium-grained rock), and gneisses or laterites (Pellant 1992).

15.3.2 Material, Methods, and Subjects

Before the field surveys, and because one of the priorities was to convert all data into digital format, a database was created (Access software) to process Geographical Information System (GIS – GeoMedia software) data. Existent topographic maps were digitized and inserted into the GIS program, allowing for the direct transference of the Global Positioning System (GPS – Garmin Map 60 CS) points to the database. For the archeological and primatological records, I used GPS coordinates

(WGs 84), archeological maps with daily plain views (1:20), transferred to digital drawing with AutoCad software, and photographic (Fuji Digital FinePix S5600) and video recordings of sites (Sony Digital Handicam, DCR-VX 1000 and Sony Digital Handicam, DCR-PC 110). For all nut-cracking sites, Geo references and surveyed areas were then introduced into our GIS database.

I carried out surveys in two areas in southern Guinea, namely Bossou and Diécké. The Bossou survey was carried out between March and May 2006. I conducted an archeological reconnaissance survey of the 5–6 km^2 area to become familiar with the local ecosystem and geology (Banning 2002; Wilkinson 2006). As proximity of oil-palm trees (*Elaeis guineensis*) increased the possibility of finding nut-cracking sites, a selected field walking survey was oriented toward oil-palm tree areas all over the Bossou forest (Carvalho et al. 2008). To have a *representative real sample*, the identified nut-cracking sites were selected from different hills around Bossou, at different altitudes and presenting tools of different raw materials. These selected nut-cracking sites were intensively surveyed to check the tools' spatial distribution (Renfrew and Bahn 1998). During the course of this survey, 13 nut-cracking sites were selected to monitor daily tool movements. However, two of these sites were finally dropped from the sample for reasons of their close proximity to humans who also used them for cracking nuts. The monitoring of nut-cracking sites was based on nonrandom sampling, in which the researcher selects units of study based on the potential productivity of the chosen areas (Roskam 2001). The selected 11 nut-cracking sites were scrutinized intensively and monitored daily to record their use by the chimpanzees (Renfrew and Bahn 1998). Before the first record at each site, a radiocentric census was done in the area around the oil-palm tree (~15-m radius) to verify the existence of tools. This method allowed confirmation of trans-port of new tools to the site area. Cartesian coordinates (X/Y/Z) were given to each site with the aid of GPS. During each monitoring visit, alterations in the position and orientation of the archeological materials were noted and plain views were drawn (a map with all the significant features drawn for each visit of the chimpan-zees to the nut-cracking site).

Because the chimpanzees using these sites could appear at any time, for tool movement recording I had to adapt a method needing a minimum of equipment, so as to be able to leave the site as soon as an individual ape appeared. Therefore, the topographic triangulation technique was applied based on two fixed points: mag-netic north and the site datum point (tree point: site datum point is one precise location in the space that is used as a reference point to record all horizontal and vertical measurements). Finally, I used a string level, a plumb bob, measuring tapes, and a compass (Suunto MC-2, with clinometer) (Fig. 15.2).

Based on indirect evidence, the classification of tool *function* relied on the examination of fresh traces of nut-cracking on the stones (e.g., shell pieces). If the stone was positioned on top of the hard-shelled debris it was considered to be a hammer; if the nuts were on top of the stone it was considered to be an anvil. In the case of wedges, their presence was confirmed only if the stone was placed under the anvil. Flake extraction was registered only if the extracted flake could be refitted to a tool previously registered. Although the presence of other debris around the

Fig. 15.2 Researcher records stone tool transport using the triangulation technique and draws a plain-view map (photograph by Tetsuro Matsuzawa)

Fig. 15.3 Bossou: two examples of different nut-crackers found in situ after being used at the Mobli site (SA 13) (photographs by Susana Carvalho)

area was noted, it was left in the same place for future studies, as this was not the main purpose of the initial study. A sequential number, a function indicator, and an area marker were assigned, with permanent ink, to all the detected tools (e.g., 2 H M, hammer number two from Mobli). This procedure allowed the tools to be recognized if they were transported and later found elsewhere (Fig. 15.3).

I surveyed the Diécké forest between January and December 2006 (see Chap. 30). All the foregoing preparations necessary to properly record tool movements and tool characteristics, including refitting trials, were also done in Diécké, using the same methods as in Bossou. When a nut-cracking site was found I conducted a survey, covering 1 km² around the site, to record the occurrence of raw material sources, water, and chimpanzee shelters (nests or beds).

During our study, the total area of Bossou (Sector A) was the home range of one group with 12 chimpanzees (see Chaps. 1 and 3 for details on group composition). The forest of Diécké is home to an unknown number of chimpanzee communities that have been rarely sighted (see Chap. 30 for details about the research at Diécké).

15.4 Results

After monitoring the 11 nut-cracking sites in the forest of Bossou, tool movements were recorded once at one site (Breton, SA 14) and seven times at another site (Mobli, SA 13). At the latter, I witnessed nut-cracking twice for a total of 132 min (Table 15.1).

All the procedures and methods, such as site preparation for recording tool movements, as well all tool characteristics, including refitting trials, were repeated at all 11 sites. In the future, this will allow a continuation of the study and the enlargement of the database from this area.

Although the results from the stone-tool analysis and their *chaîne opératoire* are reported and discussed in detail elsewhere in this book (see Chap. 7), here I report the results concerning the spatial analyses of the most visited site, Mobli (SA 13). Our spatial analysis indicated the existence of three types of strategies of resource exploitation: (1) most commonly, individuals exploit the nearest resources and optimize the task (nut-cracking occurs under the nut-bearing tree using tools that are next to the nuts); or (2) individuals transport the stone tools

Table 15.1 Summary of nut-cracking behavior and tools used at Bossou forest

Site	Date of visit	No. of hammers	No. of anvils	No. of wedges	No. of unknown	No. of flakes	Direct/indirect observation
Mobli	04.03.2006	Video recording while following the chimpanzees					Dir. (52 m)
Mobli	06.03.2006	5	4	0	0	0	Ind.
Mobli	19.03.2006	5	4	0	0	0	Ind.
Mobli	31.03.2006	8	5	2	0	0	Dir. (80 m)
Mobli	03.04.2006	7	6	1	1	0	Ind.
Breton	04.04.2006	4	5	0	0	2	Ind.
Mobli	08.04.2006	8	5	0	0	0	Ind.
Mobli	28.04.2006	10	7	0	1	1	Ind.
Total		47	36	3	2	3	

Dir. direct, *Ind.* indirect

Table 15.2 Seven nut-cracking sites in Diécké forest

Date	Site	Survey area (Diécké)	Latitude N	Longitude W	Altitude (m)	Nut species	Material and movable vs. nonmovable
22.01.06	SB1	Nonah	07°33′57.4	09°02′07.1	424	*Panda oleosa*	StoneMov. hammer Non-M. anvil
11.03.06	SB2	Nonah	07°33′54.1	09°02′01.7	366	*Panda oleosa*	StoneMov. hammer Non-M. anvil
16.04.06	SB3	Korohouan	07°27′53.7	08°55′40.5	337	*Panda oleosa*	StoneMov. hammer Non-M. anvil
17.04.06	SB4	Korohouan	07°29′04.8	08°54′23.2	386	*Coula edulis*	StoneMov. hammer Non-M. anvil
17.04.06	SB5	Korohouan	07°29′04.4	08°54′22.9	377	*Coula edulis*	StoneMov. hammer Non-M. anvil
19.04.06	SB6	Korohouan (Mont Medou)	07°24′05.5	08°58′55.0	334	*Panda oleosa*	StoneMov. hammer Non-M. anvil
10.12.06	SB7	Nonah	07°34′07.9	09°02′20.8	–	*Panda oleosa*	Stone/woodMov. stone hammer Non-M. root anvil

Mov movable, *Non-M* nonmovable

to another place where nuts are available; or (3) individuals transport both nuts and stones to another place. These three strategies follow the same pattern of those recorded at the outdoor laboratory during experimental sessions of nut-cracking (see Chap. 7), may reflect the optimization of time and management of energetic expenditure, and may be influenced by social dynamics or party composition (Carvalho et al. 2008). Furthermore, introduction of new tools into the monitored nut-cracking site often occurred (see Table 15.1). The Mobli site originally had 9 tools, but by the end of our recording there were 19. This transport is not explained by the lack of raw materials as potential tools are readily available in all areas of Bossou.

During pilot surveys in Diécké, seven nut-cracking sites were recorded (Table 15.2; see also Chap. 30). In between surveys, two of these sites, Diécké SB1 and SB2, were visited by the chimpanzees.

The analyzed nut-cracking site (SB1) was near a *Panda oleosa* tree and near the Lilaya watercourse. Nuts were dispersed around the area, and nut-cracking remnants were concentrated around outcropping anvils (Fig. 15.4). The tools originally present were represented in the plain-view drawing, according to the function recognized for each tool: hammers (H) and anvils (A), showing the movements of the used tools (Fig. 15.5). Diécké surveying is only in its initial phase, and spatial analyses have yet to be carried out.

Fig. 15.4 First nut-cracking site found in Diécké (SB 1). Note the hammer on the *left-hand* side; the anvil is a rock outcrop on the *right-hand* side. Nut remains can be seen on top and around the anvil (photograph by Susana Carvalho)

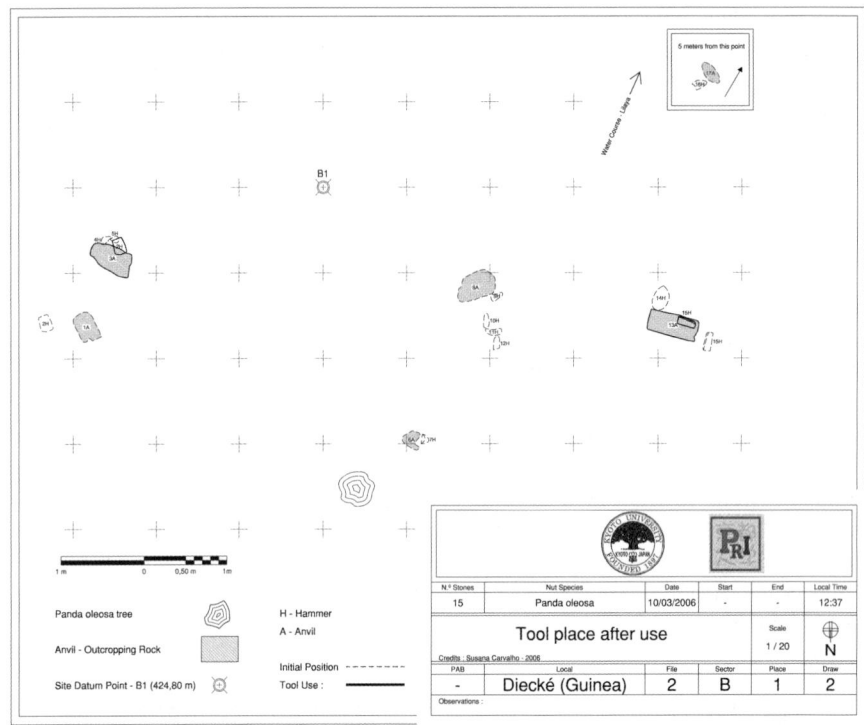

Fig. 15.5 Nut-cracking site at Diécké (SB1) after being visited by the chimpanzees. Original plain-view map was digitally gridded to define sample units on the ground and to allow spatial examination by quadrat. Note the new location of the used tools (drawing by Susana Carvalho and Pedro Gonçalves)

15.5 Discussion

By following each day the tool movements that occur at one site, I uncovered clues to help interpret the widespread frequency of the cumulative palimpsests in the archeological record (Schiffer 1985): "Many archeological deposits are very obvious palimpsests in this sense. The stone tools in a layer of a Paleolithic cave, for example, usually represent the aggregation of many different episodes of knapping, use, and discard that have become compressed into a single layer or surface and cannot be resolved into the individual episodes of activity" (Bailey 2007:7). Archeology has a coarse temporal definition that prevents us from knowing how long one population (or more) made use of a certain place and of certain objects. Because of its natural constraints, archeology examines temporal and spatial variables with accuracy comparable to that of a telescope, whereas an etho-archeological approach functions similar to a magnifying glass. To survey chimpanzee nut-cracking sites in action is, for the archeologist, to unite in practice the anthropology of the object and the method of archeology. This monitoring covers the short-scale hiatus of the actions that happen in each place, to obtain microtemporal frames that are not possible to reach through conventional archeological methods. The researcher is able to record the objects without interfering with the tool-users and to collect several sequences that are repeated over time.

This aspect adds yet another benefit to the etho-archeological method, as these sequences of behavior change the composition of the tool-use scenario but they do not change its structure. Thus, the researcher can analyze the changes in tool-use in light of stable ecological variables present in the home range of the chimpanzees/ tool-users.

Moreover, by taking this approach and observing directly individuals in action at any of the selected sites being monitored, data obtained before, during, and after tool-use can provide additional information on (1) raw material selection and tool function and durability; (2) real frequency of transport (instead of only the last location where the tool was discarded); (3) individual preferences; (4) reutilization of tools by the same or by different individuals; (5) frequency of reuse of sites; (6) synchrony of use of the home range of a group with the "home range" of its specific technology; (7) correlation of the target items consumed with particular tool types if matched to particular food items; and (8) patterns of landscape exploitation and use.

Finally, this novel etho-archeological approach gives the unique opportunity to study nonhuman primate populations in the present, while laying the groundwork for their future study.

Acknowledgments I would like to thank the Direction National de la Recherche Scientifique et Technologique, République de Guinée, and Dr. Kourouma Makan, Director of the Institut de Recherche Environnementale de Bossou for permission to conduct field work at Diécké. I would also like to thank Dr. Papa Cecé Condé from the Centre Forestier de N'Zérékoré, Prof. Tetsuro Matsuzawa and Dr. Werner Grimmelman (Progerfor) for support and materials. The research was supported by Grants-in-Aid for scientific research from the Ministry of Education, Science, Sports, and Culture of Japan: MEXT-16002001, JSPS-HOPE, JSPS-21COE-Kyoto-Biodiversity, and F-06-61 of the Ministry of Environment, Japan, to Tetsuro Matsuzawa. S.C. was supported by Municipality of Leiria, Portugal; Cambridge European Trust (RIB 00107), FCT-Portugal (SFRH/BD/36169/2007), Research Centre for Anthropology and Health (CIAS – University of Coimbra), The Wenner-Gren Foundation, Queens College Cambridge, and the Leakey Trust (UK). I am grateful to B. Zogbila, H. Gbéregbé, J. Doré, P. Goumy, M. Doré, C. Goumy, J.M. Kolié, J. Malamu, L. Tokpa, A. Kbokmo, C. Koti, and O. Mamy for field support. I am also grateful to C. Sousa for guidance during the beginning of the 2006 research; to L. Pinela (GIS configuration and training); and to P. Gonçalves for unconditional support. I thank P. Kelmendi and V. Carvalho for fruitful discussions and S. Koski, T. Humle, and W.C. McGrew for comments on the manuscript.

Chapter 16
Field Experiments of Tool-Use

Tetsuro Matsuzawa

16.1 The Logic Behind Field Experiments

To better understand chimpanzees, I initiated the parallel study of chimpanzees in both the field and the laboratory. Before field experiments at Bossou, fieldwork depended on observations of individuals in their natural habitat during the course of their daily activities, while experiments were primarily conducted in controlled laboratory settings. My research approach has focused on synthesizing these two different approaches (observation and experiment) and these two different study environments (natural habitat and laboratory setting) (Table 16.1). I thus developed two paradigms: (1) *field experiments* in the wild and (2) *participation observation* in the laboratory. In my view, this holistic approach is the most suited in providing us with a truer and deeper understanding of chimpanzees as a whole (Matsuzawa et al. 2006).

Field experiments consist of experimental manipulations in the natural habitat. My idea of running field experiments at Bossou was inspired by the work of pioneering ethologists. For example, Adriaan Kortlandt (1918–2009), a Dutch ethologist, carried out his early studies of chimpanzees in the wild using a field experimental approach. In one of his field experiments, he presented a stuffed leopard with a mobile head to wild chimpanzees (see also Chap. 4). The chimpanzees mobbed it with a stick. Kortlandt filmed the behavior and took still photos for further analysis. This field experiment potentially illustrated how early hominids may have relied on a similar tactic using sticks for defense against predators before inventing spears, bow and arrows, and other weapons. Such kinds of field experiments emerge from the tradition of ethology whose founders include Konrad Lorenz (1903–1989, Austrian ethologist), Nikolaas Tinbergen (1907–1988, Dutch ethologist), and Karl von Frisch (1886–1982, Austrian ethologist), who shared the Nobel Prize in Physiology or Medicine in 1973.

Jane Goodall (1934–), the pioneer of the long-term study of wild chimpanzees, also utilized field experiments during the early years of her research at Gombe in Tanzania

T. Matsuzawa (✉)
Primate Research Institute, Kyoto University, 41-2 Kanrin, Inuyama, Aichi 484-8506, Japan
e-mail: matsuzaw@pri.kyoto-u.ac.jp

T. Matsuzawa et al. (eds.), *The Chimpanzees of Bossou and Nimba*,
DOI 10.1007/978-4-431-53921-6_17, © Springer 2011

158   T. Matsuzawa

Table 16.1 Synthesizing laboratory work and fieldwork

Place	Method	
	Observation	Experiment
Field	Observation	Field experiment
Laboratory	Participation observation	Experiment

Fig. 16.1 Field experiment in the outdoor laboratory known as the "Bureau." The researchers and the local assistants waited for the arrival of the wild chimpanzees so they could display their tool-use skills. The distance between the chimpanzees and the researchers is about 15 m (photograph by Tetsuro Matsuzawa)

(Goodall 1986). For example, she introduced a mirror to explore mirror self-recognition in wild chimpanzees. She also introduced an apparatus for remotely opening the lid of a baited box. In general, many fieldworkers do not necessarily simply rely on following and observing wild chimpanzees. The recent advance of camera trapping equipped with an infrared censor can today also be considered as a sort of field experiment because the researcher carefully sets up the camera to film details of the situation or target behavior without the required presence of human observers.

However, with clear intention and motivation, since 1988, my colleagues and I have been systematically employing field experiments to investigate in detail tool use and tool manufacture in wild chimpanzees (Figs. 16.1 and 16.2). We clearly followed the logic of laboratory experiments as far as we could by paying attention to establish an experimental setup, controlling the situation to facilitate the comparison across different tool-use behaviors in the same location, manipulating the availability of tool materials, and so on. Because no one had previously ever clearly defined this methodology, the term *field experiment* was coined to label our collective effort (Matsuzawa 1994).

Fig. 16.2 Wide-angle view of the outdoor laboratory. A group of chimpanzees are using stones to crack open oil-palm nuts. A chimpanzee is also using leaves for drinking water from the hollow of a tree trunk (photograph by Tetsuro Matsuzawa)

16.2   The Advantage of Field Experiments

Field experiments aim to stimulate the performance of the chimpanzees' natural behavior. It is thus possible to observe and record behavioral details and to collect photographic and video archives. We first started to use battery-operated video cameras in December 1987, which must have been among the first attempts to do so in the wild because SONY first started to produce and commercialize dry-battery-operated video cameras in the same year. We have, ever since, been using and relying on behavioral video-recording of wild chimpanzees at Bossou.

Video-recording clearly presents a huge advantage. In the wild, each single event is unique. Therefore, field observations theoretically consist of a string of anecdotes if a single observer witnessed these events. However, the accumulation of observations allows us to understand the pattern of the behavior. Once the pattern of a chimpanzee's behavior is recognized, one can more readily predict chimpanzee behavior. This kind of prediction is the corroboration of our understanding of the behavior. However, if a behavior is rarely observed, it is difficult to gather any clear insight into the behavior; the single observation in such a case is known as an anecdote.

Fieldwork is filled with anecdotes in which a researcher claims that he or she witnessed a behavior. No one can deny it: the observer claims that he/she saw it. However, such kind of observations cannot strictly be considered scientific. For example, stories about chimpanzees recounted by local people are rich in anecdotes. How can we get rid of this kind of nonscientific description based on human observation? One plausible solution is to use video-recording. Even if it is only a single event, video-recording is the only way to reliably generate scientifically sound data. Video-records can be repeatedly visualized, verified, and shared among people, thus allowing for scientific and public validation of the observation.

In addition, field experiments present three major advantages. First, field experiments dramatically multiply the number of observation opportunities of behaviors of interest. Suppose that you follow wild chimpanzees at Bossou. If you are lucky enough, you may be able to witness stone-tool use. However, you may also fail to observe stone-tool use even if you follow the chimpanzees from dawn to dusk, every day for weeks. By introducing the field experiment of stone-tool use, you can observe the behavior almost every day: one to two times a day on average during the past 22 years. In our recent experience, 3 days suffice to observe stone-tool use in all able nut-crackers of the community, which is in clear contrast with the more traditional approach that relies solely on natural observations. It is extremely challenging to witness nut-cracking of all able community members if you simply follow them day after day.

Second, field experiments allow us to manipulate the availability of tool materials or target food. For example, during nut-cracking of oil-palm nuts, the stones and the nuts are provided by the experimenters, so that you can easily manipulate the availability of those resources. When you limit the number of available stones, you may be able to witness interesting behaviors such as deception. One of the best examples of deception in the wild was video-recorded at Bossou in the outdoor laboratory: a mother named Fana deceived her 9-year-old son, Foaf, to obtain the stones that he was using (Matsuzawa 1999; see the video clips attached to this volume). You may also be able to observe the flexible intelligence of young chimpanzees. For example, a young female chimpanzee, who required a hammer stone, started using a wooden club as a hammer, a type of hammer favored among chimpanzees in the Taï Forest in Côte d'Ivoire. Another young chimpanzee used a large wooden branch as an anvil because she had no access to anvil stones. A young chimpanzee climbed up a tree with a hammer stone and tried to crack open a nut using the tree trunk as an anvil. All these behaviors were observed because we manipulated stone availability.

We also manipulated the target nuts (Biro et al. 2003; see Chap. 17). The species of nuts cracked at Bossou is the oil-palm nut (*Elaeis guineensis*). We also tested different species of nuts: coula nuts (*Coula edulis*) and panda nuts (*Panda oleosa*). Our study revealed that only one female, Yo, knew that coula nuts were edible. The other chimpanzees were clearly unfamiliar with this nut species. However, none of the chimpanzees knew panda nuts. Based on these findings, we hypothesized that immigrant females are vehicles of culture to neighboring communities, potentially generating cultural zones that share common features of tool repertoire.

Third, field experiments allow us to observe different kinds of tool-use behaviors at the same time in the same location (Fig. 16.3). In 1996, we started the field experiment of using leaves for drinking water in the outdoor laboratory that we named the "Bureau" (see Chap. 8). We drilled a hole in a huge tree at the left far side of the Bureau. The experimental manipulation was so simple: adding fresh water. The researchers filled the hole with water in advance, and then waited for the chimpanzees to visit. The chimpanzees arrived at the outdoor laboratory and used leaves for

a   b

Fig. 16.3 Field experiments provide us with the opportunity to compare different tool-use behaviors, stone-tool use for nut cracking and use of leaves for drinking water, simultaneously in the same locale: (**a**) a chimpanzee at the rear, named Foaf, uses leaves for drinking water in the tree hollow, while another chimpanzee in front, named Velu, uses a pair of stones to crack open nuts; (**b**) the local assistants measure how much water was ingested by the chimpanzee by refilling the tree hole with water (photographs by Tetsuro Matsuzawa)

drinking the water contained in the artificial hollow. After the departure of the chimpanzees, we went to the tree to fill it with water again. By measuring how much water was necessary to refill the hole, we could actually measure how much water the chimpanzees had drank. Combining this kind of measurement with the video-records, we were able to estimate that adult chimpanzees drank on average about 30 ml per dip using a leaf tool. Individual adults often drank as much as 800 ml per session.

The clear advantage of these field experiments is that we can observe these two different kinds of tool uses taking place in the same location. For example, we could demonstrate that young chimpanzees start using drinking tools before cracking tools. This developmental difference might result from the complexity of the tool-use behavior: nut cracking requires three objects, that is, the hammer, the anvil, and the nut. In contrast, water drinking involved only a leaf or a bunch of leaves, that is, a single object (for further details on the level theory of tools, see Chap. 18).

Finally, the longitudinal record of tool-use in the outdoor laboratory can generate fixed-point observations. At the top of the hill in the outdoor laboratory, we observe Bossou chimpanzees in the same setting year after year, several weeks per year. The observations have been ongoing now for more than 20 years. We have accumulated an incredible longitudinal video archive of the wild chimpanzees' use of stone tools. We have the rare opportunity to perform a sort of behavioral time-sampling of all members of the community, once per annum, thus highlighting the ontogeny of the acquisition process and also the influence of age on performance in the same setting.

16.3   How did We Start Field Experiments at Bossou?

The first field experiment attempt was carried out in February 1988. The original outdoor laboratory was in the middle of the Mont Gban. Matsuzawa in collaboration with Sakura ran this first field experiment of stone-tool use, providing oil-palm nuts and stones to investigate how chimpanzees crack nuts (Sakura and Matsuzawa 1991). Matsuzawa waited behind a grass screen and succeeded in video-recording a young male chimpanzee, named Pru, using stones as a tool to crack open oil-palm nuts. Then, the following year, Fushimi in collaboration with Sugiyama moved the outdoor laboratory to its current place at the top of Mont Gban (Fushimi et al. 1991; Sugiyama et al. 1993a). The place is now known as the "Bureau," which means "office" in French.

Since then, we continued video-recording the chimpanzees every year. We usually open the outdoor laboratory for several weeks once a year in the dry season any time between November and February. Our experience has indicated that the chimpanzees do not often visit this part of their core area during the rainy season because of the low availability of fruit during this period in this part of their range. Huge fruiting trees such as *Ficus mucuso* and *Antiaris africana* located at the summit of Mont Gban often serve to attract the chimpanzees to the Bureau during the dry season.

16.4   Future Perspective

In December 2009, Matsuzawa opened a new outdoor laboratory for field experiments. This new site has been named the "Salon" and is located on Mont Guein (see Chap. 2 for location). The idea is the same as the "Bureau" on Mont Gban. However, this site provides natural palm trees and a natural tree hollow. In this sense, experimental manipulation has been reduced to a minimum level: just adding extra palm nuts in addition to the naturally fallen ones, and adding pure water in the natural tree hollow. As a result, the chimpanzees often started visiting the Salon (Fig. 16.4a–d) soon after this new outdoor laboratory was opened.

The Salon has a clear advantage: it is located in a place known as the "colline de concassage" (the nut-cracking hill) and is only about a 4-min walk from the KUPRI-IREB laboratory. When the chimpanzees cross the traffic road from Gban to Guein and vice versa, they often pass by the Salon. The field experiment of tool-use in the Salon looks very promising.

Stone-tool use was the same in both places, that is, Bureau and Salon. We also confirmed that Bossou chimpanzees select the same broad leaves (*Hypophrynium braunianum*) for drinking water. Interestingly, they drink water in two ways at the Salon, directly drinking from the hole or using leaves.

The field experiment of stone-tool use was also employed in the Nimba Mountains (see Chap. 28). Kathelijne Koops set up a camera trap and provided stones and oil-palm nuts in selected locations in Seringbara. She succeeded in video-recording Seringbara chimpanzees. Although extensive field surveys had already suggested the absence of stone-tool use in Seringbara (Humle and Matsuzawa 2001; Humle 2003b), the

Fig. 16.4 Field experiments continue in the newly opened outdoor laboratory known as the "Salon." The site is located under an oil-palm tree and provides access to a natural tree hollow. The chimpanzees have already visited the site and displayed their tool-use skills: (**a**) overview; (**b**) leftover leaves used for drinking water; (**c**) stone-tool use; (**d**) on the *left*, a chimpanzee is drinking water directly from the tree hole and, on the *right*, one is using stones for cracking nuts (photographs by Tetsuro Matsuzawa)

video-recording confirmed that Nimba chimpanzees in Seringbara do not know how to crack open oil-palm nuts. The reality of the video image of Seringbara chimpanzees manipulating stones and nuts but failing to attempt to crack them is highly impressive. Susana Carvalho also set up camera traps at naturally occurring nut-cracking sites in the Diécké forest (see Chap. 30). She also succeeded in filming the Diécké chimpanzees. Finally, the use of camera traps combined with an experimental approach such as providing tool materials may yield invaluable data on unhabituated chimpanzees, data not otherwise readily obtainable via direct observation.

The field experiment paradigm clearly demonstrates that the experimental techniques refined in the laboratory can be very useful to improve our observations in the natural habitat and complement those ad libitum observations acquired while following wild chimpanzees. In return, these field experiments can also help us improve the ecological and social validity of cognitive experiments in the laboratory. This kind of inverse translational approach gave us the idea of "participation observation," which relies on the daily-life, direct face-to-face observation of captive chimpanzees (Matsuzawa et al. 2006). Taken together, the conjoined parallel efforts in the field and in the laboratory elegantly complement one another and help us further our understanding of the chimpanzee mind.

Acknowledgments The present study was financially supported by the following grants: MEXT #16002001, #20002001, JSPS-HOPE, gCOE-A06-D07. Thanks are due to all my colleagues who have been conducting field experiments in the outdoor laboratory for decades: Drs. Noriko Inoue-Nakamura, Dora Biro, Claudia Sousa, Misato Hayashi, and Susana Carvalho. Special thanks are due to the late Takao Fushimi (1960–2009), who first succeeded in video-recording stone-tool use in the "Bureau," our outdoor laboratory, in 1989. Thanks to his effort, our colleagues and students may continue field experiments at Bossou.

Chapter 17
Clues to Culture? The Coula- and Panda-Nut Experiments

Dora Biro

17.1 Introduction

Recent interest in the evolutionary origins of culture has ignited a lively debate about the taxonomic distribution of the phenomenon within the animal world. Although definitional issues ("What *is* culture?") muddy the waters somewhat, evidence for group-typical, socially propagated behavioral traits has come forward from a variety of different species (see McGrew 2004 for a recent review). Nevertheless, it is now generally agreed that besides humans, chimpanzees (*Pan troglodytes*) are the species with the most prolific cultural repertoires (McGrew 1992; Whiten et al. 1999). There exists extensive regional variation in postnatally acquired behaviors across different chimpanzee communities: such variation, when neither genetic nor ecological factors can be shown to be clear determinants, is often considered a hallmark of culture (see Lycett et al. 2007 for a recent treatment of genetic factors being insufficient to explain intercommunity differences in behaviors). Within a community, behavioral traditions are thought to be passed on from one generation to the next through – at least in part – some form of social learning and can be transmitted from one community to another through the exchange of migrants. Through the presence of knowledgeable individuals in the group, naïve conspecifics are provided with a model to observe and from whom to acquire new skills. In this way, each community develops and maintains its own unique culture, the components of which extend across the tool-using, social, and self-maintenance domains. Chapter 6 of the current volume provides, for example, an overview of Bossou chimpanzees' tool-use repertoire (with additional detail in Chaps. 7–12). Some of the behaviors reported are unique to Bossou, some are restricted to certain parts of Africa, and at least one behavior seems to be ubiquitous across the continent. For comparison, Chaps. 27 through 31 provide information on nearby communities, highlighting clear differences even among those groups that are in relative geographic proximity to each other.

D. Biro (✉)
Department of Zoology, University of Oxford, South Parks Road, Oxford OX1 3PS, UK
e-mail: dora.biro@zoo.ox.ac.uk

T. Matsuzawa et al. (eds.), *The Chimpanzees of Bossou and Nimba*,
DOI 10.1007/978-4-431-53921-6_18, © Springer 2011

An oft-cited example of the principle of regional variation concerns the distribution of nut-cracking across Africa (see also Chap. 6 for further details on the distribution of nut-cracking). This behavior involves the opening of hard-shelled nuts with the aid of a hammer and an anvil (see Chap. 7 for details on the technique employed by Bossou chimpanzees). Restricted to populations inhabiting the western part of the continent, the behavior is absent in both central and east African communities despite the obvious availability of the raw materials involved (McGrew et al. 1997). It has been hypothesized that underlying this distribution pattern was a local (west African) innovation in food processing techniques, which subsequently spread to other communities throughout the region through the exchange of migrants. However, communities that lie beyond a geographic boundary precluding the movement of such migrants (the N'Zo-Sassandra River: Boesch et al. 1994) did not come to perform the behavior: the innovation never reached them. Thus, movement patterns of migrants – channels through whom knowledge spreads between communities – are thought to contribute to the creation of cultural zones. Nevertheless, a recent report (Morgan and Abwe 2006) on nut-cracking in a community lying outside the boundary proposed by Boesch et al. (1994) now hints toward the possibility of multiple origins for the behavior and calls for a reassessment of both innovation and extinction of skills as phenomena contributing to observed patterns in chimpanzee cultures (Wrangham 2006).

The present chapter summarizes a series of experiments aimed at addressing the mechanisms of behavioral innovation, within-community transmission, and between-community propagation in wild chimpanzees. Although these topics are currently receiving rigorous empirical treatment in studies with captive chimpanzees (Horner et al. 2006; Hopper et al. 2007; Whiten et al. 2007), under natural conditions the same phenomena have remained elusive and little studied (Gruber et al. 2009). Nevertheless, as may be seen, our field experimental approach provides data that allow us to speculate meaningfully about the processes behind the maintenance of chimpanzee cultures.

17.2   Methods: Unfamiliar Nuts at the Outdoor Laboratory

Chapter 16 of this volume introduces the "outdoor laboratory" at Bossou. Established in 1988, this facility has provided the setting for a long-running intensive study of the Bossou chimpanzees' tool-use. In this natural clearing at the top of a hill known as Gban, researchers have been providing raw materials for nut-cracking (nuts, and locally collected stones) as a way to create opportunities for reliably frequent close-range observations of individual tool-using ability and technique across many consecutive field seasons. Consequently, we have accumulated valuable insights into learning processes underlying the acquisition of nut-cracking in young chimpanzees and the ways in which individuals handle the cognitive demands of the task (see Chaps. 16, 18, 21).

The species of nut cracked at Bossou is the oil palm (*Elaeis guineensis*), which is the only nut naturally available at the site, although the nut of *Parinari excelsa*, cracked

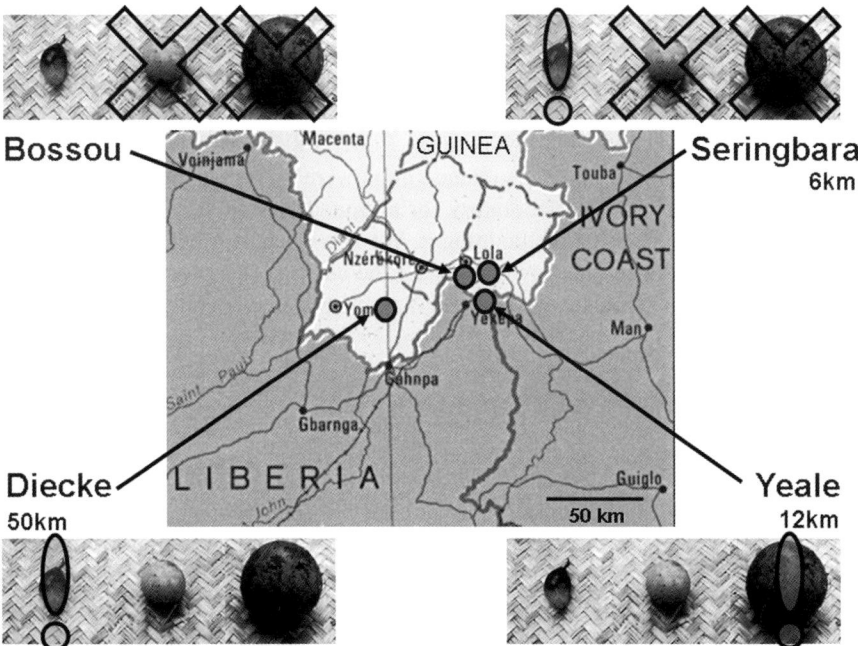

Fig. 17.1 Distribution of three different species of nuts and their utilization by chimpanzees at four sites surveyed: Bossou, Seringbara, Yealé, and Diécké. Distances correspond to distance from Bossou. Nuts, *left to right:* oil palm, coula, panda. *Crosses* indicate that the nuts are not naturally available at the site; *exclamation marks* indicate that the nuts are present but are not cracked by chimpanzees; *unmarked nuts* are available and cracked

at Taï in Côte d'Ivoire, is also available at the less-frequented extreme edges of the Bossou chimpanzees' home range (Ohashi, personal communication). Communities at nearby sites, however, have access to different species, which they utilize in some, but not all, cases (Fig. 17.1; see also Humle and Matsuzawa 2001). This variation on a theme (the use of different target items for the same behavior) coupled with the imperfect correlation between availability and utilization suggests two points: (1) that regional variation in the details of nut-cracking behavior is evident even when dealing with communities that are geographically nearby, and (2) that ecological factors are not in themselves sufficient to explain intercommunity differences. In effect, different communities seem to possess different cultures of nut-cracking. Compare, for example, the sites of Diécké and Yealé in Fig. 17.1: of the same three nut species available at both locations, chimpanzees at the two sites target a different pair.

How can we explain such distinctive patterns of raw material utilization? Why do not all communities crack all the nuts available to them? In an extension of our oil-palm nut-cracking observations at the outdoor laboratory, we presented chimpanzees at Bossou with two species of nuts found (and cracked) at nearby sites but unavailable locally: the coula nut (*Coula edulis*) and the panda nut (*Panda oleosa*). Given such artificially created availability, would the nuts' use be established within the community? If so, can we speculate about likely sources of behavioral innovations and channels of subsequent spread within the community? Further, can we generalize from these to cultural processes in wild chimpanzee communities? Or, if the novel nuts are ignored, then why so?

In the dry seasons of 1993, 1996, 2000, 2002, 2005, and 2006, small piles of coula nuts (and, only in 2000, also panda nuts) were presented alongside familiar oil-palm nuts at the outdoor laboratory for a limited period (see Biro et al. 2003, 2006, 2010 for the detailed methodology). We continued to provide the nuts until every member of the community had visited the outdoor laboratory a minimum of four times and recorded each chimpanzee's reaction to the novel food items.

17.3 Results: Innovators and Observers

Culture is built on innovation. For novel behaviors – or even just variations on existing behaviors – to appear, invention by at least one particularly enterprising individual is necessary. Thereafter, the behavior may spread among fellow community members through some form of socially mediated learning, and to neighboring groups through the arrival of a migrant bringing with her knowledge acquired in her natal community. At Bossou, the coula and panda experiments provided us with a window to both these phenomena.

17.3.1 Responses to Coula and Panda

Table 17.1 summarizes the reactions of the Bossou chimpanzees to the introduction of coula and panda nuts over all the years of presentation. These responses were categorized into three possibilities: crack (individual placed nut on an anvil stone and pounded it with a hammer, whether shell was broken or not), explore (individual made no cracking attempts, but handled, sniffed, or mouthed nuts, or ate a leftover kernel of nut cracked by others), and ignore (individual directed no cracking or exploratory behavior whatsoever toward the nuts). In addition, each individual present in the group at the time of the experiment was assigned to one of three age groups: infant (0–4 years), juvenile (5–8 years), and adult (9 years and older, based on the age of first parturition at Bossou).

Let us first examine the initial presentation of coula (in 1993) and panda (in 2000). A number of observations are of interest, common to the two types of nut. (1) Both species of nut attracted cracking attempts from some group members. Thus, importantly, the cracking of oil-palm nuts was successfully transferred by at least some individuals to the novel, locally unavailable species. (2) Not all individuals who cracked oil-palm nuts attempted to crack the unfamiliar species. (3) The age group that showed most interest toward the nuts (the highest proportion either cracking or exploring) was juveniles.

However, we also noted differences in the group's treatment of coula and panda nuts. Although in the case of coula all juveniles and about half the adults either

Table 17.1 Responses of individuals in the three age groups to coula and panda nuts across the six separate years when these nuts were presented

Age group	Nut/year	n	Crack	Explore	Ignore
Adult	Coula/1993	9	1 (11%)	4 (44%)	4 (44%)
	Coula/1996	9	3 (33%)	0 (0%)	6 (67%)
	Coula/2000	10	4 (40%)	3 (30%)	3 (30%)
	Coula/2002	9	6 (67%)	1 (11%)	2 (22%)
	Coula/2005	7[a]	5 (72%)	1 (14%)	1 (14%)
	Coula/2006	7[a]	5 (72%)	0 (0%)	2 (28%)
	Panda/2000	10	2 (20%)	0 (0%)	8 (80%)
Juvenile	Coula/1993	4	2 (50%)	2 (50%)	0 (0%)
	Coula/1996	5	3 (60%)	0 (0%)	2 (40%)
	Coula/2000	6	4 (66%)	2 (33%)	0 (0%)
	Coula/2002	4	3 (75%)	0 (0%)	1 (25%)
	Coula/2005	3	2 (67%)	1 (33%)	0 (0%)
	Coula/2006	3	3 (100%)	0 (0%)	0 (0%)
	Panda/2000	6	2 (33%)	2 (33%)	2 (33%)
Infant	Coula/1993	4	0 (0%)	0 (0%)	4 (100%)
	Coula/1996	4	0 (0%)	0 (0%)	4 (100%)
	Coula/2000	4	0 (0%)	4 (100%)	0 (0%)
	Coula/2002	5	1 (20%)	2 (40%)	2 (40%)
	Coula/2005	1	0 (0%)	1 (100%)	0 (0%)
	Coula/2006	1	0 (0%)	1 (100%)	0 (0%)
	Panda/2000	4	0 (0%)	1 (25%)	3 (75%)

[a]The eighth adult (Velu) could not be tested in 2005 and 2006 as she did not visit the outdoor laboratory during the period of coula nut presentation; this individual had cracked coula nuts in previous years of presentation

explored or cracked the nuts, with panda these figures were lower: only two-thirds of juveniles and a fifth of adults showed any interest in these novel food items. Of those individuals who cracked panda (two adults and two juveniles), adults lost interest in them after a single successful bout of cracking, while juveniles continued with their attempts on several more occasions (although they also eventually stopped). The cracking of panda nuts was in every individual preceded by extensive sniffing and mouthing; this was true also of coula nuts during the initial series of presentations for all except one individual. This individual, an adult female named Yo, proceeded, remarkably, to crack coula nuts with no prior attempts at exploration. In addition, she correctly selected ripe (dark) coula nuts over unripe (green) ones for cracking, even though neither showed obvious signs of either containing something edible inside or requiring the use of hammer and anvil. Her behavior suggested that she already possessed a familiarity with the nuts (Matsuzawa and Yamakoshi 1996); why this may have been the case and what are the implications of this possibility are discussed below. Apart from Yo, the only two other individuals who cracked coula nuts in the first round of presentation were both juveniles.

Over the subsequent years of coula nut presentation, we noted a gradual increase in the proportion of individuals in the juvenile and adult age groups who cracked coula (Fig. 17.2). At present, these levels are comparable to those found in oil-palm nut-cracking, partly because juvenile crackers have reached adulthood, and partly because

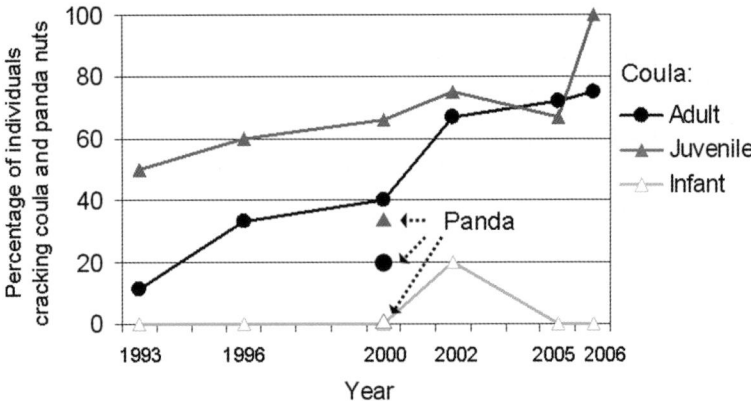

Fig. 17.2 Percentage of individuals in the three age classes who cracked coula and panda nuts in the different years of presentation at the outdoor laboratory

more and more individuals adopted the behavior. Furthermore, exploratory behaviors directed toward coula nuts have waned: almost all crackers of coula now do so without any prior sniffing or mouthing of the nuts, much as they do with oil-palm nuts.

17.3.2   Conspecific Observation

The introduction of unfamiliar nuts also precipitated a striking behavior among group members. Those individuals engaged in some form of interaction with the nuts (particularly cracking) often attracted the attention of many group members (Fig. 17.3a). We searched for patterns in such instances of conspecific observation (defined as one individual approaching another to within a distance of 1 m, with the former fixing their gaze on the latter's face or hands for at least 3 s) by noting, for each bout of observation, the identity of the observer and of the target. Rates of conspecific observation were then calculated as the number of observation bouts per hour of nuts being handled or cracked by at least one of two or more individuals present in the outdoor laboratory.

Figure 17.3b–d summarizes gross patterns in our data regarding instances of conspecific observation, from which three key points emerge. First, adult members of the community were the most likely to serve as targets of observation, which was true irrespective of the species of nut handled or cracked. Juveniles were rarely observed, and infants never, even though both these age groups contained individuals who engaged in the cracking or handling of nuts for extended periods. Second, most of the observing was performed by juveniles, followed by infants, with relatively few instances of observation of fellow group members by adults. Third, in any given bout of observation, individuals were more likely to observe those group members who were in the same age group or older, but not younger, than themselves. For infant observers, in about half the recorded instances the individual

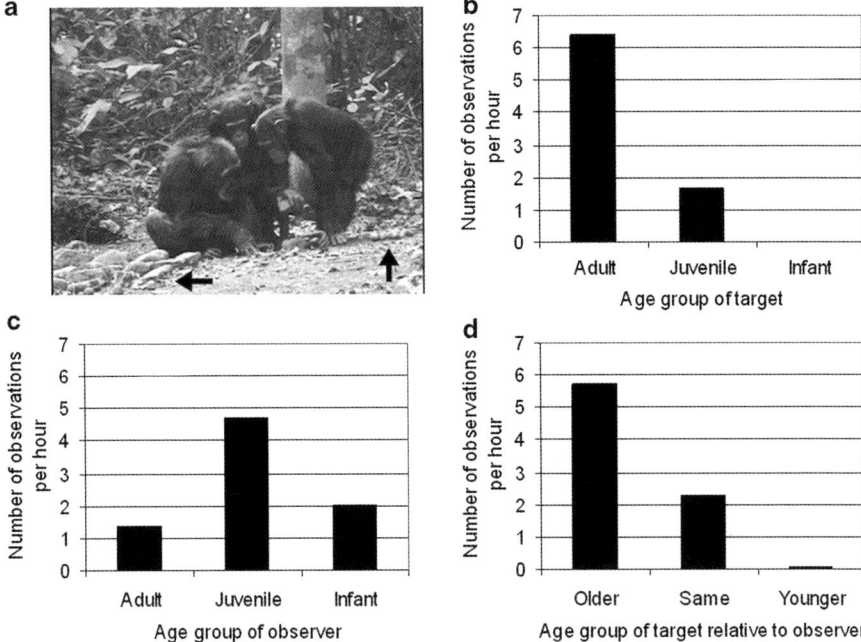

Fig. 17.3 Observation of fellow group members during the presentation of unfamiliar nuts at the outdoor laboratory. (**a**) A typical scene: adult female Yo is being observed during her cracking of coula nuts by two juvenile females, Fotaiu and Nto. Coula (*up arrow*) and panda (*left arrow*) nuts were presented to the chimpanzees simultaneously; however, all three individuals are ignoring panda nuts on this occasion. Graphs show rate of conspecific observation as a function of the age group to which the individual being observed belongs (**b**), the age group to which the observer belongs (**c**), and the relative ages of target and observer in any given bout of conspecific observation (**d**). Data are from 2000 and show averages across all three types of nut presented (oil palm, coula, panda)

attended to was the mother, while observations by adults were directed primarily toward unrelated individuals (other adults). Juveniles divided their attention among the mother, unrelated juveniles, and unrelated adults. They were occasionally waved away by the latter; in contrast, infants were tolerated by all adult targets of their observation.

17.4   Discussion: Implications for Culture

The introduction of coula and panda nuts at Bossou's outdoor laboratory represents the first truly manipulative field experiment aimed at elucidating the sources of cultural variation across different wild chimpanzee communities and the mechanisms behind its maintenance. As we are concerned with a small wild community, strict control conditions that make captive studies particularly informative (Whiten et al. 2007) have not been practical to conduct. Nevertheless, our work does possess an unparalleled ecological validity by using wild individuals, focusing on a naturally occurring behavior and on the differential utilization of naturally occurring materials constituting cultural variants in the region. So what do the coula and panda experiments tell us?

The finding that communities in relative geographic proximity to Bossou do not all crack every species of nut available to them suggests, when interpreted in cultural terms, either (1) that even when some nuts are cracked, no individual has generalized to other edible species, or (2) that individuals may in the past have innovated by cracking other nuts but that the behavior did not spread within the community (or it did spread but then subsequently died out again). On the other hand, those instances in which the same nuts are utilized in multiple communities suggest that either (1) the behavior was invented several times independently or (2) a single innovation spread across communities through the exchange of migrants. The differential fates of coula and panda nuts at Bossou may illustrate some of these alternative scenarios.

Coula, initially cracked only by a limited number of individuals, has over the years become assimilated to such an extent in the Bossou chimpanzee repertoire that the proportion of individuals cracking it now mirrors that cracking oil palm, and almost all individuals now crack without the kind of exploratory sniffing and mouthing that characterized all except Yo's early interactions with the nuts. In contrast, panda cracking was short lived within the group; although this nut was provided in only a single series of presentations, already at that stage we saw a complete vanishing of interest in the few individuals who did attempt to crack them. Thus, although innovation did occur in the case of panda as well, in contrast to coula it failed to gain a foothold within the group beyond an initial, transient interest. For a possible explanation underlying these differential outcomes let us consider sources of innovation and channels of subsequent within-community spread for behavioral variants.

For both species of novel nuts, the age group with the highest proportion of individuals who showed an interest in the unfamiliar food items was juveniles. We may therefore speculate that innovations in natural settings are most likely to originate in this group (rather than the more neophobic adults). Then, bearing in mind the patterns of conspecific observation that were evident in our data, and assuming that such observations contribute to the kind of socially mediated learning that is responsible for the diffusion of novel behavioral variants within a community, innovations in the juvenile age class will spread primarily to other juveniles and to infants, but rarely to adults. Adults themselves, on the other hand, make the most attractive models for observation and could therefore effect the highest diffusion rate within the group, with mothers having a particularly strong influence on their infant offspring (see also Chaps. 10 and 21). Adult female Yo's unhesitant and persistent cracking of coula nuts may have been responsible for the eventual spread of the behavior – she provided a reliable model for observation to individuals from all age classes. The fascinating possibility that Yo is an immigrant from a community, such as Yealé, where coula nut-cracking is habitual (see Shimada et al. 2004 for genetic analyses corroborating this hypothesis) would therefore lend support to the occurrence of intercommunity spread of behaviors through the exchange of migrants. Because no such knowledgeable adult was available in the case of panda (the nearest site where panda nut-cracking has been confirmed is Diécké, about 50 km away), and as even young innovators eventually abandoned their cracking attempts, panda nut-cracking died out at Bossou within the relatively short time-frame of our experiment. Nevertheless, it is worth noting that panda nuts are by far the hardest of the three species (Boesch and Boesch 1983), and, given the ready availability of the softer oil-palm and coula nuts at the time of presentation, panda nuts represented a relatively unattractive food item. On a related note, the notion that local ecological variables can influence the likelihood of spread of innovations has been proposed as a possible explanation behind the differential usage patterns of the oil-palm tree at sites surveyed by Humle and Matsuzawa (2004); analogously, in the present case the "ecology" of the outdoor laboratory may indeed have favored the spread of coula nut-cracking over that of panda.

Taken together, our results may thus illustrate several key phenomena in discussions of chimpanzee cultures: innovation within a group through novel invention, innovation within a group through the arrival of a knowledgeable migrant, propagation among members of a community through socially mediated learning, and (as a combination of the last two) between-community transmission. The rates at which these processes take place under natural conditions are therefore likely to influence how quickly novel cultural traditions are assimilated by wild chimpanzee communities. In turn, "cultural zones," that is, sets of neighboring communities that come to develop similar but not necessarily identical behavioral traditions, can then emerge from the complex interplay between local ecology, intercommunity migration, and within-community propagation.

Acknowledgments   The following researchers contributed to the experiments described in this chapter: Tetsuro Matsuzawa, Gen Yamakoshi, Noriko Inoue-Nakamura, Rikako Tonooka, Claudia Sousa, Misato Hayashi, and Susana Carvalho. Local guides at Bossou offered invaluable help. The author thanks the Royal Society for financial support.

Chapter 18
From Handling Stones and Nuts to Tool-Use

Misato Hayashi and Noriko Inoue-Nakamura

18.1 Characteristics of Stone-Tool Use in Terms of Object Manipulation

Chimpanzees at Bossou (*Pan troglodytes verus*) are known to use a pair of stones to crack open the oil-palm (*Elaeis guineensis*) nut to eat the edible kernel contained inside the hard shell. Because nut-cracking has only been observed in a limited number of communities in West Africa, it is considered a good example of "cultural" behavior (Whiten et al. 1999). The chimpanzees in a non-nut-cracking community do not exhibit stone-tool use even if both stones and nuts are available in their habitat. The absence of this skill at these sites shows the difficulty in the spontaneous emergence of this complex stone-tool use in naïve individuals. It may also indicate the importance of social learning through observation of other members of the community in acquiring the nut-cracking skill. Even in a nut-cracking community, an infant requires a long time to master stone-tool use during development. This chapter follows the developmental course of the nut-cracking skill in the wild chimpanzees of Bossou.

Tool-use is based on object-manipulation skill because it inevitably requires the manipulation of an object (the tool) to achieve a goal. In this sense, primates have an advantage as they typically demonstrate dexterous manipulation skills using both hands that evolved to grab branches for their arboreal lives. The skill of object manipulation develops as a function of age of the individual. In the case of chimpanzees, the hands of a newborn are exclusively used to cling to the mother. Then, they gradually start touching objects with one hand while the mother is resting, maintaining body contact with their mother with the other hand. When infants develop, they show more dexterous manipulation with a variety of actions using both hands until they finally start relating an object to another. This kind of manipulation,

M. Hayashi (✉)
Primate Research Institute, Kyoto University, 41-2 Kanrin, Inuyama, Aichi 484-8506, Japan
e-mail: misato@pri.kyoto-u.ac.jp

N. Inoue-Nakamura
Showa Women's University, 1-7 Taishido, Setagaya-ku, Tokyo 154-8533, Japan

relating one object to another, is categorized as "combinatory manipulation," and it is a precursor of tool-using behavior. Combinatory manipulation has also been used as an indicator of cognitive development in previous studies including those in humans (Connolly and Dalgleish 1989; Fragaszy and Adams-Curtis 1991; Hayashi and Matsuzawa 2003; Takeshita 2001).

In terms of combination of objects, nut-cracking behavior has a unique character-istic. Three objects, that is, a nut, an anvil stone, and a hammer stone, must be com-bined in a proper way to ensure a successful outcome. Studies in captive chimpanzees showed that the combination of three objects is difficult for naïve chimpanzees and hinders their acquisition of the skill (Hayashi et al. 2005). Infant chimpanzees in a nut-cracking community also require a long period of practice before they actually acquire complex stone-tool use. Now, the question is: How do chimpanzees in a nut-cracking community learn this complex tool-use skill during development?

18.2   Early Development of Stone-Tool Use

Nut-cracking behavior in the Bossou community has been intensively studied since 1987 during field experiments in an outdoor laboratory (see Chaps. 16 and 17). These long-term studies revealed that chimpanzees at Bossou start cracking open nuts at 3.5 years of age or older (Matsuzawa 1994). Inoue-Nakamura and Matsuzawa (1997) reported in detail the early developmental pathways in the acquisition of stone-tool use in three Bossou infant chimpanzees. The longitudinal data were grouped into four age classes ranging from 0.5 to 3.5 years of age. The older infants manipulated objects longer during their stay in the outdoor laboratory. The older infants also spent more time in manipulating objects while not in physical contact with their mother.

Inoue-Nakamura and Matsuzawa (1997) pointed out several features concerning stone–nut manipulation. The frequency of manipulating both stones and nuts in one sequence of behavior increased as the infant matured. The frequency and duration of manipulation in hierarchically higher classes also increased with age. A single action on a single object (only stone or only nut) developed into multiple actions on multiple objects (some stones, some nuts, or both stones and nuts) as the young matured. The older infants had a tendency to manipulate both stones and nuts successively or simul-taneously and to physically combine nuts with stones in a manipulative sequence. The type of hitting action was further analyzed focusing on the combination of which object (nut or stone) in which condition (on ground or on stone) was hit by what (hand, nut, or stone). The infants showed many variations of hits (Fig. 18.1), except the com-bination of hitting a nut placed on the ground with a stone, which was observed in human children (Matsuzawa 1994). It may be difficult for chimpanzees to hold a stone and hit a nut with the stone, in other words, to use a stone as a percussive tool.

There are five basic actions for nut-cracking: picking up a nut (A), putting the nut on an anvil stone (B), holding a hammer stone (C), hitting the nut on the anvil stone with the hammer stone (D), and picking up the kernel and eating it (E). Inoue-Nakamura and Matsuzawa (1997) reported that infant chimpanzees showed the basic actions

a One object involved

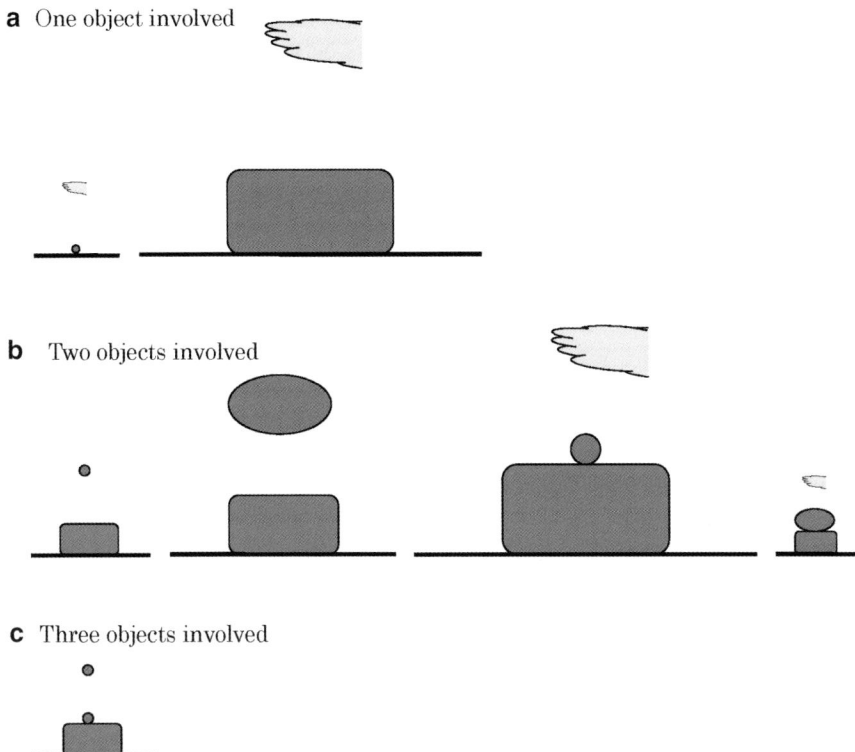

b Two objects involved

c Three objects involved

Fig. 18.1 Schematic view of the hitting patterns observed in chimpanzee infants. The relative size of the figures represents the frequency of observation of the various combinatory actions observed between 0.5 and 3.5 years of age. (**a**) Hitting pattern involving a single object; (**b**) hitting pattern involving two objects; (**c**) hitting pattern involving three objects [based on data in Inoue-Nakamura and Matsuzawa (1997)]

necessary for cracking nuts at the age of 1.5 years, much earlier than the actual success in nut-cracking. They analyzed the conditional probabilities of occurrence of each basic action. At the age of 1.5 years, the sequence from Action A (Pick) to Action E (Eat) was most dominant, and it decreased as a function of age, whereas the sequence from Action A (Pick) to Action B (Put) increased as a function of age. Many instances of "reverse" and "short-cut" sequences were observed in all age groups. The results indicated that it was difficult for the infants to combine the basic actions in an appropriate sequence to perform actual nut-cracking before the age of 3.5 years.

It should be noted that chimpanzee infants at 1.5 years of age also started to put a nut on an anvil stone and to hit the nut. This pattern differs from what is known about nut-cracking by capuchin monkeys. In capuchin monkeys, the hitting action appears at an early age and the main difficulty for young is to place and release a nut onto the anvil (Resende et al. 2008). This developmental distinction may underlie differences in the cognitive characteristics between chimpanzees and capuchin monkeys, even if the actual outcome is the same in both species.

At Bossou, chimpanzees started successfully to crack open nuts at 3.5 years of age. Individuals who did not begin to crack nuts before the age of about 7 years never acquired the skill. This observation led to the idea of a "critical period" for learning the skill (Matsuzawa 1994, 1999). A lengthy exposure to this complex tool-using skill during the early stages of development is required as well as opportunities for manipulating objects on their own.

18.3   Social Factors Promoting the Acquisition of Nut-Cracking

Inoue-Nakamura and Matsuzawa (1997) pointed out several social factors involved in the acquisition of the nut-cracking. The infant chimpanzees observed the performance of their mother from an early age. Older infants observed other chimpanzees more often than their own mother. Other than their mother, the infants observed other adult females (64%), juveniles (21%), an alpha male (9%), and older siblings (7%). Thus, the mother was not the only model for the infant in acquiring the nut-cracking skill. Actually, the offspring of two mothers who do not crack nuts also learned the nut-cracking skill just as the other infants did.

The infants continued observing the adult's performances even after succeeding in performing nut-cracking. The infants also continued to steal kernels from their mother even after succeeding in performing nut-cracking on their own. The high levels of tolerance of experienced adults may facilitate social transmission of the nut-cracking skill in wild chimpanzees. Because the infants learned the general functional relationships of stones and nuts and also learned the goal obtained by the demonstrator, the authors concluded that "emulation" could be a possible social learning process involved in the acquisition of nut-cracking behavior.

18.4   Changes After the First Nut-Cracking Success

Even after the infant chimpanzee first succeeds in cracking a nut, he or she still requires many years to achieve the level of efficiency of the skillful adults. Biro et al. (2006) reported long-term changes in the efficiency of nut-cracking. The skilled adults requires only a few hits for cracking open a nut. In contrast, juvenile chimpanzees required many more hits to achieve the same purpose.

Adult chimpanzees showed perfect laterality, as well as tool preference during nut-cracking, while young individuals showed lower fidelity to tools and techniques (Fig. 18.2). Juveniles often changed the combination of hammers and anvils. They also tended to use the same set of stones that had previously been abandoned by adult individuals. Figure 18.3 shows an example of the nut-cracking sequence recorded in an adult female, Jire, and a juvenile male, Jeje. Although the adults showed consistent bimanual coordination, young individuals used the same hand sequentially for putting a nut on an anvil stone and for using a hammer stone.

Fig. 18.2 An infant chimpanzee (Joya, 3.4 years old) is trying to crack a nut with a hammer stone held in one hand while moving irrelevant body parts, the other hand and a foot, at the same time

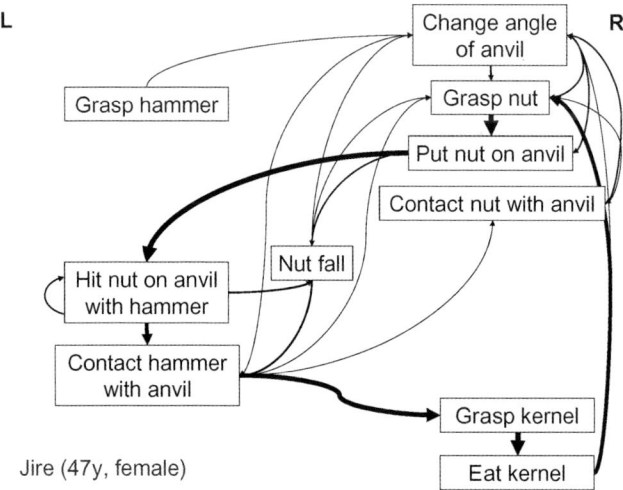

Fig. 18.3 Manipulation flow in nut-cracking sequence: *thickness* of a line indicates the frequency of occurrence; laterality is also indicated with the position of action patterns and *arrows*

Young individuals also often switched hands for hammering upon failure (see Chap. 20 for a similar pattern in captivity). These observations indicate that, even after a first success, many steps are required to reach the sophisticated level seen in adults.

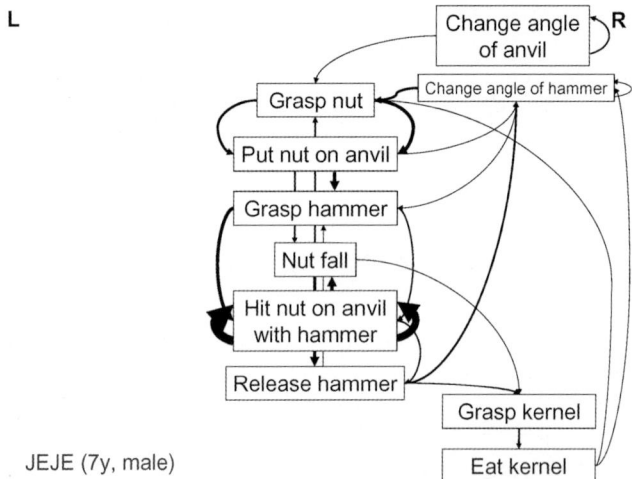

Fig. 18.3 (continued)

18.5 Hierarchical Complexity of Object Manipulation During Nut-Cracking

Nut-cracking is more complex than the other types of tool uses reported among wild chimpanzees with respect to the necessary combination of multiple detached objects. Matsuzawa (1996) used a "tree-structure analysis" to indicate the hierarchical complexity of tool use by focusing on the number of objects used and their combinatorial relationship. Most of the tool use in chimpanzees falls in the "level 1" category, which requires only one relationship between two objects, such as using "a stick" to fish "termites." The nut-cracking behavior requires the appropriate combination of three objects; an individual has to place "a nut" on "an anvil stone" and then hit the nut with "a hammer stone." Hirata and Celli (2003) reported that honey fishing through a transparent apparatus, a "level 1" tool-use, started at 1.5 years of age in captive chimpanzees. Lonsdorf et al. (2004) reported that termite fishing, another type of "level 1" tool-use where the function of the tool is obscured by the opaque termite mound, started at 2.5 years of age or onward in the wild chimpanzees of Gombe, Tanzania. Nut-cracking behavior, a "level 2" tool-use, first appeared at 3.5 years of age or older at Bossou (Matsuzawa 1994).

During the course of long-term observations of nut-cracking at Bossou, instances of using a wedge stone by chimpanzees were recorded. Use of a wedge stone as a "meta-tool" is the most complex tool-use ever reported in wild chimpanzees as it represents a "level 3" tool-use behavior based on Matsuzawa (1996)'s "tree-structure analysis". The youngest age at which wedge use during nut-cracking was observed was in a 6.5-year-old juvenile male.

Matsuzawa (1994) reported data on the development of the nut-cracking skill in human children. Twenty-eight children in the village of Bossou (1–13 years old) were tested in the same situation as the chimpanzees. Children younger than 2.5 years old failed to crack open the nuts. Human children reached a refined level of efficiency at around 10 years of age. Those ages were comparable to that of chimpanzees. Humans started to use a wedge stone to stabilize a slanted anvil stone at 6 years and 9 months of age.

18.6   Physical Causal Understanding in Nut-Cracking

Both chimpanzees and human children were tested in a situation known as the three-stone test (Matsuzawa 1994). They were given a set of three stones of different sizes. The large stone had slanted surfaces and was the anvil stone that had actually been used with a wedge by the chimpanzees during prior field experiments. The medium-sized stone was useful as a hammer and the smaller stone was useful as a wedge. A variety of solutions was observed in both species (Table 18.1). Some of the behaviors failed to reflect any physical causal understanding, such as throwing multiple nuts on the slanted anvil stone. Some others were more meaningful in the context of problem solving such as rotating the slanted anvil stone. However, those behaviors do not necessarily require a full understanding of physical causality as they can simply reflect basic behavioral strategies adopted by each individual to solve the problem.

The use of a wedge stone is a good example of the high level of cognitive abilities of chimpanzees. However, if physical causal understanding is considered, there are two distinct levels in the use of a wedge stone. The first one may correspond to the classic view of wedge-stone use: an individual inserts a wedge stone to render

Table 18.1   Behaviors observed during the three-stone test

Behavior	Chimpanzees	Humans	Causal understanding
Keep holding a nut on the slanted anvil stone while trying to hit the nut with a stone using the other hand	X	X	N
Move or rotate the slanted anvil stone	X	X	A
Turn the slanted anvil stone upside-down	X	X	A
Use a wedge stone as a meta-tool	X	X	A/Y
Throw multiple nuts onto the anvil stone	X	–	N
Support the anvil stone with the foot to keep the surface flat	X	–	A
Adjust the position of the nut or place the nut on the tip of the slanted anvil stone	–	X	Y

X, observed; –, not observed; N, not related to causal understanding; A, ambiguous relationship to causal understanding; Y, requires causal understanding
Source: Adapted from Matsuzawa (1994)

the slanted anvil flat. This active solution requires the understanding of physical causality. The second one is more passive. A skilled individual may display several behavioral strategies in solving the nut-cracking task such as rotating the anvil stone when it is not working appropriately. That behavior may sometimes lead to the appearance of wedge-stone use as a result of a single or multiple rotations of the anvil stone. Thus, future studies should focus more on the manipulative sequences involved and the intentionality underlying the use of a wedge stone in chimpanzees to clarify the cognitive abilities involved.

Analyzing object manipulation can be a useful means to assess individual cognitive development. The precise analysis of the sequential patterns of object manipulation during nut-cracking may shed further light on physical causal understanding in wild chimpanzees.

Acknowledgments The study was supported by grants from the Ministry of Education, Science, and Culture in Japan (#19700245 to Misato Hayashi and #16002001 to Tetsuro Matsuzawa) and from the Benesse Corporation. Special thanks are offered to Tetsuro Matsuzawa, Dora Biro, Claudia Sousa, Tatyana Humle, Susana Carbalho, Etsuko Nogami, Mari Hirosawa, and the guides of Bossou for their great advice and support in collecting data at the outdoor laboratory of Bossou.

Chapter 19
The Emergence of Stone-Tool Use in Captive Chimpanzees

Satoshi Hirata and Misato Hayashi

19.1 The Use of Stone Tools to Crack Nuts by Chimpanzees in Captivity

Researchers have accumulated knowledge about wild chimpanzees' use of tools to crack nuts through longitudinal observations, as described in several chapters of this volume. These investigations are a rich source of information on the implications of this type of complex tool-use from ecological, behavioral, and comparative cognitive perspectives (Matsuzawa 2001a). However, although research has expanded, with field experiments in Bossou, Guinea (Matsuzawa 1994), there are limitations to the study of chimpanzee behavior in the wild. For example, it is virtually impossible to follow the entire process by which infant chimpanzees acquire nut-cracking skills. Moreover, researchers do not know how stone-tool use emerges in wild chimpanzee communities. The study of chimpanzees in captivity may help researchers answer these questions and learn more about this type of stone-tool use.

Based on the distribution of tool-use among wild chimpanzee populations in Africa, nut-cracking likely developed in West Africa and was then transmitted socially from one generation to the next, with expansion based on individual emigration and immigration (Biro et al. 2003, 2006; Whiten et al. 2001). In this chapter, we outline studies that focused on the emergence and learning of stone-tool use to crack nuts in captive chimpanzees.

S. Hirata (✉)
Great Ape Research Institute, Hayashibara Biochemical Laboratories, Inc.,
952-2 Nu, Tamano, Okayama 706-0316, Japan
e-mail: hirata@gari.be.to

M. Hayashi
Primate Research Institute, Kyoto University, 41-2 Kanrin, Inuyama,
Aichi 484-8506, Japan

19.2   Challenges for Chimpanzees

To clarify the challenges for chimpanzees in using tools to crack nuts, Hayashi et al. (2005) investigated the behavior of three human-raised captive chimpanzees who observed a human model performing the correct sequence of nut-cracking behavior. The subjects were three adolescent-to-adult female chimpanzees living in a group at the Kyoto University Primate Research Institute, Japan. For testing, each of the chimpanzees was brought separately to a playroom in which were placed macadamia nuts and seven natural stones (400 g–2 kg) used by wild chimpanzees at Bossou. A human tester remained in the room and successfully cracked a nut in front of the chimpanzee. The tester then handed the chimpanzee several nuts, and the chimpanzee's behavior was recorded.

Two of the three chimpanzees placed a nut on an anvil stone and hit it with a hammer stone after a single observation of the human model. One of these chimpanzees successfully cracked nuts during the first session. The second chimpanzee succeeded in cracking nuts in the subsequent session, during which some degree of human active teaching took place. The third remaining chimpanzee did not succeed in cracking nuts during the first test session or subsequent training sessions.

The chimpanzees' manipulation pattern was analyzed in detail to elucidate the cognitive capabilities necessary for the emergence of nut-cracking behavior. The focus was on the difference in behavioral characteristics between the two chimpanzees that were successful and the one who had failed. First, the two successful chimpanzees showed hitting action spontaneously and increased their hitting action approximately 10 min into the test session, whereas the unsuccessful chimpanzee did not. Second, when investigators analyzed the items handled by the chimpanzees, they found that the two successful chimpanzees manipulated the stones in various ways, but the unsuccessful chimpanzee rarely handled the stones. Third, the chimpanzees combined two or three objects, and the frequency of manipulating two objects together was higher than that of three objects for all three chimpanzees.

These results indicate three main difficulties that may underlie the failure to acquire this tool-use skill: lack of hitting action, lack of stone manipulation, and difficulty in combining three objects. The action of hitting is observed in spontaneous manipulation patterns in chimpanzees, but it appears less common compared to capuchin monkeys (Izawa and Mizuno 1977; Visalberghi 1987). In fact, the capuchin monkey is the only other primate that cracks nuts in the wild (Fragaszy et al. 2004). Although capuchin monkeys crack nuts by using a familiar action, performing a hitting action is the first challenge that the chimpanzees face when cracking nuts. The goal of nut-cracking is to obtain the edible kernel contained inside a nut; thus, it is reasonable for chimpanzees to pay attention to the nut. However, to succeed in using tools to crack nuts, chimpanzees have to shift their attention from the nut to the stone tools. This shift in attention is the second challenge with respect to nut-cracking.

According to Matsuzawa (1996), the use of tools to crack nuts requires the relationship of three objects, that is, a nut, a hammer, and an anvil, which necessitates greater cognitive demands than relating two objects, as seen in many other forms of tool-use in chimpanzees and other primates such as drawing a banana with a stick, inserting a grass stem into a hole, pulling a rake to draw something closer, or stepping on a box to reach a banana hanging from the ceiling (e.g., Köhler 1925; see Tomasello and Call 1997 for a review). Therefore, combining three objects is the third challenge for chimpanzees.

19.3   Behavior Acquisition in Juvenile Chimpanzees

Hirata et al. (2009) investigated the acquisition of tool-use for cracking nuts in a group of five young captive chimpanzees aged 4–7 years at the Hayashibara Great Ape Research Institute, Japan. They clarified what types of understanding of tools and actions would lead to the acquisition of nut-cracking skills in the presence of a skilled model.

The most dominant male of the group was selected as the model and was trained separately to use tools to crack nuts through human modeling and teaching. After intensive training, which was carried out in an indoor experimental room, this individual was eventually able to crack open macadamia nuts using a hammer stone and an anvil stone. The model chimpanzee was then placed with four naïve chimpanzees in an outside enclosure approximately 7,400 m² in area in which nut-cracking sites with abundant nuts, hammer stones, and anvil stones were available. The anvil stones were fixed into the ground to prevent the chimpanzees from carrying them to unobservable areas. Although the anvil stones were not detached objects, they were coded as objects in the analysis. This group test situation allowed the researchers to track the entire history of the acquisition of stone-tool use by the four naïve individuals' observation of the skilled conspecific model (Fig. 19.1). A session of the group test lasted for 30 min, and two or three sessions were conduced per week.

All four naïve chimpanzees acquired the nut-cracking behavior during the group test. The individuals first succeeded in the 8th, 11th, 13th, and 15th sessions, respectively. Each chimpanzee made many relevant and irrelevant manipulations of the nuts and stones before his or her first success. To clarify the acquisition process, the behaviors of the chimpanzees before the first success were analyzed in detail. Combinatory manipulations of three or more objects were initially rare, but began to increase one or two sessions before the first success, whereas combinatory manipulations of two objects were observed from the first session. Also, the hitting action was initially rare in all subjects. Stones were more frequently handled when the chimpanzees manipulated a combination of two objects than a single object, and they were more frequently handled when the chimpanzees manipulated a combination of three objects than two objects. These results were consistent with Hayashi et al.'s (2005) three challenges for chimpanzees: combinatory manipulation of three objects, hitting action, and shifting of attention to stones.

As the sessions proceeded, the chimpanzees began to combine three objects, but initially their manipulations were often inappropriate (e.g., three stones or one stone and two nuts). Gradually, however, the frequency of combining the correct set of three objects (i.e., nut, hammer, and anvil) increased. When the chimpanzees did combine the correct three objects, they sometimes used these objects in the incorrect

Fig. 19.1 A naïve chimpanzee observing a skilled model crack nuts successfully using stone tools (photograph by Satoshi Hirata)

order (e.g., placing the nut on the hammer stone on the anvil stone). The frequency of combining objects in the correct order (i.e., the anvil at the bottom, the nut in the middle, and the hammer on top) increased during the last stages of the process.

Another key factor in acquiring nut-cracking skills is the action of hitting. At first, the chimpanzees sometimes stepped on the nut to apply pressure. All subjects stepped more frequently than they hit in the first half of the process but began to hit more frequently in the second half. The fact that the chimpanzees stepped on the nuts in the early stage indicates that they recognized the goal (i.e., to put pressure on the nut shell to crack it open), but did not understand the action required to achieve this goal, even though they had observed a model successfully cracking nuts by hitting them with a hammer stone.

More detailed analysis of the chimpanzees' hitting action revealed that they initially used their hands to hit the target. Each subject's first hitting actions were performed with empty hands, and hitting with a hammer stone came to predominate later on. Therefore, even after the chimpanzees recognized the necessary action (i.e., hitting), they did not understand how to use a tool (i.e., a stone hammer) to hit the target. In other words, understanding of the use of tools emerged after understanding of the hitting action as one of the necessary components to crack nuts. Further analysis of the hitting action indicated that the target of the hitting was not always a nut. Chimpanzees sometimes hit an empty anvil stone, or they hit a hammer stone on an anvil stone (Fig. 19.2). This observation shows that the goal of the action (i.e., cracking open a nut) was separate from the actual action (i.e., hitting) in the acquisition process. Although the goal was to crack open a nut, the chimpanzees sometimes disregarded the nuts and performed the hitting action using another target altogether.

Fig. 19.2 A chimpanzee hitting a stone on an anvil stone with a hammer stone (photograph by Satoshi Hirata)

In this situation, the anvil stones were fixed to the ground, which differs from the situation of wild chimpanzees in Bossou, Guinea, who use loose stones as anvils. Rather, the situation is similar to that of chimpanzees in Diécké (see Chap. 15) or capuchin monkeys in Brazil (Fragaszy et al. 2004), which use tree roots or outcrops as anvils to crack nuts. This type of behavior may be easier because the task involves relating two detached objects (a nut and a hammer) at a certain spot (a rock), rather than three detached objects (a nut, a hammer, and an anvil). However, the results would not have differed markedly if loose stones had been used as anvils because the chimpanzees in the study described here readily used loose stones as anvils in another situation after their first success (Foucart et al. 2005; see Chap. 20).

This study shows how chimpanzees came to succeed in using stone tools to crack nuts. The process can be broken down into several steps. First is the recognition of the goal, which is to apply pressure on a nut to crack it open. Second is the emergence of the combinatory manipulation of three objects. Third is the emergence of the hitting action. Fourth is the use of a tool (i.e., a hammer stone) for hitting, rather than an empty hand. Last is the aiming of the hitting action at the nut. The chimpanzees grasped these ideas at different stages, and they gradually united these factors in their behavior, leading to their first success. In other words, success was not brought about by random trial and error, but arose from the chimpanzees' systematic understanding of this type of tool-use.

Regarding the effect of observing a model, the chimpanzees did not show evidence of immediate true imitation (Whiten and Ham 1992). That is, their behavior did not markedly improve immediately after they observed successful nut-cracking by a skilled conspecific peer. Nevertheless, the influence of observing a model is apparent when a longer time span is considered: The observation of a model might have had a penetrative effect over a few days, during which time the subjects also engaged in their own trials and errors. In sum, the results illustrate the step-by-step learning of several components of nut-cracking, which might be practiced through intermittent observation of a model and direct handling of related objects between observations.

19.4   Infant's Behavior Acquisition Before Age 2 Years

Of the five subjects in the study just described, one gave birth to a female on July 8, 2005. When the infant was 2 months old, she became the subject of an observational study of the acquisition process of using stone tools to crack nuts. The study began in October 2005 and consisted of two sessions per month. Each session comprised approximately 30 min of observation in the presence of nuts, hammer stones, and anvil stones in a large outdoor enclosure.

Initially, the infant clung to her mother's back or abdomen most of the time. Sometimes she watched her mother using tools to crack nuts, and at other times she reached her hand toward the nuts and stones on the ground while holding on to her mother's body with her opposite hand. At approximately 3 months old, she gradually began to move away from her mother and touched nuts or mouthed stones. At 7 months old, she spent more than 90% of the observation time away from her mother, exploring her surroundings on her own.

When the infant was 9 months old, she was observed hitting a hammer stone with her hand. This was her first use of the hitting action, one of the fundamental elements of cracking nuts. However, the hitting action was not observed regularly after this. Approximately 3 months later, when she was 1 year and 2 months old, she was observed performing the hitting action again. She picked up a nut, placed it on the ground, and hit it repeatedly with her hand. Her hitting behavior became more frequent after this. During that same month, she was observed placing a small stone on an anvil stone and then hitting it with her hand. Subsequently, she put a nut on an anvil stone and hit it with her hand. The placement of a nut or a stone on an anvil stone is a type of combinatory manipulation. Thus, the combinatory manipulation of placing one object on to another emerged when the infant was 1 year and 2 months old.

Another milestone occurred when the infant was 1 year and 9 months old. She picked up a hammer stone and used it to repeatedly hit an anvil stone. Although there was no nut on the anvil stone, she was using a hammer stone to hit an object, which is one step closer to the correct sequence to crack nuts. This event indicated that she could perform the three necessary elements to crack nuts: placing a nut on an anvil stone, hitting the nut, and hitting with a hammer stone. However, these elements did not come together at the same time. The infant repeatedly hit the nut on the anvil stone with her empty hand, but lacked the hammer stone. She also hit the empty anvil stone with the hammer stone, but lacked the nut. That is, she did not combine three objects (i.e., a nut, an anvil, and a hammer), but only two.

Further progress was observed when the infant was 1 year and 11 months old. She brought a small stone, put it on an anvil stone, held a hammer stone, and hit the small stone on the anvil using the hammer stone. The target of this hitting action was the small stone, not a nut, so this was not technically nut-cracking, but she successfully related three objects.

During the following session, which was also conducted when she was 1 year and 11 months old, she finally succeeded in cracking her first nut. She picked up a nut by herself, placed it on an anvil stone, hit it several times using a hammer stone, and cracked it open. The age of first acquisition, 1 year and 11 months, was considerably earlier than in cases of wild chimpanzees, which acquire this skill between the ages 3 to 7 years. The anvil stones in this study were fixed to the ground; thus, the number of detached objects that the chimpanzee had to relate was two (i.e., the nut and the hammer stone), which may have facilitated the learning process. To address this, another test situation was introduced immediately after the infant's first success. The infant was observed in an indoor experimental room with a detached, movable anvil stone. She quickly succeeded in cracking a nut by putting the nut on an anvil stone and hitting it with a hammer stone (Fig. 19.3). Therefore, after this infant had first succeeded in cracking nuts by using an anvil stone fixed to the ground, she readily transferred this skill to a situation in which the anvil was not fixed.

Three factors possibly contributed to the early acquisition of nut-cracking by this infant. The first was the presence of a skilled mother and other skilled group members. Chimpanzee infants have a strong motivation to do the same thing as their mother and other adults. This infant grew up watching her mother and other peers repeatedly cracking nuts. The second factor was the place in which the study was conducted. Most studies of tool-use in captivity are conducted in an experimental

Fig. 19.3 An infant chimpanzee who succeeded in using tools to crack nuts before 2 years of age (photograph by Satoshi Hirata)

room that is prepared specifically for the test. Such an environment is unusual for chimpanzees. By contrast, the observations of Hirata et al. (2009) were conducted in the outdoor enclosure in which the infant and other chimpanzees spent most of their time. The infant was already familiar with the environment, including the stones, and this may have made it easier for a new behavioral pattern to emerge. The third factor was the infant's early independence from her mother. By the age of 7 months, she was already spending more than 90% of her time away from her mother. According to Inoue-Nakamura and Matsuzawa (1997), wild infant chimpanzees at Bossou spend none of their time away from their mothers at the age of 0.5 years, one-third of their time at age 1.5 years, one-half at age 2.5 years, and two-thirds at age 3.5 years. The 7-month-old infant in the present study was spending more time away from her mother than do normal 3.5-year-old wild infants, which may have been partly because there were few lethal dangers in her captive environment. When an infant clings to the mother, there is little opportunity for him or her to manipulate objects. By contrast, this infant had many opportunities to manually explore her environment, which might have facilitated her acquisition of tool-use at such an early age.

In short, this observational study clearly shows that an infant chimpanzee is able to acquire the skills to use stone tools to crack nuts before 2 years of age when there is a model, when opportunities are provided in a familiar environment, and when manual exploration of the surrounding objects is possible from an early age.

19.5   Conclusion

Studies of captive chimpanzees have illuminated the following challenges for chimpanzees in mastering the use of stone tools to crack nuts: the use of the hitting action, the shift of attention from nuts to stones, and the combinatory manipulation of three objects. The acquisition process can be broken down into several steps, including the recognition of the goal, the emergence of the use of a combination of three objects, the emergence of the hitting action, the use of a tool for hitting, and the hitting of the nut. Chimpanzees recognize these different components separately and practice them separately. Success is not brought about by random trial and error, but arises from the systematic understanding of this type of tool-use. The example of the acquisition of this behavior by an infant illustrates the fact that in favorable conditions, chimpanzees have the potential to acquire the skills to use stone tools to crack nuts before the age of 2 years, which is much earlier than indicated by studies in the wild.

Acknowledgments This study was financially supported by the Ministry of Education, Culture, Sports, Science and Technology, Japan (#07102010, 12002009, 16002001, 1870266, and 20680015) and JSPS-HOPE. We thank all the collaborators at the Primate Research Institute of Kyoto University and Hayashibara Great Ape Research Institute for assistance during the study.

Chapter 20
A Gibsonian Motor Analysis of the Nut-Cracking Technique

Blandine Bril, Gilles Dietrich, and Satoshi Hirata

20.1 Nut-Cracking as a Goal-Directed Action

Nut-cracking can be considered as a goal-directed action aimed at removing a kernel from a shell without smashing the kernel. Once the tools have been chosen, the behavior involves taking the hammer stone with one hand, positioning the nut on an anvil with the other hand, and hitting the nut with ballistic movements until the shell cracks open.

The foregoing is how a witness might describe the activity of chimpanzees cracking open nuts before eating their contents. However, this description says nothing about the physical nature of the interaction (in terms of forces) that causes the cracking of the shell. In this chapter, we discuss the Gibsonian concept of *affordance* viewed as relating to the animal's capacity for using energy transfer to realize the task.

Field studies in West Africa, especially in Bossou, have clearly described the nut-cracking behavior of chimpanzees, as well as the learning processes involved in its acquisition (Biro et al. 2006) (see Chaps. 18 and 21 for more details). In the wild, chimpanzees have been observed selecting the appropriate tools, that is, hammers and anvils of particular sizes, shapes, and materials, suggesting that they understand the functional properties of the nut-cracking task (Sakura and Matsuzawa 1991; Sugiyama 1981a; Sugiyama and Koman 1979b). Sugiyama and Koman (1979b) suggested that the stone hammers selected by Bossou chimpanzees are most probably the best adapted to the hand of the chimpanzees and to the shape and dimensions of the nuts. Observations concerning the choice of an anvil show that

B. Bril (✉)
École des Hautes Études en Sciences Sociales, 54 Bd Raspail, 75006 Paris, France
e-mail: blandine.bril@ehess.fr

G. Dietrich
Université Paris Descartes, 1 rue Lacretelle, 75015 Paris, France
e-mail: gilles.dietrich@univ-paris5.fr

S. Hirata
Great Ape Research Institute, Hayashibara Biochemical Laboratories, Inc., 952-2 Nu, Tamano, Okayama 706-0316, Japan

T. Matsuzawa et al. (eds.), *The Chimpanzees of Bossou and Nimba*,
DOI 10.1007/978-4-431-53921-6_21, © Springer 2011

chimpanzees have a clear understanding of its function in nut-cracking: hammers were systematically smaller than the anvils (Sakura and Matsuzawa 1991), variations in hammer size were systematically smaller than the variations among the anvils, and the horizontality of a stone was a criterion for anvil selection (Sugiyama and Koman 1979b). Fushimi et al. (1991) have observed adult subjects manipulating the anvil stone on the ground to get the useful surface into a horizontal position. The other major feature of the anvil's surface is related to the immobilization of the nut: Sugiyama and Koman (1979b) noted that chimpanzees use depressions on the surface to prevent the nut from shifting when it is hit by the hammer.

The observations summarized above indicate that chimpanzees are able to apprehend a priori the relationship between potential tools (the stone hammer or anvil) and the referent (the nut); that is, they perceive the properties of the tools that are directly relevant to successfully reaching the goal, that is, cracking open the nut. In other words, they recognize the function of the tool (Sakura and Matsuzawa 1991).

Gibson invented the term *affordance* to provide an account of the features of the environment that are directly relevant to behavior (Gibson 1977, 1979; Turvey 1992; Reed 1996). For Gibson, "the affordances of the environment are what it offers animals, what it provides or furnishes, for good or ill. (…) Moreover, the objects of the environment afford activities like manipulation and tool using." (1977: 68). In more general terms, what makes an affordance is having the right properties to support a behavioral process of a species (Reed 1996: 40). Furthermore, Reed describes a skill as constituted by an organism's ability to detect special relationships among affordances and to organize these relationships to achieve a particular functional outcome (Reed 1996).

Data on nut-cracking in the wild undoubtedly show that chimpanzees perceive the affordances of stones as potential tools to achieve nut-cracking. Now, considering that action is the realization of affordances, we may hypothesize that chimpanzees are able to adapt their striking movement to the tools, the hammer and/or anvil. When chimpanzees choose an anvil with or without depressions on the surface, their striking movement should specify the characteristics of the anvil surface. Along the same line, if the hammer stone is too small or too big, they will either choose another stone hammer or adapt their striking movement to produce the right functional properties for the strike, that is, the appropriate kinetic energy (see following).

In the following section, we report some results from experiments conducted with five juvenile captive chimpanzees from the Hayashibara Great Ape Research Institute (see Chap. 22). The nut-cracking behavior was examined with reference to two points: the choice of the best tool that fits the goal and the capacity of the animal to adapt his or her behavior to the properties of the tools and of the nuts. Two series of experiments were performed. The first series was dedicated to investigating the perception of affordances of objects with different properties: what were the properties of the object that make a suitable hammer or anvil? In the second series, we analyzed how the chimpanzees adapted their movement to meet the demands of the task. It is important to emphasize that what is of interest here is not the movement per se, but what the movement produces on the environment, that is, the satisfactory amount of kinetic energy.

20.2   Affordances of Tools

All five chimpanzees had previously been trained to crack open nuts with stone hammers. One of them, Loi, had substantially greater experience in nut-cracking, having been taught to crack nuts by humans. The other four chimpanzees had learned how to crack nuts through social learning sessions. Loi was considered the model who would elicit nut-cracking behavior in the other four chimpanzees (for further details, see Chap. 19).

In the first experiment, the chimpanzees had to crack 15 nuts of two different species (macadamia and Brazil nuts), and artificial nuts. Artificial nuts were made of two plastic parts joined by a metal belt. To make them as attractive as real nuts, they contained a piece of fruit within as a reward. The chimpanzees had to choose a hammer among eight stones differing either in shape (regular or irregular) or in weight (100, 300, 600, and 1,000 g; Fig. 20.1a). Once the nut was cracked open, they put the stone hammer back in its tray. No constraints were imposed on the choice of the hammer. If a chimpanzee was willing to change his/her hammer during the striking sequence he/she could freely do so. The irregular stones were never chosen (whatever their weight) except by one chimpanzee. When weight is considered,

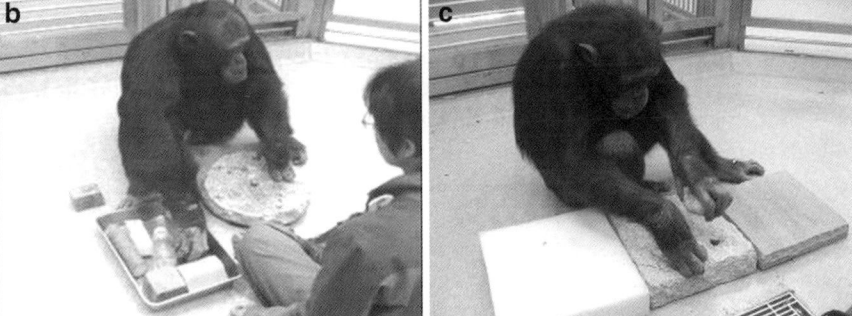

Fig. 20.1 (**a**) Affordances of stone tools varying in shape (regular and irregular) and weight. (**b**) Affordances of non-stone objects varying in their appropriateness and efficiency as tools. (**c**) Affordances of anvils (stone anvil with cavities, stone anvil with a smooth surface, and foam anvil)

the 100-g stone was never used as a hammer; a majority of choices involved the 600-g stone and, to a lesser extent, the 300-g one. One chimpanzee only, Mizuki, chose the 1,000-g stone (including the irregular one). The interesting point here is that she was the only one to fail frequently in holding the nut steady, which shifted away a greater number of times (1.5 times per nut on average, while for the other chimpanzees the figure was less than 0.1 time per nut).

In the second experiment, the same number of objects was offered; none of them was made of stone and they differed in their degree of efficiency [they consisted of two wooden blocks that were parallelepiped shaped, one small branch and one bracket, a primate toy, two pieces of foam, and one plastic bottle (Fig. 20.1b)]. As these objects were unknown to the chimpanzee, two habituation sessions were conducted before the affordance experiment. A session consisted in cracking 15 nuts in succession. The eight objects were introduced to each chimpanzee following the same procedure as in the previous experiment with stone hammers. He or she was able to change hammer as many times as he or she wanted. The most functional objects were the two wooden blocks and the primate toy. They were chosen more than 80% of the time by the five chimpanzees.

Quite a few characteristics of the behavior of the chimpanzees indicated some kind of understanding of the functionality of the tool. When the nut was "rebellious" to the hits, it was not rare to observe the reorientation of the wooden hammer in the hand. In addition, when using a "nonfunctional" tool, the chimpanzees kept trying to a lesser degree than when a more efficient tool was used. In other words, they gave up when they understood that the tool was not efficient, without trying again and again. In addition, when the hammers were less proficient, and consequently the strikes less efficient, the chimpanzees tried a greater number of hammers and changed hammers more frequently. Although the number of changes of stone hammers during a session was quite small, less than 8 times, it could reach up to 25 times per session with non-stone hammers. This observation indicates clearly that not only did the chimpanzees base their choice of hammer tool on the functional properties of the hammer, but when they failed to crack the nut, they engaged in an exploration process with other objects. Surprisingly, in contrast to what Boesch and Boesch (1983) described in the wild, no systematic difference appeared in hammer choice depending on the species of nut being cracked. This lack of differences in behavior is not quite clear: perhaps the difference in the hardness of the nut shell was not great enough. Alternatively, the sample size of individuals taking part in the experiment may have been too small for such a relationship to emerge. This question is worth further investigation.

In a third experiment, different anvils (a stone anvil with a smooth surface, a stone anvil with cavities, and a foam anvil) were provided simultaneously. Here again, the chimpanzees chose the most efficient anvil, regardless of its location (the three anvils were arranged randomly in a row in front of the chimpanzee, see Fig. 20.1c). Three sessions were conducted, and the layout of the anvils was changed each time to minimize the possibility of the layout influencing the choice. Each time there was a different anvil in the central position. In addition, again to avoid influencing the chimpanzee's choice, the nut was presented from the side of the least efficient anvil. The hammer was a familiar one, that is, a stone weighing approximately 600 g, the most frequent choice in the hammer affordance experiment. Four of the five chimpanzees chose the anvil with depressions for more than 95% of the nuts being cracked. Only one chimpanzee used the anvil with a smooth surface 20% of the time. When a chimpanzee changed the anvil used, most of the time it was to go from a less amenable one (with a smooth surface or made of foam) to the amenable one (the anvil with cavities). This change of anvil occurred following a small number of unsuccessful strikes (fewer than 7). In the rare cases when the anvil with depressions was abandoned for the smooth one, it was always following a very high number of unsuccessful strikes (more than 25).

Other adaptive behaviors are worth mentioning. There was a change in the position or orientation of the nut on the anvil (especially with the artificial nuts). However, this behavior was more frequent when the tool was less functional (the wooden hammer), and it was observed twice as often with Loi who had the greatest experience in cracking nuts. Another strategy that appeared when the chimpanzee had difficulties in cracking nuts, especially with the less efficient hammers, was a switch from the preferred hand to the nonpreferred one. This behavior was observed in three chimpanzees (two right-handed and one left-handed); the most experienced chimpanzee, Loi, only used his preferred hand. However, in most cases, the sequence of strikes was initiated with the preferred hand. This result corroborates McGrew and Marchant (1999), who observed that in termite fishing more skilful chimpanzees were more lateralized than less skillful ones.

To summarize, these results appeared:

1. When offered various unknown objects as tools, the chimpanzees preferentially chose the most efficient ones.
2. When they failed to crack the nut, they tried another tool, or repositioned the tool in the hand or the nut on the anvil, or even switched hands.
3. Except in cases where there was obviously low motivation, after a series of unsuccessful strikes, the chimpanzee typically discarded the tool and chose another one; there was almost no continuation of the ineffective action.

20.3   Generating the Functional Requirements to Reach the Goal

In the brief review of the observation of nut-cracking in the wild, and in the experiments described earlier in this chapter, we have seen that chimpanzees are able to perceive the properties of objects directly relevant to the successful cracking of nuts.

If, as Smitsman emphasizes, *affordances* are the properties (of the environment) that are perceived through actions (1997: 303), or in other words, if to perceive an affordance is to perceive what actions are possible (Bongers 2001; Turvey 1992; van Leeuwen et al. 1994), the counterpart should be the capacity to adapt one's actions to the functional demand of the task. In the case of cracking nuts, this means that a chimpanzee is able to adapt his or her striking behavior according to the tools and the types of nut.

How can we define nut-cracking in functional terms? Nut-cracking is defined as a task that consists in delivering a blow to a nut in such a way that the shell cracks open leaving the kernel intact. To meet this demand of the task, the blow must be elastic, that is, a blow in which the total energy is conserved (the sum of potential and kinetic energies). This requirement means that the total impulse is constant before and after the blow, that is, all forces are used to modify the velocity of the object, or to generate its deformation.

To reach this goal, the right amount of kinetic energy, which depends on the hardness of the shell, must be delivered to the nut to produce an adequate deformation of the shell so that it breaks. Therefore, the kinetic energy, which depends on the weight of the hammer (m) and its velocity (v), will be the main parameter to be controlled. Consequently, the way in which the action must be carried out depends on several factors: the weight of the hammer, the properties of the support surface and of the object to be hit, the velocity of the hammer, the orientation of the trajectory, etc. Conversely, in the case of a nonelastic blow (either plastic or viscoelastic depending on the characteristics of the hammer and the anvil), part or all of the forces are dissipated, and it will be difficult to crack open the nut. For example, if the nut is lying on a soft anvil, or if the hammer is not hard enough, the energy will be absorbed by the support or the hammer, and it will be impossible, or at least quite difficult, to crack the nut (for more details, see Bril et al. 2008).

To illustrate this necessary complementarity between perception and action to reach a goal, we analyzed how chimpanzees adapted their striking actions to meet the functional properties of the task, when either the weight of the hammer or the hardness of the shell varied.

In other words, regardless of the hammer, when a given type of nut is considered, what must be produced is the right amount of kinetic energy. Kinetic energy is defined as $\frac{1}{2} mv^2$, where m is the mass of the effector and v the velocity. Consequently, there is an infinite number of combinations between mass and velocity that end up in the same amount of kinetic energy: the smaller the mass, the higher the velocity, and conversely, the greater the mass of the hammer, the lower the velocity. The amount of kinetic energy to be produced must, however, be adjusted to the hardness of the nut. In a recent study (see Bril et al. 2008 for details), we analyzed how chimpanzees adapt their movement, or more precisely what their movements produce in terms of amplitude and velocity of the displacement of the hammer.

Figure 20.2 shows the amount of kinetic energy for two values of the weight of the stone hammer: 600 and 1,000 g. In theory and for a given type of nut, there should be no difference regardless of the hammer used. When macadamia nuts are considered, there was no difference in the amount of kinetic energy produced for three of the five chimpanzees, whereas with the other two chimpanzees, the amount of kinetic energy was lower in the case of the heavier hammer. Now, if we compare the value of kinetic energy produced for the Brazil nuts, two chimpanzees produced the same amount of kinetic energy whatever the hammer's weight, while the other three displayed a slightly smaller value with the heavier hammer. It is worth noting

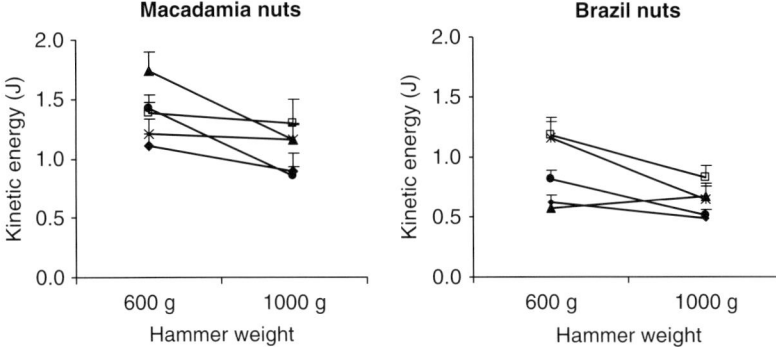

Fig. 20.2 Value of kinetic energy produced by each of the five chimpanzees, depending on the type of the nut (macadamia or Brazil nuts) and weight of the hammer (600 and 1,000 g)

that the more-experienced chimpanzee, Loi, was the only one to produce the same amount of kinetic energy for the two nut species whatever the weight of the hammer.

Now, do all chimpanzees use the same motor strategy to deliver the right amount of kinetic energy to the nut? If we consider that a biological system can use conservative mechanics, one extreme solution would be to rely only on potential energy (potential energy: $Ep=mgh$, where m is the mass of the hammer, g is the value of the gravitational acceleration, and h is the gradient of the vertical position of the hammer during the movement). Thus, once the tool has been chosen, the chimpanzee has only to control the tool's initial position because the subsequent movement is passive. However, a biological system can add a new kind of energy based on muscular activity: in terms of movement, this means the chimpanzee can modify the kinetic energy, hence velocity, by using muscular energy. Consequently, an infinite number of solutions can be used to produce this total energy, which must meet the task constraints, depending on the two independent parameters that have to be selected: the initial position of the tool and the muscular force produced.

Of the five chimpanzees, only one, Mizuki, produced a significantly larger amplitude of movement (a longer trajectory of the hammer during the strike), meaning that she relied less on muscular force; this was obvious when we computed the ratio of the kinetic energy over the potential energy (Ek/Ep), which indicates the additional energy applied by the chimpanzee during the movement, that is, the amount of muscular energy added (Foucart et al. 2005; Bril et al. 2008). Mizuki showed significantly lower values than all the other chimpanzees except Loi. Yet, in addition to the hammer's weight, there is another factor that plays a role in the amount of kinetic energy produced: the weight of the animal's hand–arm system. Here, Loi weighed almost one-third more than the other animals, so he had a heavier hand. Consequently, if he displayed the same amplitude of the hammer, then, consequent to his heavier hand, the total potential energy, for a given amplitude of movement, would be higher than that of the other chimpanzees. To produce the same amount of kinetic energy, this chimpanzee needed less additional muscular energy.

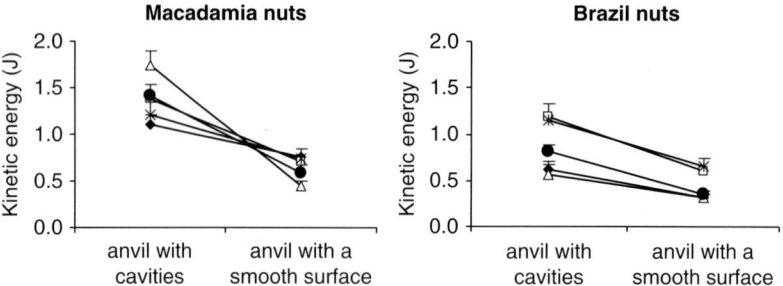

Fig. 20.3 Value of kinetic energy produced by each of the five chimpanzees depending on the type of the nuts (macadamia or Brazil nuts) and type of anvil surfaces (with cavities or smooth). Stone hammer of 600 g

We have seen in the previous section that the anvil with a smooth surface was rarely chosen as a functional anvil. When the kinetic energy produced during the cracking of nuts on this smooth anvil was compared with that produced on the more functional anvil with cavities, all the chimpanzees displayed smaller amounts of kinetic energy (Fig. 20.3). Indeed, a slight deviation of the hammer trajectory from the vertical at contact could shift the nut from its actual position and the nut would fall off. This result clearly suggests that the slippery surface of the anvil was perceived as being more hazardous and difficult. To avoid the possibility of the nut shifting, a tendency to use a slightly more vertical strike was observed for all the chimpanzees. A specific type of behavior has been observed in the more-experienced chimpanzee. When cracking macadamia nuts on the flat-surfaced anvil, Loi produced small irregular strikes preceding large-amplitude strikes, as if he were trying to immobilize the nut.

20.4   The Benefit of the Ecological Framework in Studying the Nut-Cracking Technique

In this chapter, we proposed an ecological Gibsonian framework to study the use of tools among captive chimpanzees. The Gibsonian approach emphasizes the fact that the control of actions does not lie in animals nor in the environment, but in the animal–environment system. As such, the approach provides an ideal framework in which to study what may be considered as the core of tool-use, that is, the functional coupling between an actor and a tool. *Affordances* appear as a key concept in expressing "what perceptual capacities are needed to perceive an object as a tool and what kind of information is available in a situation that specifies the affordance of a tool in terms of complementary relationships among actor, tool and target" (van Leeuwen et al. 1994: 175).

Following on from this, the choice of objects as tools is seen as depending on the action capacities of the agent. In this chapter, we emphasized the validity of taking a functional approach to tool-use, and to nut-cracking in particular, to better understand the actor–tool coupling: in other words, the animal's adaptive capacity for action.

Acknowledgments This research has been supported by the Action Concertée Incitative TTT P7802 n° 02 2 0440 from the French Ministère Délégué à la Recherche et aux Nouvelles Technologies and the Ministry of Education Culture, Sports Science and Technology of Japan (grant for the Biodiversity Research of the twenty-first century COE, A14). A fraction of this research was part of the PhD of Julie Foucart, who has left the academic world for a private position. We thank Tetsushi Nonaka for his helpful comments on an earlier version of this chapter.

Chapter 21
Education by Master-Apprenticeship

Tetsuro Matsuzawa

21.1 Introduction

Humans and chimpanzees shared a last common ancestor about six million years ago: we are evolutionary neighbors. We continuously seeks answers to "what makes us human?," "how and why we have evolved?," and "where will we go?" through understanding brain function, analyzing the genome, excavating fossils, etc. However, the mind and the brain cannot be excavated from the ground. To appreciate the evolutionary history of the human mind, we have to query the minds of other living creatures. The human body is an evolutionary product, and so are the human mind, education, culture, society, mother–infant relationships, etc. This chapter aims to answer the question of the origins of mother–infant relationships and education through a comparative and developmental approach. Not many people recognize that brain size triples from birth to adulthood in both humans and chimpanzees. We reach adulthood after many years of education and long-term learning. There is no schooling in chimpanzees. However, they have their own way of education, known as "education by master-apprenticeship."

21.2 The Chimpanzee Way of Rearing: A Single Working Mother

A recent study reported 534 births in the past 46 years in wild chimpanzees (Emery-Thompson et al. 2007). The study was based on the collaboration among six long-term research sites in Africa. These data served to explore female longevity and reproduction. Infant mortality was high: about 30% of infants died before the age of 5 years. The average life expectancy was 15 years. The longest record of

T. Matsuzawa (✉)
Primate Research Institute, Kyoto University, 41-2 Kanrin, Inuyama, Aichi 484-8506, Japan
e-mail: matsuzaw@pri.kyoto-u.ac.jp

T. Matsuzawa et al. (eds.), *The Chimpanzees of Bossou and Nimba*,
DOI 10.1007/978-4-431-53921-6_22, © Springer 2011

longevity of females in the wild was about 50 years old. The average interbirth interval was about 5 years. In other words, an infant chimpanzee monopolizes his or her mother until the age of 5.

Female chimpanzees usually start giving birth at around 12 years of age. However, there are community differences. At Bossou, we have had four cases of females giving birth at the age of 9 (see Chap. 3). We also recorded cases of females giving birth at an estimated age exceeding 50 years. Bossou female chimpanzees, in a sense, can be distinguished by their long reproductive period. Jire, an old female, last gave birth to a female infant, named Jodoamon, on November 18, 2009; this is her tenth offspring since Sugiyama and his colleagues started observations in 1976. Let us explore Jire's reproductive data in detail. Jire is the most reproductively successful chimpanzee in the history of Bossou. Jire, who was estimated by Sugiyama, in 1976, to have been born in 1958, has so far given birth to ten infants, five females and five males. The first infant was named Jima (male, estimated to have been born in 1972, and who disappeared in 1980 at the age of 8 years), JI (male, estimated to have been born in 1975, and who disappeared in 1977; he was presumed dead because he was only 2 years old when he disappeared), Jieza (male, estimated to have been born in 1978, and who disappeared in 1988 at the age of 10), Ja (female, born in 1983, and who disappeared in 1993 at the age of 10), Jokro (female, born in 1989, and who succumbed to a respiratory disease in 1992 at the age of 2.5 years), Juru (female, born in 1993, and who disappeared in 2001 at the age of 8), Jeje (male, born in 1997, and still in the community at the age of 13 years), Jimato (male, born in 2002, and who died in 2003 from a flu-like epidemic), Joya (female, born in 2004, and who still is in the community at the age of 6), Jodamon (female, born in 2009, and who passed away at the age of 0.5 years). Sugiyama estimated the age of Jire's first two offspring in 1976: he noticed these dependent offspring but had not actually witnessed their birth so their estimated age could be incorrect. However, it is very clear that Jire was already a mother when she was first observed in 1976. Among her ten offspring, four passed away before the age of 3 years (Matsuzawa 1997b; Biro et al. 2010; see Chaps. 13, 25, 32), which means that 40% of her offspring died before being weaned: this statistic closely fits the overall dataset compiled by Emery-Thompson et al. (2007).

In chimpanzees, it is predominantly the mother who takes care of her infant whereas the biological father does not directly invest in infant care. When the infant reaches about 5 years of age, the mother starts caring for her next offspring. In contrast to humans, no chimpanzee siblings are as close as 2–3 years apart in age; mothers take care of their infants one at a time.

There is, of course, a biological father who most often is one of the males of the community, especially at Bossou where there are no overlapping neighboring communities around. Chimpanzee society is so-called patrilineal: males are typically the philopatric sex and females generally emigrate from their natal community to join a neighboring one. The males often form alliances and patrol the territory of their community, thus protecting mother–infant pairs, and monopolizing females in estrus within their community. Intercommunity interactions are most often agonistic.

Therefore although there is no obvious direct paternal care, all the adult males typically make a collaborative effort to protect the multiple mother–infant pairs within their community (Hockings et al. 2007; see Chap. 23). From the infant's perspective, he or she has one mother and, in a sense, multiple fathers. From the adult male's perspective, an infant in the community is likely to be his own son or daughter, his younger full- or half-sibling fathered by his father, or his nephew or niece fathered by his brother. The likelihood of kin-relatedness between adult males and offspring within the community is fairly high as a consequence of male philopatry. This is the nature of chimpanzee society.

21.3 The Human Way of Rearing: Collaborative Breeding and the Role of Grandmothers

Once you understand chimpanzee society, you can identify the unique features of human society. The physiology of the two species is comparable. The gestation period in humans is about 280 days while it is 240 days on average in chimpanzees. At birth, human infants weigh approximately 3 kg and chimpanzee infants a little less than 2 kg. It is often not well appreciated that the age difference between surviving chimpanzee siblings necessarily exceeds 5 years, which is also true in gorillas and orangutans. In contrast, humans may have brothers and sisters 2–3 years apart in age, even sometimes only 1 year. The rate of twin birth is also higher in humans than among the other three genera of Hominidae. Humans are unique in terms of being able to rear multiple children at the same time.

Humans have invented special foods for babies to facilitate weaning and thus shorten their interbirth interval. Once the infant stops suckling, the sexual cycle of the mother resumes, allowing her to conceive again: this is a reproductive strategy aimed at maximizing the number of offspring during a limited period of fertility counterbalanced by a longer investment in rearing infants.

The disadvantage of this human strategy is clear: multiple children are being reared at the same time. This can be costly and energetically expensive. A single working mother cannot easily raise multiple children at the same time. Rearing multiple offspring simultaneously is often facilitated by having a spouse or husband. Paternal care in this sense helps in rearing multiple offspring.

In the case of chimpanzees, females advertise their period of ovulation. Females in estrus present a large pinkish swelling of their perineal region. This signal attracts the sexual attention of males in the community. In general, females mate with multiple males but rear their offspring on their own.

In contrast to chimpanzees, humans evolved to conceal their estrous. It is difficult to tell whether a woman is ovulating simply based on her outer appearance. This is a human female strategy for securing the long-term commitment of a single male. Females are indeed constrained by a long gestation period, breastfeeding, and their attachment to their offspring. Males, on the other hand, can maximize their reproductive success by producing multiple offspring with different females, just like chimpanzees. Concealed ovulation is therefore a human female counterstrategy to the reproductive strategy of males. Paternity certainty favors human males' investment in rearing their partner's offspring. Although male chimpanzees sometimes mate-guard females for several days or even weeks to maximize their paternity certainty, the relationship between a woman and a man is typically far more prolonged and enduring, sometimes lasting a lifetime. Among primates, humans have a clear tendency to maintain a strong male–female pair bond. In this sense, humans are alleged collaborative breeders. This system has evolved from the necessity of rearing multiple children at the same time.

Humans have not only evolved a pair-bonded mating system, but have also favored cooperative breeding. Grandmothers often act as caregiver to her offspring's offspring (Hawkes et al. 1990). The grandfather may also have a role. Human grandparents do not typically require as high an energy intake as younger, more active and reproducing adults. However, they often contribute to the collaborative rearing of multiple children, which has been an argument for the evolution of the prolonged postreproductive life span of human females. In humans, close kin also sometimes serve as helpers. This cooperative rearing system may explain why community-based mutual support has developed in humans. The burden of having multiple children at the same time may be one of the factors that facilitated group living and mutual support and cooperation in early hominids.

The evolutionary origins of the human family is not a nuclear one based on a single male–female pair bond that defends its territory and practices collaborative breeding with the aid of its newest generation of offspring. This kind of cooperative breeding system is not uncommon in the animal kingdom and is observed among different species of birds and mammals (Wilson 1975). However, the collaborative breeding system of the human family is unique in terms of involving individuals across at least three generations: grandparents, parents, and grandchildren. Breeding and rearing in this sense transcends several generations. For that purpose, exogamy, the marriage between different unrelated communities, is essential in preventing inbreeding.

Relationships between neighboring chimpanzees communities are typically highly agonistic (e.g., Goodall 1986) but are quite peaceful in bonobos (e.g., Kano 1992; Idani 1991). Different bonobo communities may feed together in the same fruiting tree and engage in sexual interaction with one another. These social nonaggressive encounters may provide young females with the opportunity to transfer from their natal community to the neighboring one. These immigrant females typically maintain a special bond with a specific resident older female (SSF, specific senior female). The SSF often accompanies her sons, and the sons can thus develop a new consort relationship with the newly immigrated female; this could be a mechanism for generating a nuclear family within the community. Although we still poorly understand the evolutionary origins of the human family, further studies of the intercommunity relationships of chimpanzees and bonobos may shed further light on the emergence of the three-generation family that is unique to humans.

Humans typically rear multiple children with the collaboration of individuals spanning three generations and their extended networks. In a sense, natural selection has favored mutual and reciprocal support in the way humans rear their offspring.

In short, humans are collaborative breeders consisting of three generations of both sexes that depend on social communication for survival and reproductive success.

21.4   The Stable Supine Posture Makes Us Human

It is not the bipedal upright posture and bipedal locomotion, but the stable supine posture, which distinguishes humans from other animals (Takeshita et al. 2009; Matsuzawa 2007, 2010). Let us suppose that you lay a primate infant down on its back, that is, in a supine position. Macaque infants immediately show the so-called righting reflex and succeed in turning over to adopt a prone posture. I have tested both chimpanzee and orangutan infants. They are incapable of turning over: they usually slowly raise one arm and the contralateral leg. Several seconds later, the infants then slowly lower their arm and leg, and proceed to gradually raise the opposite arm and leg. This movement alternates spontaneously. This behavior means that great ape infants cannot sustain a stable supine posture and always struggle to cling. Chimpanzee infants until 3 months of age always cling onto their mother and are never separated. Only human infants can adopt and maintain a stable supine posture.

Bipedalism is often thought of as the impetus for human evolution. Accordingly, human ancestors stood up on their hindlimbs and freed their forelimbs from supporting the body. Many contend that the freed forelimbs favored their ability to manipulate objects, which stimulated the brain, which in turn further facilitated object manipulation, and tool-use and manufacture. However, I will argue instead that the ability to maintain a stable supine posture has been the primary impetus for promoting unique features of social and physical intelligence in humans. In contrast to infant chimpanzees, human infants do not need to cling onto their mother, can remain physically separated, and can maintain a stable supine posture. This posture facilitates face-to-face communication (Mizuno et al. 2006), vocal exchange, and the manipulation of objects from an early stage of development soon after birth (Hayashi and Matsuzawa 2003). It is in fact the supine posture that has freed our hands. Human infants can stand up on their feet at around the age of 1 year. However, before even being able to stand bipedally, the stable supine posture frees the infant's hands, enabling him or her to manipulate various objects such as rattles from a very young age. This early onset of object manipulation is a precursor to tool technology.

21.5   Education by Master-Apprenticeship

The absolute brain size of humans is almost three times as large as that of chimpanzees. However, it is not well recognized that the brain size triples from birth to adulthood in both species (3.26 times in humans versus 3.20 in chimpanzees; Matsuzawa 2007).

This point highlights the importance of postnatal development, a period during which young need to acquire an array of skills and behaviors for their survival. This long learning phase is channeled via the maternal bond in chimpanzees and via the parental and grandparental bond in humans.

There is of course no schooling in chimpanzees; however, they have their own form of education. Each chimpanzee community develops its own set of cultural traditions based on observational learning. Knowledge and techniques are passed on from one generation to the next, such as tool-uses and ways of greeting (McGrew 2004; Nakamura and Nishida 2006; Whiten et al. 1999). For example, chimpanzees of Gombe use a twig to fish termites from mounds (Goodall 1986). However, chimpanzees at Bossou do not perform termite fishing, although they feed on other termite species without the need for a tool. In contrast, chimpanzees at Bossou use a pair of stones to crack open the hard shell of oil-palm nuts to get the edible kernel within (Fig. 21.1). However, chimpanzees at Gombe do not perform this stone-tool use behavior, although they eat the outer red soft tissue of the oil-palm fruit and stones are readily available in their habitat. This example reminds us of human culture. For example, Japanese use chopsticks to eat sashimi, but not all human groups use a pair of sticks as a tool to eat raw fish. Just like humans, each chimpanzee community has its own unique set of cultural traditions.

The chimpanzee way of education is known as "education by master-apprenticeship" (Matsuzawa et al. 2001), or "bonding and identification-based observational learning" (BIOL) (de Waal 2001). The most important "teacher" of the infant is the mother (Boesch 1991b). The mother and the infant are always together at least for the first 4 years of life until the infant is weaned. Chimpanzee mothers do not teach per se; they do not provide their young with good stone tools, nor do they mold their infant's hands when he or she tries to use stone tools.

The foundation of education by master-apprenticeship is the strong and long-lasting mother–infant bond. In this context, (1) chimpanzee mothers serve as good models for their young, (2) the infants have an intrinsic motivation to copy their mothers through intensive observation, and (3) the mothers are highly tolerant of their infants.

The infants learn a lot from their mothers, and then start paying attention to other older members of the community. Young individuals carefully watch older individuals, but never do older individuals watch the younger ones, when it comes to nut-cracking (Biro et al. 2003; see Chap. 17). In the case of human society, elders may carefully watch and learn from the younger generation. However, this kind of oblique transmission of behavior from the younger generation to the older generation is indeed not observed among wild chimpanzees. Young chimpanzees learn how to use stone tools

---

Fig. 21.1 (continued) (c) An adult female, named Yo, is cracking open coula nuts, not available in Bossou, as a part of a field experiment. Two young chimpanzees are attentively watching her crack this new species of nuts: one is her son, Yolo, on the *right*, and the other is his playmate, a young female, named Fotaiu. These two young chimpanzees have already acquired the skill of stone-tool use. However, they are always curious about novel behaviors performed by older members of the community. They have learned a lot from watching older members of the community, but never the reverse (photographs by Tetsuro Matsuzawa)

Fig. 21.1 This is an illustration of behaviors involved in "education by master-apprentice-ship" in wild chimpanzees. (**a**) Joya, a female infant, is attentively observing her mother, Jire, crack open nuts. Jire's son, Jeje, situated in the background, just learned to crack nuts by himself. (**b**) Jire is cracking open nuts, which are then scrounged by her daughter, Ja. Pama, the adult female chimpanzee in the background, does not know how to crack open nuts. Her son, Poni, is stamping on the nut placed on the stone anvil in an attempt to crack it open.

via long-term observation and practice of the behavior. It takes them about 4 years to learn how to use a hammer and anvil stone to crack nuts (Matsuzawa 1994).

An appreciation for the way chimpanzees educate can help us better identify the unique aspects of human education. The human way of education is founded on the collaborative breeding system composed of multiple elder individuals in addition to the mother. From the start, the mother is not the only teacher. Others include the father, the grandparents, elder brothers and sisters, and uncles and aunts. This elaborate social network based on a collaborative breeding and rearing system implies that each community member may take on the role of the teacher.

Active teaching is, of course, a unique feature of human education. Teaching in humans also entails other unique behaviors, such as social praise, scolding, verbal instruction, and molding, in addition to more subtle mechanisms of social praise such as nodding and smiling. Human children seem to have a strong desire to be socially praised.

Let us imagine the situation of a mother–child pair who for the first time goes to the park to play in the sand pit. Before starting to play, the child looks up at the mother and the mother smiles back. Suppose that the child successfully scoops up sand and puts it into a bucket – I would bet that the child then looks up at the mother again. The mother then smiles, nods, or even claps her hands to praise her child's success. The English word "educate" originates from the corresponding Greek word that means "extract." Education extracts something already within the children. Subtle praise and an intrinsic motivation for receiving social praise, therefore, play an important role in human education.

In summary, chimpanzees have their own unique way of education; the coined term is "education by master-apprenticeship." This form of education is based on the long-lasting mother–infant bond. This bond is cemented in the intensive care that the mother provides her infant during his or her first 5 years of life. Understanding this process in chimpanzees provides us with important insights into the unique evolution and characteristics of human education. Human education is based on an elaborate social network of parents, grandparents, kin, and non-kin. The human baby can adopt a stable supine posture, which facilitates face-to-face communication between the baby and other community members. Collaborative breeding and rearing in humans might have evolved as an adaptation to a change in environment and cognitive niche when our ancestors shifted from living in the forest to a milieu dominated by savanna and woodland.

Acknowledgments The present study was financially supported by the following grants: MEXT #16002001, #20002001, JSPS-HOPE, gCOE-A06-D07. Thanks are due to the colleagues, the students, and the Guinean collaborators who helped me to understand the importance of education by master-apprenticeship. Special thanks are due to our late two first field assistants: Mr. Tino Camara and Mr. Guanou Goumi. Without their help and guidance, I could not have acquainted myself with the life of wild chimpanzees at Bossou.

Part V
Social Life and Social Intelligence

Chapter 22
The Crop-Raiders of the Sacred Hill

Kimberley Jane Hockings

22.1 Introduction

Humans (*Homo sapiens*) and wildlife have interacted for thousands of years, coexisting in many different ways (Fuentes and Hockings 2010). As our closest phylogenetic relatives, chimpanzees (*Pan troglodytes*) in particular occupy a special importance in terms of their complex relationship with humans: this is especially valid in many parts of West Africa, where chimpanzees (*P. t. verus*) form an integral part of human myths. For example, the village of Bossou in Guinea is home to the Manon people, who hold the neighboring chimpanzees sacred as the reincarnation of their ancestors and believe that their ancestors' souls rest on the hill of Gban (Kortlandt 1986). As the chimpanzee is a totem of the most influential family of Bossou, it is strictly forbidden for anyone to hunt or eat the chimpanzee. Humans and chimpanzees in Bossou are not only neighbors – their coexistence is preserved by a delicate balance of wild and cultivated resource use (Yamakoshi 2005; see Chap. 4). Similarly, in neighboring Liberia, the local Manon people preserve their coexistence with chimpanzees according to similar local beliefs (see Chap. 31). In Cantanhez National Park, south-eastern Guinea-Bissau, humans and chimpanzees are able to coexist by reason of a strong local taboo against the hunting of chimpanzees, despite regular crop-raiding reports (Gippoliti and Sousa 2004). Chimpanzees and humans in Fongoli, South-East Senegal, generally coexist peacefully thanks to a cultural taboo against hunting of chimpanzees, but concerns over chimpanzees discovering human crops might threaten such relationships (Pruetz 2002).

Chimpanzees have a highly flexible social system and a very mixed diet, and seem able to adapt to areas of secondary vegetation and human agriculture that are

*Electronic supplementary material The online version of this chapter (doi: 10.1007/978-4-431-53921-6_23) contains supplementary material, which is available to authorized users.

K.J. Hockings (✉)
Department of Anthropology, Faculty of Social and Human Sciences, New University of Lisbon, Avenida de Berna, 26-C, 1069-061 Lisbon, Portugal
and
Department of Psychology, Stirling University, Stirling, UK
e-mail: hock@fcsh.unl.pt

impinging on their natural habitat (Yamakoshi 2005). Crop-raiding is probably an adaptation by wildlife to both a loss of natural habitat and wild foods and an increase in access to new energy-rich food resources. As more areas are being cultivated in direct proximity to the forest edge, the geographic ranges of many species shrink and fragment, causing human and nonhuman primate species to increasingly compete for resources. Although in some parts of Africa humans and chimpanzees are able to live as neighbors, in many parts crop-raiding is not tolerated, with often fatal consequences when the apes "trespass into human land."

Around the Budongo Forest Reserve, different chimpanzee communities vary in their crop-raiding propensities (Hill 1997; Reynolds 2005a). Individuals from the Sonso community in the main Budongo Forest block occasionally raid crops, namely mango from surrounding orchards and sugarcane from commercial fields on the forest edge, with sometimes fatal consequences (personal observation, 2003). However, the Nyakafunjo community, which lives closer to human settlements, exhibits higher crop-raiding levels than their Sonso neighbors, possibly because of increases in human cultivation and consequent reductions in home range size. The Kasokwa chimpanzees, a small community inhabiting a riverine strip of forest to the south of Budongo, subsists mainly on forest foods, but occasionally feed on papaya, mango, and sugarcane during forest food shortages (Reynolds 2005a). Furthermore, chimpanzees at Bulindi in the Hoima District of Uganda, approximately 30 km south of Budongo, live in a fragmented farm-forest-woodland mosaic and regularly raid human crops including sugarcane, mango, cocoa, guava, papaya, banana, and jackfruit (McLennan 2007, personal communication). The chimpanzees of Kibale National Park, Uganda, occasionally feed upon maize and were reported to cause significant damage in banana plantations (Naughton-Treves et al. 1998). Although quantitative data are lacking, the loss of forest habitat in the southern region of Gombe National Park in Tanzania has driven one community of chimpanzees to raid crops such as bananas, mango, and oil-palm fruits at the forest edge (Greengrass 2000). In the Mahale Mountains, Tanzania, the M-group feed on different agricultural species, including guava, mango, lemon, and oil-palm fruit (Takahata et al. 1985). However, recent human interventions have affected numbers of available cultivated resources (Nishida 2008). In Cantanhez National Park in Guinea-Bissau, chimpanzees regularly crop-raid orange, pineapple, and cashew fruit. Chimpanzees in this area also raid human-installed beehives for honey and compete with local people for access to water wells in the dry season (Sousa 2008, personal communication). In the village of Yealé, in the Nimba Mountains in Côte d'Ivoire, chimpanzees have been reported to raid cultivated foods, in particular, cacao, papaya, pineapple, oranges, and cassava, during periods of wild fruit scarcity (Humle 2003a).

Although much is known about crop-raiding from the human perspective, very little research has been done on the underlying reasons why the chimpanzees choose to crop-raid. Understanding the behavior of primates whose home ranges border agricultural land and human settlements is central to answering questions about how they perceive and adjust to such environments. The aim of this chapter is to provide an overview of the feeding, behavioral, and social adaptations demonstrated by the chimpanzees of Bossou in response to living in a heavily human influenced environment, and to discuss how these findings might help us to better understand resource conflict situations throughout Africa.

22.2 Adopting an Interdisciplinary Approach

Human utilization in forested ecosystems in the Republic of Guinea is extensive, and the area surrounding the village of Bossou is no exception. The majority of people living at Bossou are subsistence farmers practicing swidden ("slash-and-burn") agriculture and rely heavily on rice and cassava for carbohydrate intake, which are cultivated in surrounding forested, savanna, and mangrove areas. Local people also produce a wide variety of fruits in orchards and next to their houses, including pineapple, papaya, orange, mandarin, and avocado, for their own consumption and for local market sales. Consequently, the hills (70–150 m high) that constitute the chimpanzees' home range are covered in primary and secondary forest, cultivated and abandoned fields, and orchards (Hockings et al. 2006, 2007). Primary forest accounts for just 1 km² of their 15 km² home range and is predominantly located at the summit of the largest and most sacred hill (Gban). The main body of forest is mostly characterized by secondary and scrub forest consequent to abandoned cultivation (see Chap. 2 for further description of habitat).

To monitor forest fruit availability, transect lines that pass through the chimpanzees' core area (a total of six transect lines; total distance, 4,739 m) were monitored twice monthly, and villagers who owned farmland or orchards within or around Bossou ($n=39$) completed verbally presented questionnaires on the planting patterns and monthly harvest of the foods they produced for the preceding year (2005).

A focal adult chimpanzee (12 years or older; Sugiyama 2004) was randomly selected daily ($n=8–9$ individuals), and feeding patterns, associated behaviors, and the presence of party members (for details of definition of a party, see Sakura 1994) were recorded in 5-min instantaneous samples. All-occurrence sampling was also employed to record all incidents of crop-raiding and rough self-scratching, a self-directed behavioral pattern exhibited by chimpanzees possibly in response to anxiety (van Lawick-Goodall 1972; Aureli and de Waal 1997).

Some cultivated species, such as mango fruit, were only consumed by chimpanzees in abandoned orchards or fields. As these areas were never guarded, acquiring these foods was not considered to represent crop-raiding. Accordingly, cultivated foods were divided into two groups: abandoned, that is, crops that were not guarded by humans; and guarded, crops which were at least intermittently guarded by humans. Even though the chimpanzees are totemic to local people and are therefore not killed, they were often chased away with noise and sometimes with the use of stones. The chimpanzees probably associated certain areas as higher risk than others, but were likely to fear local human presence to a certain extent in any exposed area.

A crop-raiding "event" was defined as any foray by an individual to obtain guarded cultivated food, from time of exit from to return to natural vegetation (Naughton-Treves et al. 1998). Party compositions were categorized as adult male-only, mixed (at least one adult male, adult female, and immature present), adult male and other (at least one adult male and at least one adult female or immature), or no male (adult females and immatures only). For each group of cultivated food, presence or absence (either auditory or visual) of local people and the location of the field or orchard were recorded. All instances of food sharing (defined as an individual holding a food item but allowing another individual to consume part of that item) were recorded (see Hockings et al. 2007). Females were classified as "of reproductive age," "cycling," or more specifically, "maximally swollen." Consortships, in which an adult female and an adult male move together to the periphery of their community range so that the male gains exclusive mating access, were also recorded.

Data were collected over 12 months (observations were recorded during each month of the year in three periods, from May 2004 to December 2005), and 187 focal samples were recorded, totaling 1,673 h of focal observation. During this study the Bossou chimpanzee community size ranged from 12 to 14 individuals, always with the same 3 adult males (Matsuzawa 2006a). The social rank of the Bossou males varied over the years, but during this study the relative status of the alpha male (Yolo), the second-ranking male (Foaf), and the third-ranking male (Tua) was stable (Biro et al. 2003; Sugiyama 2004). Throughout this study, infants or juveniles less than 8 years old were classified as immature.

22.3   Results and Discussion

22.3.1   Crop-Raiding: An Ecological Approach

Human cultivation provides chimpanzees at Bossou with easy access to a range of different cultivated foods, which they exploit more frequently than any other chimpanzee community. The Bossou chimpanzees feed on 17 varieties of cultivated foods: in particular, simple-sugar fruits such as papaya, orange, pineapple, mango, and banana, but also complex carbohydrates and proteins such as cassava, maize and papaya leaf, are frequently consumed (Table 22.1).

Significant variations exist in the importance of various cultivated foods in the chimpanzees' diet. In particular, simple-sugar fruits were taken during months of low wild fruit availability. However, when mango fruits, which are abundantly found in abandoned orchards within the forest, were available to the chimpanzees, crop-raiding rates of most cultivated foods including simple-sugar fruits decreased, with mango fruits being preferentially consumed. The high rate of mango consumption in May therefore effectively resembles a month of wild fruit abundance. Consequently, access to other foods might at least partly explain temporal variations in cultivated food consumption. Following this point, chimpanzees at Bossou consumed the tuberous root of cultivated cassava, a spatially abundant and continuously available plant, especially during periods of wild fruit scarcity and lower availability of other cultivated foods. Cassava appears to represent a "filler" fallback food (Laden and

Table 22.1 Guarded and abandoned crop-feeding event frequencies for each cultivated food and part during the study period

Cultivated food			Event frequency	
Common name	Scientific name	Part	Guarded	Abandoned
Papaya	*Carica papaya*	FT	126 (4)+	0
		LF	69+	0
Banana	*Musa sinensis*	FT	62 (3)+	23 (2)*
		PI	63+	26*
Orange	*Citrus sinensis*	FT	86	3
Mandarin	*Citrus reticulata*	FT	18	0
Pineapple	*Ananasa comosus*	FT	21	3
Mango	*Mangifera indica*	FT	0	149
Rice	*Oryza* sp.	PI	81	0
Maize	*Zea mays*	FT	48	0
Cassava	*Manihot esculenta*	TB	74	26
Cacao	*Theobroma cacao*	FT	34	0
Oil palm	*Elaeis guineensis*	FT	57 (1)+	24 (14)*
		NT	8	53*
		FL	1	0
		PI	0+	0
Coula	*Coula edulis*	NT	0	5
Okra	*Hibiscus esculentus*	LF	18	0
		FL	1	0
Raphia palm	*Raphia gracilis*	GM	10	0
Sugarcane	*Saccharum officinarum*	PI	1	0
Avocado	*Persea americana*	LF	0	2
Yam	*Dioscorea*	TB	0	1
Grapefruit	*Citrus grandis*	FT	0	1

FT fruit, *LF* leaf, *PI* pith, *TB* tuber, *NT* nut, *FL* flower, *GM* gum
Numbers in brackets indicate the frequency where more than one crop part was consumed within one crop-raiding event, and + (guarded), * (abandoned) indicate the parts. Please note that oil-palm fruits and nuts, and coula nuts, were supplied during the study period for nut-cracking experiments (see Chaps. 16 and 17 for details); these are included in the abandoned section. In addition to the results presented, there were five events where multiple simple-sugar crops were raided and two in which multiple carbohydrate, protein, and lipid cultivated foods were raided

Wrangham 2005; Marshall and Wrangham 2007), never constituting the entire diet and not sufficient on its own to sustain chimpanzees at Bossou (Hockings et al. 2010b). This observation shows that when available, underground storage organs can become an important food for chimpanzees inhabiting tropical wet forests, as well those inhabiting drier environments (Hernandez-Aguilar et al. 2007).

Certain crops were raided in direct response to wild fruit scarcity, whereas others were raided according to their availability (Hockings et al. 2009); this illustrates the significance of analyses of individual and specific groups of cultivated foods. Such variations also illustrate the importance of crop choice by farmers when establishing land management strategies for alleviating human–primate conflict, and how simple measures can sometimes be adopted by local people to reduce human–ape contact, especially when preferred chimpanzee foods are involved. For example, the presence of papaya trees in Bossou village frequently brings chimpanzees into proximity with people's houses, and local people perceive themselves, especially children, to be at increased risk of chimpanzee attacks (Hockings et al. 2010a). This fear has resulted in some people cutting down papaya trees located near the forest edge in an attempt to reduce human–chimpanzee contact (Hockings 2007).

Certain cultivated foods were preferred by the Bossou chimpanzees and others were not, which raises the possibility that different chimpanzee populations exhibit material and behavioral cultural patterns in their usage of cultivated foods. For example, during the study period the Bossou chimpanzees were not observed feeding on guava or avocado fruits, even though these are readily eaten by captive chimpanzees (de Nijs 1995; Matsuzawa 2007, personal communication); avocado trees were widely available and a guava tree was present in a regularly visited orchard. In contrast, sugarcane was rare throughout the chimpanzees' home range but, when encountered, was preferentially raided compared to other cultivated foods.

22.3.2   Crop-Raiding: A Behavioral Approach

Fluctuations in the composition of temporary parties should reflect differences in costs and benefits of association. For example, Boesch (1991a: 236) reports that "Taï chimpanzees are mostly found in parties with the best defense capacities (mixed and all-male) that allow both sexes to profit from others' support." Party compositions during crop-raids at Bossou vary depending on factors that influence the degree of risk associated with raiding, including location and the presence of people. More specifically, simple-sugar fruit-raiding by adult male-only parties is associated with greater exposure and degree of risk: male-only parties are again more likely than other party compositions to raid in the village, to raid further from the forest edge, and to raid in more highly guarded situations (Fig. 22.1). Despite adult male chimpanzees exhibiting elevated levels of rough self-scratching, which is a potential indicator of anxiety, when raiding in the presence compared to absence of local people, male-only parties are more likely than any other party composition to crop-raid when local people were present. Interestingly, in a comparison restricted to forested locations, a higher proportion of "no-male" parties are found feeding on abandoned mango fruits compared to raided simple-sugar fruits. In general, male-only parties are more common when the degree of risk increases, suggesting a perception of the need for greater security, whereas females and immatures rarely raid in exposed areas in the absence of an adult male. Although further research is required, observations suggest that it is often possible to predict when chimpanzees at Bossou will crop-raid according to their focused directional movements and location in the forest. If party members perceive the degree of risk during crop-raiding to be too high, they will often wait in the forest next to where others are raiding. Chimpanzees at Bossou appear to evaluate the degree of risk posed by different cultivated foods and within different party compositions in an attempt to decide whether to crop-raid.

Fig. 22.1   An adult male chimpanzee raiding pineapple fruit (photograph by Laura Martinez)

A positive association exists between the number of adult males and the proportion of time that they spend feeding on raided compared to wild foods. When all three males were present, a greater proportion of time was spent feeding on raided foods. Conversely, when only one or two males were present, more time was spent feeding on wild foods. Additionally, as the number of males in a raiding party increased, so did the likelihood that they would raid in the presence of people and further from the forest edge. The perceived risk associated with crop-raiding may be reduced by having more adult males present, and the presence of cooperative partners might increase the readiness to crop-raid. Although there is some debate over the extent to which male chimpanzees hunt cooperatively (see Boesch and Boesch-Achermann 2000; Mitani and Watts 2001; Gilby 2006), it is worth remembering that they may decide to hunt for social as well as simple energetic reasons (Teleki 1973; Mitani and Watts 2001). Crop-raiding certainly provides energetic benefits, but as has been proposed for hunting, it might also provide males with opportunities to "show off" their boldness (Teleki 1973), especially when other individuals get access to some of the food.

As the conversion of forested habitats to agricultural fields continues; the potential for the acquisition and propagation of new feeding habits within primate populations increases (Takasaki 1983). In the case of the Bossou chimpanzees, such adaptation includes behaviors to cope with the risks associated with crop-raiding. By transporting food from a risky environment to one of relative safety, the chimpanzees reduce the amount of time spent in an exposed area and thus reduce the likelihood of detection by people. In general, when food was obtained in wild, abandoned, and supplied conditions, it was rarely transported to another place for feeding. However, chimpanzees commonly transported guarded foods back to the safety of the forest

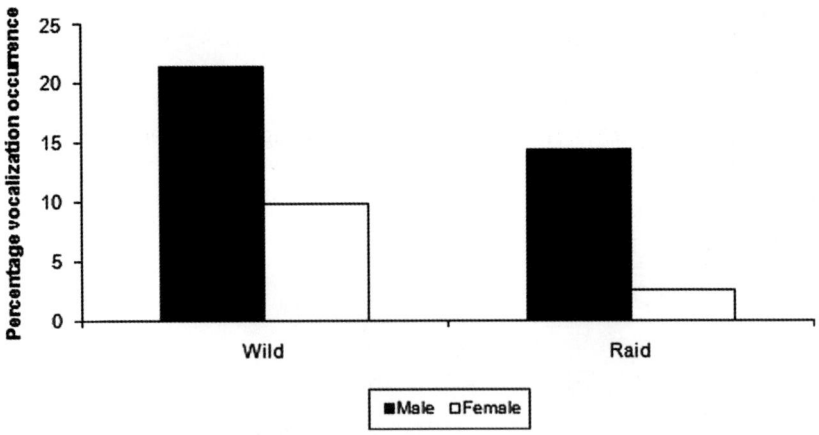

Fig. 22.2 The percentage of events during which males and females vocalized during wild fruit- and raid fruit-feeding situations

before feeding, especially when the crops were obtained at a greater distance from the forest edge (Hockings et al. 2007). The Sonso community of chimpanzees at Budongo, Uganda also exhibited this transport behavior when raiding sugarcane from plantations bordering the forest (Hockings 2003, personal observation).

Past research suggests that chimpanzees may suppress vocalizations as a tactic to avoid detection by predators and other chimpanzees (Wilson et al. 2001). The Bossou chimpanzees vocalized less frequently when feeding on raided fruits compared to wild fruits, with females showing the greatest reduction (Fig. 22.2); this is consistent with females being more afraid than males when raiding. The chimpanzees did vocalize relatively more when feeding on simple-sugar cultivated foods and rarely vocalized when raiding other crop types. However vocalization rates did not vary with the degree of risk associated with simple-sugar fruit-raiding. In accordance with past research (Clark 1993), adult males were the most vocal, especially when crop-raiding. Studies suggest that male pant-hoot vocalizations are directed to particular audiences within their social group specifically to maintain contact and to recruit allies and associates (Mitani and Nishida 1993). Adult males were less likely to crop-raid alone, but when they did, they were more likely to remain quiet; vocalization rates increased when two or three males were present during crop-raids. Although this might be a function of number of individuals, adult males may simply be more confident when other males are present. However, other chimpanzee communities reportedly use pant hoots to advertise the presence of large parties in an effort to deter approach by extra-group chimpanzees (Nishida et al. 1985). It is possible that the Bossou adult males may be promoting their presence in an area and intimidating local human bystanders rather than other chimpanzees. The use of complex communicative behavior in risky situations constitutes a compelling line of inquiry regarding cooperative networks in chimpanzees.

To facilitate detection of potential risks in fields or farms when crop-raiding, the Bossou chimpanzees often visually scanned the surrounding areas intently before entering. As the chimpanzees' forest is located on hills, they are already in an elevated position to scan potential cultivated foods in the village. It would be interesting to examine how chimpanzees might use their environment to their advantage when entering fields and orchards within the forest. Scanning behaviors are sometimes enhanced by taking up bipedal positions and climbing up vegetation. In contrast, bipedal standing within the forest was less frequent than in the context of crop-raiding, and most occurrences were in a display context. Quantitative measures such as head turn rate would be useful for accurately determining the role of vigilance as a function of degree of risk.

22.3.3 Crop-Raiding: A Social Approach

In addition to nutritional reasons, crop-raiding can provide adult male chimpanzees at Bossou with highly desirable food commodities that might affect their socio-sexual behavior. Adult chimpanzees at Bossou very rarely transferred wild plant foods (excluding transfers of wild foods from mother to infant) or food from abandoned sources. In contrast, they share cultivated plant foods, raided from orchards and fields, much more frequently. Overall, papaya, the largest and most easily divisible cultivated fruit available, was the most frequently shared crop type. Other cultivated plant foods (includes pineapple and orange fruits and the cassava tuber) were generally shared but less often than papaya fruit (for detailed descriptions, see Hockings et al. 2007).

Sharing primarily consisted of adult males allowing reproductively cycling females to take food, mostly papaya fruit that they possessed. This behavior is referred to as "tolerated theft." Adult males shared particularly with one cycling female who at the time of this study took part in most of all the consortships with males. The second-ranking adult male, who shared most with this cycling female, was also her most frequent consort and grooming recipient. In contrast, the alpha male shared less frequently with this female, and in spite of his dominance, was less likely than the second-ranking male to consort with and receive grooming from her. Males shared crops with a maximally swollen female infrequently and were never observed mating with that female immediately after sharing. Crop-sharing episodes were nonaggressive, and clusters of individuals begging the possessor for a share of raided crops was rare. Although further research is required to ascertain the importance of begging intensity on sharing patterns, it appears that chimpanzees share the fruits obtained during crop-raids to enhance affiliative relationships with reproductively valuable females.

Males rarely shared cultivated foods with one another, despite failing to obtain a fruit in more than one third of all papaya raids; crop sharing at Bossou therefore does not appear to enhance cooperative raiding, as proposed for meat sharing at other sites (Boesch and Boesch-Achermann 2000). Mother–offspring sharing was relatively common at Bossou, but this sharing pattern, in addition to male–immature sharing patterns, may be explained on the basis of kin selection. Adult females never seem to share crops with unrelated adults; however, they do occasionally share with unrelated immatures (Hockings et al. 2007).

In summary, chimpanzees at Bossou exploit a high variety of cultivated foods that are fully integrated into their dietary repertoire. The chimpanzee's crop-raid in parties with protective compositions while assessing the costs and benefits of feeding on different crops. They also vocalize less and show more food transportation and specific vigilance behaviors when crop-raiding. Analyzing how chimpanzees adapt their social organization and behaviors when exploiting different environments bolsters our understanding of why chimpanzees make such successful crop-raiders. Effective management of human–chimpanzee conflict situations requires understanding how chimpanzees adjust their behavior when engaged in crop-raiding.

22.4 Conclusions

A chimpanzee's decision to crop-raid will be subjective to an intricate web of factors. Although there may be overlap between chimpanzee communities in the factors that influence choices about whether to raid, each community may be faced with different combinations of social, ecological, and cultural factors to those found at Bossou. Crop-raiding is definitely not tolerated throughout Africa to such an extent as at Bossou, and thus chimpanzees will have to evaluate a localized set of costs and benefits when deciding whether to crop-raid.

Elevated levels of crop-raiding by wildlife are a by-product of natural resources becoming less available and the nutritional riches of agricultural production becoming increasingly known to them. This change will result in most wildlife species, including chimpanzees, being considered progressively more problematic to local and commercial farmers. This pressure in turn increases the need to develop interdisciplinary conflict mitigation techniques (see SSC primate specialist group guidelines by Hockings and Humle 2009), which require data on human and great ape behavior and ecology combined with a complete understanding of local people's perceptions of the situation.

Acknowledgments I wish to thank all the local assistants and Bossou villagers who helped during this research period. This work was supported by a Stirling University studentship, a postdoctoral research grant from Fundação para a Ciência e a Tecnologia, Portugal, and MEXT grant #20002001, JSPS-HOPE, and JSPS-gCOE (A06, Biodiversity).

Chapter 23
Behavioral Flexibility and Division of Roles in Chimpanzee Road-Crossing

Kimberley Jane Hockings

23.1 Introduction

Regularities in spatial patterns are a well-known occurrence in the animal kingdom; for example, during group movements monkeys reduce the risk of predatory attacks through adaptive spatial patterning (Altmann 1979; Bicca-Marques and Calegaro-Marques 1997; DeVore and Washburn 1963; Rhine and Tilson 1987; Waser 1985). Although increased survival is their ultimate function, at a proximate level differences among age- and sex-classes in fear or confidence may result in nonrandom progression orders, whereby positioning will alter as a function of perceived or anticipated danger. Busse (1980) reported that the majority of lion attacks on baboons in Moremi Wildlife Reserve in Botswana were from the direction in which the group was traveling. In this context it is noteworthy that adult males in several savannah baboon troops tend to move toward the front of the group during progressions and to a lesser degree toward the rear (Rhine and Westlund 1981). In adult male chacma baboons, their forward tendency increases when approaching a waterhole with potential predators, with their rearward tendency also increasing when retreating from a source of danger (Rhine 1975; Rhine and Westlund 1981; Rhine and Tilson 1987). This slight flexibility of response shown by monkeys during progression orders highlights the possibility that individuals might be cooperating – defined as joint action for mutual benefit – to maximize party protection.

Chimpanzees have flexible fission–fusion social systems, and the social organization they employ in human-influenced risky situations might be comparable to some aspects of their strategies for predator avoidance (Sakura 1994; Hockings 2007). However, almost nothing is known about progression order in chimpanzees. Sakura

*Electronic supplementary material The online version of this chapter (doi: 10.1007/978-4-431-53921-6_24) contains supplementary material, which is available to authorized users.

K.J. Hockings (✉)
Department of Anthropology, Faculty of Social and Human Sciences, New University of Lisbon, Avenida de Berna, 26-C, 1069-061 Lisbon, Portugal
and
Department of Psychology, Stirling University, Stirling, UK
e-mail: hock@fcsh.unl.pt

(1994) reported that the Bossou chimpanzees formed into parties that usually included the alpha male before crossing a road; this was interpreted as resulting from heightened perceived risk. Additionally, it was reported that the first individual to visually scan and cross the road was nearly always the second-ranking male, not the alpha male (Matsuzawa and Sakura 1988). On one occasion, Matsuzawa (personal communication, 2006) observed an adult female chimpanzee, struggling to carry both her infant and fearful juvenile during a road-crossing, pass her infant to an adult male chimpanzee that was traveling besides her. Although the limited data available on the effects of such situations on chimpanzee behavior are inconclusive (Itani and Suzuki 1967), the interplay between risk and vulnerability can produce a complex set of adaptive behaviors (Miller and Treves 2006). For example, cooperative behaviors by chimpanzees might prove beneficial during road-crossings by increasing the protection of more vulnerable group members; this has implications regarding the importance of altruistic behaviors, which are often difficult to observe in wild chimpanzee communities.

The Bossou chimpanzees often hesitate to cross roads because of wariness of people and the risk of injury from vehicles traveling at speed, but they frequently need to do so to access foraging sites in their relatively constrained home range. Given that adult males are usually the most physically powerful group members, they might be most likely to enter unexplored areas. As a consequence, they may be expected to take up higher-risk positions during group movements in human-dominated environments (DeVore and Washburn 1963; Hamilton 1971; Matsuzawa and Sakura 1988). Opportunities to test the fear hypothesis of progression orders are quite rare in wild primates because naturally occurring fearful events are usually difficult to anticipate or predict (Rhine and Tilson 1987). However, road-crossing provides an excellent opportunity to analyze this aspect of sociospatial organization.

23.2 Approach

The village of Bossou and the chimpanzees' home range are dissected by one large road (approximately 12-m wide) that stretches from the Guinea–Liberia border through into the forested region of Guinea. A narrower dirt road (approximately 3-m wide) branches off from the large road and is used by pedestrians (Fig. 23.1). The chimpanzees must cross both roads to move from one forested area to the next. Study 1 was carried out between January and April 2005 following the widening of the road, and a follow-up study (study 2) was carried out from November to December 2005. Both roads had forest cover to the roadside and were separated by a middle zone of secondary forest and coffee plantations that normally takes 2–3 min to cross. Except for researchers and field assistants, local people were never observed in this area during the study period. The chimpanzees crossed going from west to east and from east to west. When moving from west to east they passed from the forest of the hill of Gban, onto the large road, then into the middle zone, across the small road, to reach the forest of Guein; the reverse itinerary applied from east to west.

Throughout these study periods the community size remained at 12 individuals (including three adult males, five adult females, three juveniles, and one infant). I

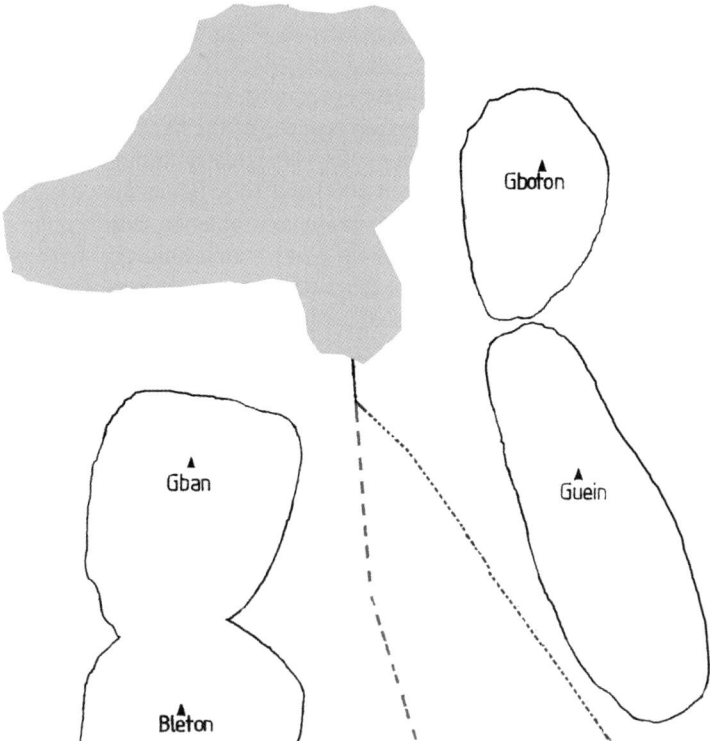

Fig. 23.1 A map of the field study site Bossou showing the main parts of the village and three main hills (forested) of Gban, Guein, and Gboton; the presence of the large and small roads is *highlighted*

recorded exact progression orders in which chimpanzees left the forest and moved onto the road at regular crossing points. It was possible for chimpanzees to visually scan the road without crossing the counting line. The first chimpanzee to scan the road was recorded; this individual was not necessarily the first to cross the counting line. The latency between arrival of the first individual to scan and the last party member to cross the forest–road edge was termed waiting time. Presence or absence (either auditory or visual) of humans and vehicles at each road-crossing event was recorded. For analyses, the expected frequency of being first to scan and to cross, or being last to cross, were calculated from the mean number of adult males per progression divided by the mean number of group members present (excluding the dependent infant). Party compositions are categorized as adult male-only, mixed (at least one adult male, adult female, and immature present), adult male and other (at least one adult male and at least one adult female or immature), or no male (adult females and immatures only). The frequency of pedestrians and vehicles present on both the roads from 0700 to 1829 hours were counted on three randomly selected nonmarket days and then averaged.

In this chapter I present data on party progression orders and associated behaviors during road-crossings, with the aim of assessing whether chimpanzees exhibit any flexibility in their responses over time and according to degree of risk when coping with this anthropogenic aspect of their environment.

23.3 Degree of Risk

Road-crossing involves leaving forest for open areas (Fig. 23.2) and is a potentially risky situation for chimpanzees at Bossou because of the likelihood of vehicle and human presence. Vehicle presence does not vary greatly throughout the day, but pedestrian numbers are lowest between 1030 and 1630 hours. However, the chimpanzees do not always time crossings to avoid confrontation with people: crossing times ranged from 0650 hours through to 1740 hours, although there were two peaks in road-crossing times, the first from 0800 to 0930 hours when pedestrian numbers were high and an additional peak in road-crossings from 1100 to 1330 hours when pedestrian numbers were low.

The elevated risk associated with crossing the large road compared with the small road during both studies is reflected in increased waiting time (study 1: median, 80 vs. 4 s; study 2: median, 40 vs. 10 s) and increased levels of rough self-scratching by chimpanzees, a potential indicator of anxiety (Aureli and de Waal 1997; Hockings et al. 2007; van Lawick-Goodall 1972). The presence of people and vehicles is probably the crucial factor in the apes' assessment of each road-crossing event. However, the reduction in waiting time at the large road in study 2 (study 1, 80 s vs. study 2, 40 s) highlights the possible influence of neophobia, defined as caution toward novel features of the environment, on the chimpanzees' perception of risk of the newly widened road in study 1: over time, the chimpanzees became used to the new condition. It is also possible that this difference results from the absence of vehicles during large road-crossings in study 2. As party size increased, waiting time on both roads also increased. Although this may simply be a case of

Fig. 23.2 Chimpanzee progression during a large road-crossing at Bossou (photograph by Kimberley Hockings)

larger parties taking more time to get organized, an alternative explanation is that as the number of vulnerable individuals present increases, the party as a whole becomes more cautious. Party waiting time on the large road was further influenced by the direction of travel: when initially leaving the protection of a forested area to move into a more open area, the chimpanzees are more cautious, indicated by longer waiting times. The larger field of observation available to the chimpanzees when crossing the large road did not reduce waiting times, which reinforces the idea that the chimpanzees are influenced by the inherent degree of risk in this environment.

23.4 Division of Roles

During dangerous excursions certain positions within a progression might be more advantageous than others, depending upon age and sex. In the case of road-crossing, the first individual to scan the road checks for potential danger, but can also be caught unaware, whereas the individual leading the progression might have to face unnoticed dangers. Additionally, the last individual may perhaps be exposed to risk or get left behind. In contrast, the individuals in the middle occupy the safer positions.

Most road-crossings include at least one adult male; but some do occur when no males are present. In this situation, females and juveniles often run speedily to cross the roads, especially the large road. This technique obviously minimizes the time spent exposed out of the forest and at potential risk from vehicles and humans. However, to analyze correctly the division of roles in a road-crossing party it is necessary to have a combination of age- and sex-classes; therefore, only mixed-party progressions were included. The analysis of road-crossing progressions during study 1 focused on 28 mixed-party progressions (17 small and 11 large road-crossings), and the second analysis of road-crossing progressions focused on the data from 25 mixed-party progressions (12 small and 13 large road-crossings); all three adult males were present in both studies.

Overall, adult males mostly take up forward positions; however, variations existed in an individual's positioning during the two study periods (Fig. 23.3). From data collected after the large road was widened (study 1), comparisons of small and large road-crossings showed that the second- and third-ranking Bossou males were often first to scan and to cross at both roads. Although the first individual to scan was usually the first individual to cross, this happened less frequently on the large road than the small road, which may be caused by greater uncertainty or risk when crossing the large road, resulting in the first individual continuing to survey the surrounding area more thoroughly. This possibility could explain the decreased frequency of the second-ranking male being first to cross but not first to scan at the large road, and the consequent increase in the elderly third male and alpha female leading the large road progressions. The alpha male was more rearward during large road-crossings, where the risk of crossing was higher, whereas the elderly alpha female showed a dramatic reduction in frequency of being last when crossing the large road compared to the small road. This female may have brought up the

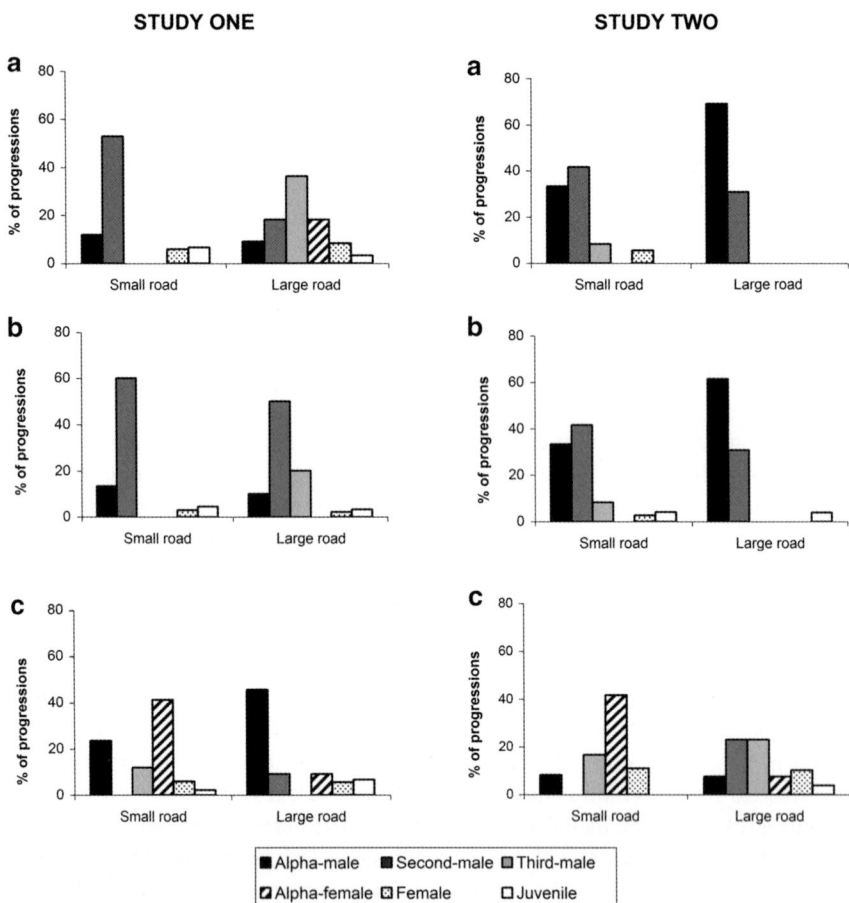

Fig. 23.3 The percentages of study 1 (small, 17; large, 17; as seen in Hockings et al. 2006) and study 2 (small, 12; large, 13) progressions in which the three males, the alpha female (determined by the direction of pant grunts by other females), and the average non-alpha female and juvenile were (**a**) first to cross, (**b**) first to scan the road, and (**c**) last in the progression on the two roads

rear more frequently than expected on the small road as she was physically slower than other members of the group. When the perceived threat increased on the large road, however, she took up a more frontward position. Excluding the alpha female, each adult female and juvenile was first or last in very few progressions. Additionally, when the alpha male was present in mixed-party progressions containing one other adult male (second-ranking male), he was first to scan and cross in half of large road-crossings and last in only one third, suggesting that his rearward position at other times was not the result of fear.

During study 2, the alpha male showed a dramatic reduction in his rearward positioning during both large and small road progressions and increased his frontward positioning particularly on the large road. In contrast to study 1, the second- and third-ranking males increased their rearward positioning, meaning that across both studies, adult males were rearward in a similar percentage of progressions. The alpha female again showed a decrease in her rearward presence when crossing the large road compared to the small road. One can speculate that this old female makes a special effort not to get left behind when crossing the large road. As in study 1, each adult female (excluding the alpha female) and each juvenile was first or last for a very low percentage of progressions.

Guarding behavior is defined as "standing in a quadrupedal posture on the road for more than 5 s without moving." The individual to exhibit this behavior was usually the first individual to cross the road and was predominantly the alpha male. Guarding appears to be a response to the presence of people on both the roads, although it does not occur every time that people are present: it is possible that when females and juveniles are present during road-crossings, the alpha male increases his protective role in the group as an opportunity to "show off" his boldness (Hockings et al. 2007; Teleki 1973). Guarding behavior has also been seen during road-crossings by eastern lowland gorillas *(Gorilla gorilla graueri)* in Kahuzi-Biega National Park in the Democratic Republic of Congo. In this instance, a silverback male displayed similar protective guarding behaviors while female and immatures hurriedly crossed a road that cuts through the forest (movie clip by Deschryver; year unknown). On occasion juvenile males at Bossou were observed joining the adult males in exhibiting guarding behaviors. Although more detailed studies are required, it supports the idea that younger members of the group are learning in a "master-apprentice" fashion (Matsuzawa 2006b) and that cultural behaviors when road-crossing are being passed down through generations. It would be interesting to examine whether the occurrence of guarding behavior varies according to the nature of the human presence, for example, age- and sex-class of the people present, and their behavior.

Figure 23.4a illustrates a hypothetical example of a large road-crossing where no observable sociospatial organization is employed by party members; adult males, adult females, and juveniles are randomly distributed within the progression. In comparison, the time chart (Fig. 23.4b) of an actual large road-crossing at Bossou visually demonstrates the complexity of the crossing, including waiting time and guarding behavior, and the protective positioning of the adult males.

23.5 Antipredator Behavior?

Miller and Treves (2006) suggested that primates may have developed specialized behaviors for avoiding human predators, but relevant data are lacking. Modern Bossou chimpanzees encounter predators infrequently (Sugiyama 2004), and although humans themselves are not "predators" of these chimpanzees, they might harass them. To adapt to more recent and current dangerous situations, the Bossou chimpanzees seem to employ a phylogenetically old mechanism that involves an ancient male tendency to stand up to predators. However, the positioning of dominant and bolder individuals is flexible over time and changes depending on both the degree of risk and number of adult males present. This variation promotes the likelihood that dominant individuals are acting cooperatively with a high level of flexibility to

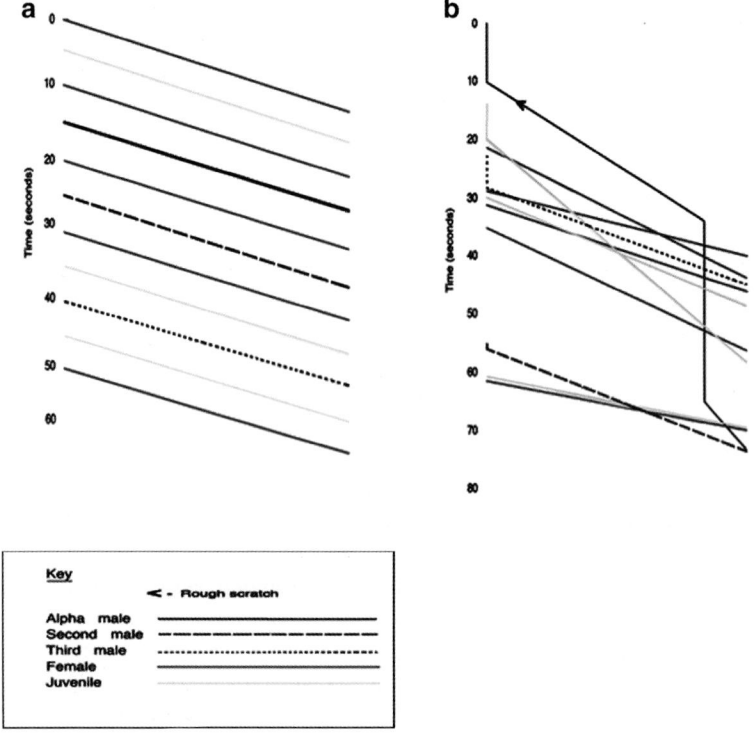

Fig. 23.4 Time chart illustrating the complexity of a large road-crossing progression: an example of a road-crossing with no sociospatial organization (**a**), and an actual large road-crossing (**b**), with crossing time in seconds on the *y*-axis and the width of the road on the *x*-axis

maximize the protection of the party, something that has not been concluded from studies of wild monkeys. To this aim they display a degree of collective intention, suggesting knowledge of party composition, relative vulnerability of members, and physical capabilities (McGrew 2004). This finding has obvious implications regarding the importance of altruistic behaviors, which are often difficult to observe in wild chimpanzee communities. At a proximate level each individual may have preferred and recognized positions; however, it remains unknown whether positioning is individual- or rank specific. A long-term study on progression orders in relationship to changes in rank would be informative in this regard.

23.6 Conclusions

In contrast to reports of high mortality rates during road-crossings in other primate species (e.g., *Colobus angolensis* in Diani Beach, Kenya), there is no evidence that chimpanzees at Bossou have been killed or injured by vehicles when road-crossing: this might be a testament to the caution and cooperative techniques that they employ.

Behavioral adaptations do not fossilize, so it is unknown how early hominins traveled in risky habitats. Data on progression orders of other great ape populations are required to help shape hypotheses about emergence of this hitherto neglected aspect of hominoid adaptive social organization. A changing human-dominated landscape provides chimpanzees throughout Africa with fresh challenges and researchers with opportunities to study behavioral flexibility in wild situations. However, a significant number of ape populations precariously reside in areas that are seriously affected by habitat alteration and human disturbance, such as mining and logging concessions (Hockings and Humle 2009; see Chaps. 5, 39, 40). In the latter case, Morgan and Sanz (2007) highlight that the conservation outlook of these endangered apes will improve significantly if forestry companies make changes in management policies to reduce logging road width. They estimate that the total width of forest cleared for primary and secondary roads, including graded portion and shoulders, should not exceed 12.5 and 8.5 m, respectively (for further details, see Morgan and Sanz 2007). In agreement, this study has shown that chimpanzees' perception of risk during road-crossing increases with road width; therefore, to reduce individual stress or possibly community isolation, roads that dissect great ape home ranges should be as narrow as possible.

Acknowledgments I am grateful to the Direction Nationale de la Recherche Scientifique et Technologique (DNRST) and the Institut de Recherche Environementale de Bossou (IREB), the Republic of Guinea, for granting permission to carry out this research. I wish to thank all the local assistants and Bossou villagers who helped during this research period. This work was supported by a Stirling University studentship, MEXT grant #16002001, and JSPS-HOPE.

Chapter 24
Play Behaviors Involving the Use of Objects in Young Chimpanzees at Bossou

Masako Myowa-Yamakoshi and Gen Yamakoshi

24.1 Play: A Step Toward Using Tools

Decades-long field studies have revealed a wide variety of tool-using behaviors in wild chimpanzees. For example, chimpanzees in Bossou, Guinea, use twigs as dipping sticks to collect ants, leaves as sponges to drink water, and pairs of stones as a hammer and anvil to crack open oil-palm nuts (Matsuzawa 1994; Sugiyama 1993; Yamakoshi and Myowa-Yamakoshi 2004; see Chaps. 6, 8, 9, 16–18). A recent extensive comparison of seven long-term wild chimpanzee study sites identified regional variations in 39 behavioral patterns that appeared unlikely to be the result of differences in local environmental conditions and have been proposed as putative examples of chimpanzee "cultures" or "traditions" (Whiten et al. 1999).

Chimpanzee infants spend several years learning local-specific tool-use techniques. During this learning process, they observe and attempt to copy others' behaviors. Interestingly, during this process, they appear to play rather than simply learn. For example, it takes 3.5 to 5 years for chimpanzees at Bossou to master nut-cracking using a pair of stones (Matsuzawa 1994; see Chaps. 18 and 21). Before mastering the technique, chimpanzee infants manipulate the nut or stone in their own ways. Chimpanzee infants start manipulating the nut or the stone when less than 1 years of age by touching, mouthing, and pushing the stone around. At around 2 years of age, they start combining objects in various ways, for example, stacking and pushing one stone on top of the other (Inoue-Nakamura and Matsuzawa 1997; see Chap. 18). Even though at this early age infant chimpanzees are unable to crack nuts to reach the edible kernel within, they eagerly continue to manipulate the nut or the stone as if enjoying the process of manipulating objects. Through repetitive

M. Myowa-Yamakoshi (✉)
Graduate School of Education, Kyoto University, Yoshida-honmachi, Sakyo, Kyoto 606-8501, Japan
e-mail: myowa@educ.kyoto-u.ac.jp

G. Yamakoshi
Graduate School of Asian and African Area Studies, Kyoto University, 46 Shimoadachi-cho, Yoshida, Sakyo-ku, Kyoto 606-8501, Japan
e-mail: yamakoshi@jambo.africa.kyoto-u.ac.jp

T. Matsuzawa et al. (eds.), *The Chimpanzees of Bossou and Nimba*,
DOI 10.1007/978-4-431-53921-6_25, © Springer 2011

Fig. 24.1 Play-face in an infant chimpanzee (photograph by ANC Production)

experience with trial and error, infant chimpanzees learn the function of, or connection between, objects, such as using a stone to crack hard-shelled nuts to gain access to the kernel within. After 3.5 years of age, infants finally master this stone-tool use technique (see Chap. 18).

It is likely that play behaviors involved in the manipulation of objects provide important opportunities for infant chimpanzees to learn various tool-using techniques. In this chapter, we focus on young chimpanzees' object manipulation in the context of play in their natural habitat.

Similarly to humans, chimpanzee infants play frequently. While their mothers eat or rest, the infants can be seen wrestling, jumping off trees, and throwing twigs, stones, and other objects. We can identify their play behaviors by detecting specific facial expressions, such as smiling. Chimpanzees' smiles are similar to those of humans; they pant and open their mouths with their eyes half open. This type of facial expression is called a "play-face" and is observed during play (Fig. 24.1). In this study, we identified chimpanzees' play behaviors whenever their behaviors were accompanied by a "play-face."

24.2 Observation of Play Behaviors in Young Chimpanzees at Bossou

During this study, Bossou chimpanzees numbered about 21 individuals. We observed the use of objects during play behavior in eight chimpanzees younger than 6 years of age (see Table 24.1 for details). Behavioral data were collected during two study periods. We conducted a 1-month study from December 1996 to January 1997 (30 days) and a 3-month study from January to March 1998 (62 days). Video-recording was used during observations during these two periods using digital

Table 24.1 Eight individuals observed during the two study periods

Name	Sex	Birth year	Age group (years)
Fotaui	F	1991	5–6
Vuavua	F	1991	5–6
Yolo	M	1991	5–6
Poni	M	1993	3–4
Nto	F	1993	3–4
Juru	F	1993	3–4
Pokuru	M	1996	0–2
Fanle	F	1997	0–2

M male, *F* female

video camcorders (Sony DCRTRV9 and Sony DCR-PC10). We recorded on an ad libitum basis all behaviors observed involving the use of objects.

A play bout started when the target individual began touching the object(s) and ended more than 30 s after the individual or playmate(s) stopped manipulating the object(s). Although the definition of play was fairly restricted, that is, it required the use of objects and did not include behaviors such as games involving chasing and wrestling (i.e., dyadic interactions), we recorded 229 play bouts involving the use of objects. These bouts were divided into two groups: (1) play bouts during which a single chimpanzee manipulated objects (*solitary* play), and (2) play bouts involving more than two individuals participating in the manipulation of the same object (*social* play). Among the 229 play bouts recorded, 112 (48.9%) were categorized as solitary play and 117 (51.1%) were categorized as social play.

The individuals who participated in play behavior involving the use of objects were categorized into three age groups: (1) 0–2 years, (2) 3–4 years, and (3) 5–6 years (see Table 24.1). We also divided all play bouts into those involving the use of *nondetached* objects such as a tree trunk, and those involving the use of *detached* objects, such as a loose branch.

24.3   Types of Play and Developmental Changes in Play Behaviors

24.3.1   Solitary Play

A total of 112 cases of solitary play were observed; 42 (37.5%) involved the use of nondetached objects and 70 (62.5%) involved the use of detached objects. The most typical objects used for play were branches, vegetation bundles (including leaves and fruit), and vines. Table 24.2 shows the list of objects used during solitary play. Age differences were found in the type of solitary play. There was a significant difference among the three age groups ($\chi^2 = 14.60$, $df = 2$, $P < 0.01$). A post hoc multiple comparison using Ryan's method revealed significant differences

Table 24.2 Instances of solitary play among young chimpanzees at Bossou

Object	Behavior
Nondetached	Hang on a bundle of vegetation or vines
	Hit or kick a bundle of vegetation or vines
	Swing on a bundle of vegetation or vines
	Pull vines
	Hit or kick a tree trunk
Detached	Swing a branch
	Have a branch/mouthing an elephant's-ear fern (*Platycerium angolense*)
	Hit a branch on the ground
	Hit a tree trunk with a branch
	Put a branch/an elephant's-ear fern/a musanga stipule (*Musanga cecropioides*)/a carapa nut (*Carapa procera*)/a pod of *Parkia bicolor* in groin pocket
	Wrap a vine/a fruit of *Parkia bicolor*/around neck
	Groom a leaf (leaf-grooming)
	Clip a leaf (leaf-clipping)
	Throw leaves/a carapa nut
	Kick a musanga stipule/a carapa nut/a pod of *Parkia bicolor*
	Hit bark/a carapa nut/a pod of *Parkia bicolor*/a stone
	Hold bark/an elephant's-ear fern/a musanga stipule/a carapa nut/a stone/a bundle of vegetation
	Place bark on abdomen
	Rub bark/a carapa nut/a leaf
	Cover face with a bundle of vegetation/bark

between the 5–6 years and 0–1 year age groups (Fisher's exact test: $\chi^2 = 6.60$, $df = 1$, $P < 0.05$, two-tailed) and the 5–6 years and 3–4 years groups ($\chi^2 = 14.60$, $df = 1$, $P < 0.01$, two-tailed). Results showed that individuals aged over 5 years played using nondetached objects more frequently than detached objects compared to the other two age groups (Fig. 24.2). The most frequently used detached object was a branch. In addition, the chimpanzees used colorful objects such as the golden-brown seeds of *Carapa procera* trees and the red stipules of *Musanga cecropioides* trees, which are approximately 15–20 cm long (Fig. 24.3). The chimpanzees carried these objects in their groin pocket or repeatedly threw them overhead, as if playing catch.

24.3.2 Social Play

A total of 117 cases of social play was observed: 78 (66.7%) involved the use of nondetached objects and 39 (33.3%) involved the use of detached objects. Most instances of social play with detached objects involved incidents in which chimpanzees broke off branches and threw them at playmates (51.3%). Social play involving the

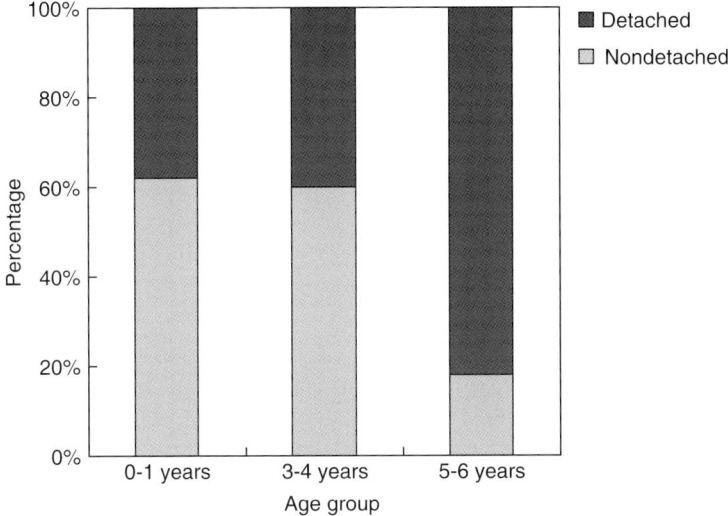

Fig. 24.2 Age differences in solitary play. Individuals aged over 5 years played using detached objects more frequently than nondetached objects

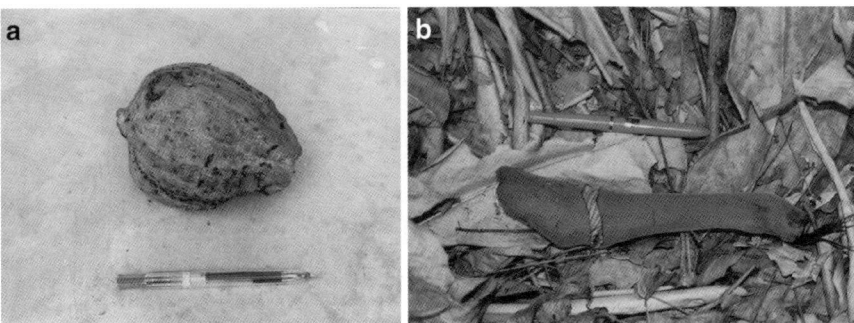

Fig. 24.3 The colorful (*vivid red* or *yellow*) objects that young chimpanzees manipulated in the context of solitary play: (**a**) carapa nut (*Carapa procera*) and (**b**) musanga stipule (*Musanga cecropioides*) (photographs by Masako Myowa-Yamakoshi)

use of detached objects was only observed in individuals 3 years old or older. No social play involving the manipulation of a detached object was observed in the 0- to 1-year-old group (Fig. 24.4). These younger chimpanzees interacted with one another using nondetached objects, such as hanging onto and swinging from bundles of vegetation or vines suspended from a tree bough.

Bouts of social play were classified into three categories: (1) *one-way* play, during which an individual manipulated an object(s) in the direction of another playmate(s); (2) *two-way* play, during which an individual manipulated an object(s) jointly with the other playmate(s) (Fig. 24.5); and (3) *object or action-role turn-taking* play, during which an individual manipulated object(s) with a playmate(s)

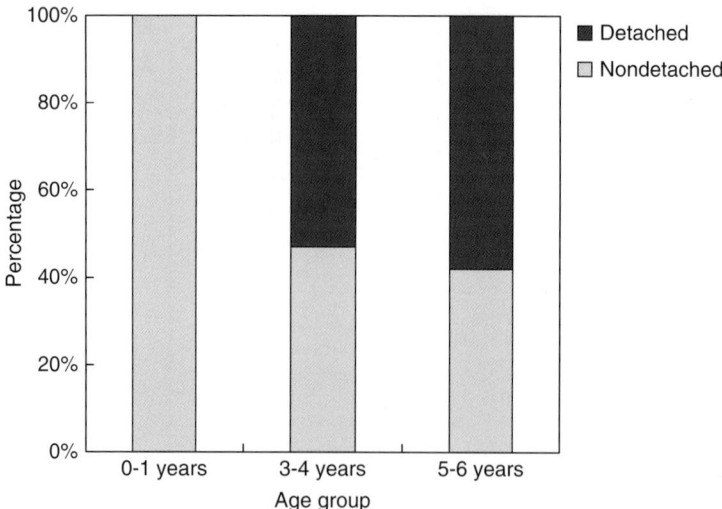

Fig. 24.4 Age differences in social play. Detached objects were not used by individuals 0–1 year of age

Fig. 24.5 An example of two-way social play. A chimpanzee manipulated an object (a branch) jointly with another chimpanzee (photograph by Masako Myowa-Yamakoshi)

while taking turns. Table 24.3 presents the types of triadic play (i.e., play including *self–object–other*) involving the use of detached objects. Most instances of social play involved one-way play. Conversely, instances of object/action-role turn-taking play were not observed.

Table 24.3 Types of social triadic (self–object–other) play with detached objects among young chimpanzees of Bossou

One-way	Two-way	Object or action-role turn-taking	Behavior
*			Break off a branch – Swing – Throw the branch
*			Break off a branch – Swing the branch
*			Break off a branch – Throw the branch
*			Throw a wad of vegetable fiber/a branch/a bundle of vegetation
*			Disrupt – Throw – a bundle of vegetation
*			Break off a branch – Poke another with branch
*			Break off a branch – Hit another with branch
	*		Hit a carapa nut (*Carapa procera*)/a bundle of vegetation
	*		Tug a branch/a vine
*			Obtain bark from a tree – Throw bark

We found an interesting instance of social play involving object modification behavior. First, a chimpanzee broke off a piece of tree bark and licked it to make the tip slippery. Then, the chimpanzee held the tip with another chimpanzee and they started to twist it together using both hands. Because neither chimpanzee fed on the tree bark, this behavior could be interpreted as *object modification during social play*. Figure 24.6 shows a scene of this behavior and Fig. 24.7 shows the finished modified object made by rotating the bark strip counter wise on both ends.

24.4   Characteristics of Play Behaviors in Chimpanzees

This study demonstrated that play behavior in wild chimpanzees involving the use of objects varied with age. They were more likely to use nondetached objects than detached objects between 0 and 4 years of age and then began using detached objects more frequently than nondetached objects after 5 years of age. Thus, during development, young chimpanzees increasingly learn about the properties of detached objects and their potential affordances as play objects. This process may play a fundamental role in the cognitive development of young wild chimpanzees and in their acquisition of local-specific tool-using techniques.

Social play behaviors in wild chimpanzees diversified at around 3 years of age, at which time young chimpanzees begin to bring detached objects into social interactions with other chimpanzees. In human cases, such interactions based on the triadic relationship of self–object–other appear in the context of social play at around 9 months of age (Trevarthen and Hubley 1978; Tomasello 1999). Human

Fig. 24.6 Object modification behavior during social ("two-way") play in young chimpanzees. A chimpanzee strips bark from tree (**a**), licks the tip to make it slippery (**b**), and starts to revolve it with another chimpanzee using both hands (**c**) (photographs by Masako Myowa-Yamakoshi)

Fig. 24.7 The finished modified object made by twisting the bark (photograph by Masako Myowa-Yamakoshi)

infants begin to look where adults are looking, to act on objects in the way adults are acting on them, and to point to request something or when they see an interesting event or object. Such action-role turn-taking is typical in human triadic interactions and is considered to help infants understand other people's mental states, such as their intentions, desires, or beliefs, that is, theory of mind (Rochat 2001; Tomasello 1999).

It is important to note, however, that action-role turn-taking interactions were not observed among the wild chimpanzees of Bossou. Similar findings have been reported from captive studies. For example, frequently instances have been observed during which an infant chimpanzee approaches its mother or another individual, who is manipulating an unfamiliar object or food, and eagerly observes and attempts to touch the object (Hirata and Celli 2003; Matsuzawa et al. 2001; Ueno and Matsuzawa 2005). However, the chimpanzee manipulating the object neither intentionally moves it to show it to the approaching individual nor vocalizes to attract the individual's attention toward it (Tomonaga et al. 2004).

Humans and chimpanzees both appear motivated to share interests or experiences with others. At the same time, a marked difference is found in the triadic play interaction between the two species. Humans, in particular, have frequent opportunities to share sensory body experiences with others through turn-taking interactions. By copying the body movements of others, human infants may learn to understand the mental states of others. In comparison to humans, young chimpanzees have few opportunities to learn about an object's function with active help from other individuals. Instead they acquire information mainly through means unique to chimpanzees, involving intense repeated observation of skilled models combined with solitary trial-and-error learning, that is, "education by master-apprenticeship" (Inoue-Nakamura and Matsuzawa 1997;

Matsuzawa et al. 2001; see Chaps. 18 and 21). Differences and similarities in play behavior between chimpanzees and humans may shed important light on the development and the evolution of the unique functioning of the human mind.

Acknowledgments This study was supported by Grants-in-Aid for Scientific Research from the Japan Society for the Promotion of Science (JSPS) and the Ministry of Education, Culture, Sports, Science and Technology (MEXT) (Nos. 12002009 and 16002001 to T. Matsuzawa, 10041168 to Y. Sugiyama, and 16683003 and 19680013 to M. Myowa-Yamakoshi), MEXT Grants-in-Aid for the 21st Century COE Program (A2 and D2 to Kyoto University), a research fellowship for Young Scientists from JSPS (No. 3642 to M. Myowa-Yamakoshi), the JSPS Core-to-Core HOPE program, and the Cooperative Research Program of the Primate Research Institute of Kyoto University. I wish to thank T. Matsuzawa, Y. Sugiyama, and S. Hirata for their help and their assistance in the field. Thanks are also owed to Direction Nationale de la Recherche Scientifique et Technologique, and Institut de Recherche Environnementale de Bossou of the Republic of Guinea, for permission to conduct this study, and G. Goumy, T. Camara, P. Goumy, P. Chérif, and D. Samy for their assistance in the field.

Chapter 25
Chimpanzee Mothers Carry the Mummified Remains of Their Dead Infants: Three Case Reports from Bossou

Dora Biro

25.1 Introduction

Maternal behavior toward recently deceased infants has been documented in a variety of primates. Goodall (1968), for example, reports having observed chimpanzee mothers as well as red colobus monkey and baboon mothers carrying and grooming the bodies of their dead infants. Warren and Williamson (2004) observed captive mountain gorillas carrying dead infants (in one case for over three weeks); interestingly, both the mothers and unrelated individuals performed the carrying, the latter possibly in response to a hormone-induced predisposition to allo-mothering. Nakamichi et al. (1996) compiled a list of existing reports on the carrying of dead or immobilized infants in different species, which include Old and New World monkeys, as well as prosimians, in both wild and captive settings. Sugiyama et al. (2009) provide the largest dataset with as many as 157 cases of dead infant carrying by the Japanese monkeys of Arashiyama over 24 years of research at the site. In the majority of all reported cases, infants were carried at most for a few days; nevertheless, the phylogenetic continuity of the behavior can be taken to demonstrate generalities of mother–infant relationships across primate taxa. Furthermore, a recent report by Anderson et al. (2010) has argued that close examination of chimpanzees' responses to the passing of conspecifics reveals potential parallels with human reactions to death.

Among Bossou chimpanzees, the various fates of infants born to mothers within the community can be summarized as follows. Since the start of the field project in 1976, 35 chimpanzee births have been recorded in the group (see also Chaps. 2 and 3 of this volume on demography and life history). Of these, 24 individuals have grown to at least juvenile age, 2 are still currently infants, 3 have disappeared together with their mothers (probably emigrated, although this has never been confirmed by subsequent direct observation), 3 have disappeared without their mothers (almost certainly died, but their remains have not been recovered), and 3 are known to have died as their deaths were observed and documented by researchers.

D. Biro (✉)
Department of Zoology, University of Oxford, South Parks Road, Oxford OX1 3PS, UK
e-mail: dora.biro@zoo.ox.ac.uk

T. Matsuzawa et al. (eds.), *The Chimpanzees of Bossou and Nimba*,
DOI 10.1007/978-4-431-53921-6_26, © Springer 2011

The present chapter focuses on these last three infants. In each case, the mothers continued to carry their dead offspring for several weeks – even months – following death, exhibiting a variety of maternal behaviors toward them, before finally abandoning the bodies (Biro et al. 2010). I explore here the likely causes of death and the factors subsequently contributing to the extended carrying by the mothers, as well as fellow group members' reactions to these events. Emerging from these observations is a clear portrait of the extremely strong mother–infant bond that exists in chimpanzees.

25.2   Three Infant Deaths at Bossou: Jokro, Jimato, and Veve

Table 25.1 introduces the three infants whose deaths were observed at Bossou. The first death, that of 2.5-year-old female Jokro, occurred in January 1992, 2 weeks after the first signs of illness were detected in the infant (see Matsuzawa 1997b for the original case report). The likely cause of death was a form of bronchitis or pneumonia, which seemed to cause a lethargy in the infant to an increasing degree in her final days: she stopped eating and refused invitations to play from fellow group members. During the night of January 24, 1992, Jokro died. The mother, Jire, continued to carry the corpse with her, much as she had done with her live infant. Over the first few days following death, Jokro's body initially swelled, then gradually dried out. All hair was lost, but body parts, apart from the lower jaw, remained intact, encased in dry leathery skin resembling that of a mummy. Jire continued to carry her infant's body for at least 27 days; at that point, field observations for that year came to an end, so that the exact length of the period of carrying could not be ascertained (Jire no longer had the body when research recommenced in the following field season).

The other two infants, 1.5-year-old male Jimato and 2.5-year-old Veve, died also of a respiratory disease during an epidemic in 2003 that claimed the lives of five chimpanzees at Bossou (see Chap. 32). Signs of the disease within the community were first detected on November 24, 2003, and Jimato was last seen alive on November 26. On December 3, his death was confirmed when the mother, Jire, was seen carrying his corpse, which she abandoned only on February 9. Jire thus carried Jimato's body for approximately 68 days (estimated, as the exact date of death was unknown).

Falling victim to the same illness, Veve was initially abandoned under unknown circumstances by her mother, Vuavua, while still alive (see also Chap. 32, this volume). Veve was suspected to have died when Vuavua was first observed without her infant on December 2. However, on December 10, Veve was found by researchers, sitting on the ground in a coffee field within the group's core area, alive but much weakened. Almost inconceivably, she had survived on her own for more than a week. It took three more days – during which time Veve was continuously monitored and provided with pieces of fruit by researchers – before the group next passed through the field and came across the abandoned infant. On seeing Veve, Vuavua immediately

Table 25.1 Infant deaths at Bossou confirmed through observation

Name of infant (sex)	Mother (year of birth)	Date of birth	Date of death	Age at death (years)	Cause of death	Body condition	Time carried after death (days)
Jokro (f)	Jire (~1957)	Early 1989	Jan 24, 1992	~2.5	Respiratory ailment	Mummified	>27*
Jimato (m)	Jire (~1957)	Oct 2002	Nov 27 to Dec 2, 2003	1.2	Respiratory ailment	Mummified	68**
Veve (f)	Vuavua (1991)	May 2001	Dec 29 to Dec 30, 2003	2.6	Respiratory ailment	Mummified	19**

*Exact length of carrying beyond 27 days unknown, as field observations concluded before Jire had abandoned Jokro's body.
**Estimated; based on state of decomposition of body when first sighted.

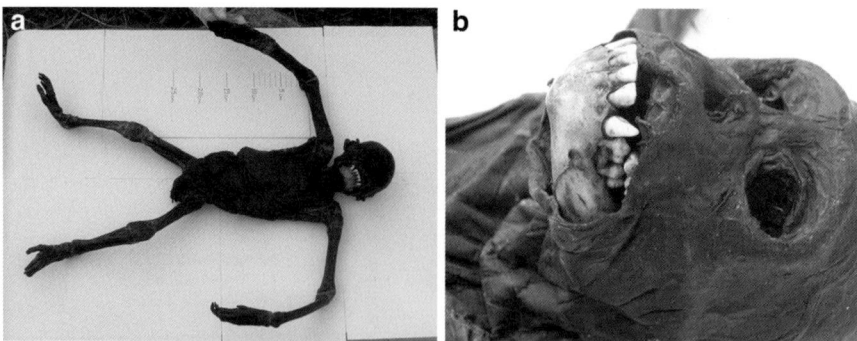

Fig. 25.1 Veve's mummified body recovered 19 days after death. The mother, Vuavua, carried the body and cared for it much as she did when her infant was alive. The mummification process set in during the first few days after death and ensured that the body stayed almost completely intact (**a**), apart from the loss of teeth in the upper jaw (**b**) (photographs by Claudia Sousa)

lifted her up and carried her off, pausing only to examine the face of her infant. Over the days that followed, Veve seemed initially to be recovering; however, from around December 25 she was rarely seen to feed and seemed to be overcome by weakness. Her breathing became shallow and fast, and she, like Jokro before her, no longer engaged in play with other chimpanzees. Veve was last seen alive on December 28, and her death was confirmed on the afternoon of December 30. As did Jire, Vuavua also continued to carry the body of her dead infant, only abandoning it 18 days later, on January 17. During this period, the community at Bossou, already at its lowest number in all the years of research, was home to *two* mothers carrying their dead infants: a sad reminder of the legacy of the 2003 epidemic.

As in Jokro's case, the bodies of both Jimato and Veve underwent complete mummification. At the time of recovery by researchers, Veve's remains were remarkably intact, the skeleton missing only a number of teeth in the upper jaw (Fig. 25.1). The consequence, most likely, of the longer period of carrying, by the time Jire abandoned Jimato's body much of the bony cranial structure had been destroyed, making most facial features unrecognizable. Nevertheless, fingers, toes, and even genitals were preserved within the layer of tough dry skin.

25.3 Why Did the Infants' Bodies Mummify?

In contrast with most other reports of dead infant carrying, Jire and Vuavua continued to care for the remains of their offspring for extended lengths of time. This attention may have been aided by the fact that the bodies mummified: the gross body structures remained almost completely intact and were relatively sturdy, which facilitated carrying the bodies while traveling both terrestrially and arboreally. Several factors are likely to have contributed to the mummification process.

First, all three deaths occurred during the dry season (see Chap. 2), when the relative lack of humidity may have contributed to the drying out of the bodies. However, it is important to note that although the 2003 epidemic at Bossou claimed the lives of five chimpanzees, only the bodies of the two infants, Jimato and Veve, mummified; those of the other victims who were recovered decomposed much more fully while lying on the forest floor (Chap. 32). Hence, it is likely that carrying itself contributed to mummification, by decreasing the bodies' contact with soil microorganisms. In addition, the chasing away of flies by the mothers would have reduced the incidence of egg-laying, and grooming would have removed fly larvae feeding on the corpses' flesh (see following). Thus, the mothers' refusal to give up the bodies in the first few days after death was likely an important factor in the degree to which the corpses were preserved and thus, in turn, facilitated extended carrying.

25.4   Mothers' Treatment of the Dead Infants

Both Jire and Vuavua adapted their techniques used for carrying their offspring's body after death: because the infants could no longer cling by themselves, they had to be gripped or appropriately balanced on the mother's body during travel. On the ground this was sometimes accomplished by holding the trunk in one hand and walking tri- or bipedally, and more often by draping the body over the mother's back where it could be additionally supported by gripping a limb between shoulder and neck (Fig. 25.2). When moving arboreally, a mother would hold the infant's arm, leg, or trunk between her shoulder and neck, or in her thigh pocket. During the night, the mothers took the corpses into their nest, as they would a live infant.

Jire and Vuavua also continued to groom their infant's body, and chased away flies that began to circle them in large numbers soon after death. Vuavua was also observed to use a tool for the latter purpose: twice she was seen to pick up a long thin twig and wave it back and forth over Veve's body on the ground (Fig. 25.3), and once to break off a small branch and do the same while resting high up on a tree.

By the age of 2.5 years (around the time when Jokro and Veve died), infant chimpanzees are highly active. Although they still travel by clinging to the mother, during times when the group is stationary they stray considerable distances from her, especially during play. A 5-min scan sampling of the distance between mother and offspring throughout the day showed that in the case of a live infant (2.5-year-old Fokaiye with mother, Fotaiu) the average distance between mother and infant was greater than that tolerated by mothers carrying dead infants (Veve and Vuavua, and Jimato and Jire). During certain relatively sedentary activities, such as nut-cracking (assessed in the outdoor laboratory; see Chap. 16), both mothers placed their dead infants on the ground next to them. When moving to a different location a short distance away, they would often leave the bodies in place, but generally did not tolerate a distance greater than about 10 m, and never left the bodies behind when the group moved on. They allowed other group members to handle the corpses, and even to transport them short distances before retrieving them (see below for descriptions of youngsters playing with the corpses). Very occasionally, the bodies would become separated from the mothers accidentally, such as by falling out of a tree, or during fights. In these cases,

Fig. 25.2 Live infants cling to the mother during travel, but mothers had to adapt their techniques for carrying the bodies of their dead offspring. Jire carries Jimato on her back by holding Jimato's arm between her shoulder and neck (**a**); Jire carries Jimato by holding his trunk in one hand (**b**); Vuavua drags Veve along the ground (**c**); and Vuavua holds Veve's trunk in her thigh pocket while climbing (**d**) (photographs by Dora Biro)

mothers appeared highly distressed, and screamed and searched until they recovered the bodies.

25.5 Reactions of Other Group Members to the Infants' Bodies

Many individuals in the group examined visually and sniffed the mummified bodies of the infants, often during greeting bouts directed towards the mother. Adult males in particular were seen also to poke the bodies, and lift and release limp extremities as if to examine them for signs of life.

Fig. 25.3 Tool-use in chasing flies from the decomposing body of an infant. Vuavua uses a long thin stalk that she repeatedly whisks over Veve, whose corpse is being circled by a large number of flies (photograph by Dora Biro)

In contrast to the gorillas in Warren and Williamson's (2004) report, none of the three chimpanzee infants' corpses were carried by any of the other group members for extended periods. Short episodes of carrying did occur, usually lasting not longer than a few minutes. These events included Tua, the alpha male at the time, dragging Jokro's body along the ground as part of a dominance display during a fight (Jire had briefly dropped the body as a result of the commotion). Jimato's body was also dragged on the ground by Fokaiye, a 2.5-year-old male infant, in play, and a juvenile female, Fanle, was briefly able to steal and carry the corpse into a tree before it was recovered by Jire (Fig. 25.4). Matsuzawa (1997b) also reports that Jokro's body was used in a form of simulated play-chase by a juvenile male, who carried the corpse into a tree, dropped it, climbed down to retrieve it, and repeated this sequence several times.

Despite the extremely pronounced smell of decay emanating from the bodies in the days following death, and the corpses' progressively more mummy-like appearance, only a single instance of what could be described as "aversion" was observed among fellow community members (Fig. 25.5). This episode occurred during Fokaiye's bout of play with Jimato's corpse mentioned above, while adult members of the group (including Jire) were engaged in nut-cracking at Bossou's outdoor laboratory (see Chap. 16). Fokaiye had been pulling Jimato's mummified remains round and round in circles, and three times she bumped into her mother, Fotaiu, sitting nearby. The first two times, Fotaiu threw her arms into the air and drew back

Fig. 25.4 An adolescent female, Fanle, steals Jimato's body (**a**) and carries it into a nearby tree (**b**) while the mother, Jire (*front center* in **a**), is engaged in nut-cracking on the ground below. Jire continued to nut-crack, then retrieved the body from Fanle 25 min later (photographs by Dora Biro)

as the corpse came into contact with her; the third time she jumped up, stepped backward, and moved away carrying her tools, withdrawing from the area where Fokaiye was playing.

Fig. 25.5 Possible case of aversion to a mummified corpse. A male infant, Fokaiye (*front center*), is playing with Jimato's body by dragging it along the ground. His mother, Fotaiu (*back center*), throws her arms in the air and jumps back when the body comes into contact with her. Jire, Jimato's mother, is watching from a few meters away (*left*) (photograph by Dora Biro)

25.6   Discussion

Chimpanzee infants go through an extended period of attachment to the mother during the first 5 years of life. They remain in close proximity to the mother at all times, the infant straying only a limited distance from her even once becoming fully mobile. Mother and infant eat the same food, always travel together through the forest, and sleep in the same nest at night. Mothers serve as the most important models whom the infants observe as they learn those skills present in the community that are propagated through social learning (see Chaps. 9, 16, 17, 21). In turn, mothers groom their offspring and defend them during aggressive encounters. The observations described here demonstrate that this extremely close social bond between mother and infant can extend beyond the individuals' lifetime. Chimpanzee infants who have lost their mothers have been reported to respond by refusing to eat and withdrawing socially (thus often themselves dying): it seems that chimpanzee mothers may also continue to experience the strength of the attachment once their infant passes away. It is an obvious and fascinating question to ponder the extent to which Jire and Vuavua "understood" that their offspring were dead. In many ways they treated the corpses as live infants (particularly in the initial phase following death); nevertheless, they may well have been aware that the bodies were completely inanimate, consequently adopting techniques of carrying never normally employed with healthy young. In effect, they refused to let their offspring go. The fact that they continued to carry the bodies for weeks, or even months, is poignant testament to the close mother–infant bond in chimpanzees.

What factors contributed to the mothers' eventual abandoning of the corpses? Besides accidental loss of the bodies and subsequent failure to recover, physiological changes in the mother associated with the death of an infant may also have played a role. Postpartum amenorrhea lasts on average 4 years (see Chap. 3, this volume) but is much shortened in the event of an infant's death. Because lactation had ceased once Jokro, Jimato, and Veve passed away, the mothers' reproductive cycle returned: Jire was in estrus 21 days after Jokro's death, and Vuavua 17 days after Veve's. These changes prepare the mother for the arrival of a new infant (normally around the time of weaning) and may have contributed to a gradual "letting go" of the previous infant's remains.

An additional note of interest concerns whether Jire's behavior was a trigger for Vuavua's carrying of Veve. At the time of Jokro's death in 1992, Vuavua was a 1.5-year-old infant, and would thus have seen Jire carrying the mummified Jokro. In addition, Jimato's death occurred about a month before Veve's, such that Vuavua once again had the opportunity to observe Jire's treatment of her dead infant. Although it is possible that such observation may have facilitated Vuavua's carrying, further data are needed to clarify this issue. However, it is hoped that the death of infants in this already-threatened community will remain a rare occurrence – leaving the question of a "tradition of dead-infant carrying" at Bossou unresolved.

Acknowledgments Several researchers contributed with field observations to the description of events herein: Tetsuro Matsuzawa, Tatyana Humle, Misato Hayashi, Claudia Sousa, Kathelijne Koops, Akino Watanabe, Yuu Mizuno, Gen Yamakoshi, and Gaku Ohashi. Local guides at Bossou offered invaluable help, in particular, Gilles Doré. The author thanks the Royal Society for financial support.

Chapter 26
Comparison of Social Behaviors

Michio Nakamura

26.1   A Comparative Look at the Social World of Chimpanzees

What are social behaviors? The answers may differ depending on how one treats sociality. In behavioral science, a social behavior is generally defined as a behavior of one individual directed to another individual whose subsequent behavior is somehow influenced by the former. Agonistic, affiliative, postconflict, and courtship behaviors are often easily accepted as being social in nature. However, in addition to such "conspicuous" social behaviors, much more subtle social behaviors permeate the social world of chimpanzees, particularly if we define the term "social" more broadly. For example, even slight movements of an arm or faint vocalizations can be significant to different individuals, although human observers may have difficulty detecting them without careful attention.

Local variation in such subtle social behaviors can often only be recognized through direct comparison by an observer well educated in the behaviors of at least one population. For example, *social scratch* (Nakamura et al. 2000) at Mahale had not been independently described until researchers from Gombe (L.F. Marchant and W.C. McGrew) visited Mahale and mentioned that this behavioral pattern was absent at Gombe. This rather simple behavioral pattern, in which one individual scratches the back or other body parts of another individual, was not novel to the Mahale researchers, and the pattern had been included in the larger category of social grooming. Similarly, the *grooming hand-clasp* at Mahale was described as a social custom only after Gombe researchers visited Mahale (McGrew and Tutin 1978). Mahale researchers were familiar with this behavior, because the pattern was very conspicuous: two chimpanzees sit face to face and raise their corresponding hands in the air to form an A-frame posture. The Mahale researchers were unaware that this pattern was absent at different sites.

M. Nakamura (✉)
Wildlife Research Center of Kyoto University, 2-24 Tanaka-Sekiden-cho, Sakyo, Kyoto 606-8203, Japan
e-mail: nakamura@wrc.kyoto-u.ac.jp

T. Matsuzawa et al. (eds.), *The Chimpanzees of Bossou and Nimba*, DOI 10.1007/978-4-431-53921-6_27, © Springer 2011

A collaborative project aimed at comparing chimpanzee behaviors across seven long-term research sites (Collaborative Chimpanzee Cultures Project; CCCP) has provided the most definitive picture of chimpanzee culture to date (Whiten et al. 1999). Updates to the CCCP are ongoing across 12 research sites across Africa (CCCP2; see Whiten 2010). When conducting such behavioral comparisons, it is more difficult to detect and confirm differences in patterns of social behavior than in patterns of tool-use, in part because researchers are more likely to observe and notice behavioral flexibility and variation in tool-use behaviors. In addition, researchers can more easily objectively describe tool-use than social behaviors because of the material nature of the behavior. Thus, tool-use behaviors observed among chimpanzees are well described in the literature. An observer of one population can therefore easily recognize a new type of tool-use that is absent from the published literature. For example, the use of leafy twigs for rain cover by Goualougo chimpanzees (Sanz and Morgan 2007) was readily identified as a novel tool type, as it had never been described elsewhere previously. These characteristics of tool-use observation differ substantially from the cases of social scratch and grooming hand-clasp described above. Therefore, "fresh eyes" that have experienced the social behaviors of different group(s) of chimpanzees may reveal subtle behavioral patterns that are unique to the chimpanzees of Bossou.

26.2 Studies of Social Behaviors at Bossou

Compared to the wealth of fine-scaled studies of tool-use, there are relatively few detailed examinations of social behaviors at Bossou. As a founder of the long-term research at Bossou, Sugiyama made initial, general descriptions of Bossou chimpanzees, including their social behavior and social relationships: for example, the number of individuals in proximity, charging displays by males, and party size (Sugiyama 1981a; Sugiyama and Koman 1979a; see Chap. 3, this volume). Sugiyama (1989a) also described some social behavioral patterns in the context of frustration, reassurance, appeasement, and invitation. An ethogram for Bossou chimpanzees that includes various social behaviors and is comparable to those for Gombe (Goodall 1989) and Mahale (Nishida et al. 1999) has not yet been compiled.

Sugiyama (1988) presented quantitative social grooming data for Bossou chimpanzees from 1976 to 1983. Bossou females groomed each other more often than expected, and males groomed each other less often than did East African chimpanzees (see Chap. 3). Muroyama and Sugiyama (1994) compiled a comparative study of social grooming among study sites of chimpanzees and bonobos (including Bossou data from Sugiyama) and also found that the frequency of male–male grooming was lower than expected only at Bossou. Unfortunately, this unique pattern of social grooming at Bossou has not been sufficiently evaluated (but see Yamakoshi 2004b; also see Nakamura 2010, for more discussion).

Sakura (1994) found that party size was not correlated with feeding ratio in Bossou chimpanzees, but chimpanzees formed larger parties in dangerous situations such as when crossing roads. Occasionally, one party would wait and pant hoot before a second party joined them (also see Chap. 23 for relevant observations). Yamakoshi (2004b) also reported that party size was relatively stable throughout the year and found no clear relationship with fluctuations in fruit availability.

Several previous studies have also described some of the other social behavioral patterns seen in Bossou chimpanzees. For example, Ohashi (2007) and Hockings et al. (2007) documented the sharing of papaya fruits (also see Chap. 22 by Hockings). Zamma and Fujita (2004) were the first to report genito-genital rubbing in wild chimpanzees, including at Bossou, a behavioral pattern often performed by female bonobos (Kitamura 1989). The acquisition of tool-use is also considered a social process (Matsuzawa et al. 2001), and numerous studies have described social interactions between immature chimpanzees and other individuals, especially their mothers, in the context of tool-use acquisition. Reviews of such social interactions can be found elsewhere in this volume (see chapters in Parts 3 and 4, this volume).

26.3   Comparing Bossou and Mahale Populations

From 1994 to the present, I have studied chimpanzees in the Mahale Mountains National Park, Tanzania, focusing primarily on social behaviors (Nakamura 2000, 2003a). I visited Bossou for a short time from mid-January to mid-March 2003. Thus, in this chapter, I cannot present a comprehensive view of the social behaviors of Bossou chimpanzees; instead, I provide an outsider's perspective of their social behaviors in comparison to those of Mahale.

As is Bossou, Mahale is a long-term research site of chimpanzees, in which research has been conducted since 1965 (see Nishida 1990b; Nishida et al. 2002 for a history of research at Mahale). The Mahale Mountains National Park is located on the eastern shore of Lake Tanganyika in the western Republic of Tanzania (Fig. 26.1), East Africa, and is inhabited by the eastern subspecies of chimpanzee (*Pan troglodytes schweinfurthii*). Compared to the group size at Bossou (19, including infants) during my visit, the size of the Mahale M group (the main study group) was three to four times larger (60, including infants, in November 2008).

At Bossou, I followed four sexually mature (adult and adolescent) males and five mature females for a total of 241 h. I recorded the behavior of target individuals as well as others in the vicinity (~5–10 m) of the targets. I took particular note of several patterns that were absent or unfamiliar at Mahale (see Nakamura and Nishida 2006).

26.4   Observations at Bossou

26.4.1   Rise in Rank of an Adolescent Male

During my brief 2-month stay at Bossou, I was fortunate to have observed an incident in which a young male (YL; Fig. 26.2; see Appendix A for explanation of abbreviated chimpanzee names) outranked an old ex-alpha adult male (TA) to become the beta

Fig. 26.1 Locations of Bossou and Mahale

male at the age of 11 (early adolescence as defined by Goodall 1983). This young male also exhibited insubordination to the alpha male (Nakamura and Ohashi 2003).

During the beginning of my stay at Bossou, YL regularly pant-grunted to TA until at least January 2003, but no pant-grunts were subsequently heard. On February 8, YL and TA displayed to each other. Neither individual pant-grunted, but TA finally grimaced, screamed, and sought reassurance from YL. During the next evening, TA began to make opportunistic displays around an estrous female who then screamed. As YL began to display and approached TA, TA pant-grunted to YL. On February 18, when TA reunited with YL, TA pant-grunted without displaying or behaving aggressively toward YL. These interactions occurred in the absence of the then-alpha male (FF). However, on March 2, TA pant-grunted to YL while FF was in a tree nearby, indicating that YL was dominant over TA and was now the beta male.

The attitude of YL toward the alpha male, FF, also changed drastically during my stay. YL was completely subordinate to FF until early February 2003, performing exaggerated pant-grunts characteristic of adolescent males. After YL outranked TA, he never pant-grunted to FF, except once on February 22. YL tended to range far from FF, but when he encountered FF, YL sometimes displayed from a distance of 10–20 m, while standing bipedally and swaying his arms and showing piloerection. Therefore, YL appeared to have begun to challenge the alpha male. After my stay at Bossou, YL obtained the alpha status and he was still the alpha male as of November 2008 (Humle, personal communication).

Fig. 26.2 Eleven-year-old male, YL, who outranked former alpha male (photograph by Michio Nakamura)

26.4.2 *Patrolling*

An additional noteworthy incident that occurred during my stay at Bossou was behaviorally quite similar to what is considered boundary patrolling at other study sites (Watts and Mitani 2001). On February 16, at approximately 8 a.m., three mature males (TA, YL, and PO) were observed together with females at the foot of Bouton Hill in the core area of the Bossou group. Another mature male, FF, was absent, because he was in consort with an estrous female. At approximately 8.15 a.m., the three males began to move quickly away from the females. After about 30 min of traveling through cultivated or abandoned fields, the males reached Go-yigba, which is about 2 km northeast of Bouton Hill and represents the approximate border of their known home range (Sugiyama 1999). Judging from the vocalizations, the females appeared to have remained in the core area. On the way to and from Go-yigba, the three males formed a compact line (Fig. 26.3). They only ate small portions of fruits, such as *Myrianthus*; thus, they did not travel to Go-yigba to raid crops. After remaining in trees at Go-yigba for approximately 1 h, the three males returned to the core area before 11 a.m.

26.4.3 *Greeting*

The conspicuous vocalization called pant-grunting is a well-known greeting behavior in chimpanzees (Hayaki 1990). However, much more subtle greeting interactions were also observed among females at Bossou. For example, one such greeting was

Fig. 26.3 Three males, TA, YL, and PO, traveling single file on a manmade trail while returning from the patrol (see Appendix A for explanation of abbreviated chimpanzee names) (photograph by Michio Nakamura)

the *mutual genital touch*. When two adult females met each other after some time apart, they approached and then closely passed by each other. They then paused, with the face of one female close to the other's hip. They then simultaneously and gently touched each other's genital area from underneath with the outer hand (Fig. 26.4). I judged that this might be a type of greeting between females, as its context was similar to peering into the face, kissing, or extending a hand. However, in contrast to noisy greetings such as pant-grunting, no conspicuous vocalizations were heard during mutual genital touching, but there were a few cases in which faint soft grunts were uttered by one individual. Mutual genital touching was performed by seven of nine sexually mature females (Fn, Jr, Ka, Nn, Pm, Yo, and Vv, but not Vl and Ft). This behavioral pattern has not been observed at Mahale.

26.4.4 Courtship and Mating

Courtship displays function to solicit copulation by a male or an estrous female from an individual of the opposite sex (Nishida 1997). Several variations of courtship displays have been documented among different sites (Whiten et al. 1999). *Leaf clipping* at Mahale is one example (Nishida 1980): a male or an estrous female clips a leaf (or leaves) to produce an audible sound that attracts the attention of a prospective mate. At Bossou, this same behavioral pattern is often exhibited to express frustration, although it is also used in courtship contexts (Sugiyama 1981a).

Fig. 26.4 Mutual genital touching by Bossou females (drawn from memory by Michio Nakamura)

I noticed an additional courtship pattern, the *heel tap*, which was frequently used by Bossou males but is completely absent at Mahale. In this behavior, a tree bough, rock, or the ground was rhythmically tapped with the heel (Fig. 26.5) to produce a conspicuous sound. During stamping behavior typically exhibited by chimpanzees, the sole makes contact with the substrate, but in heel tap the sole remains upright facing forward and only the heel makes contact. Sugiyama (1989a) had previously described this behavior as *knock branch with heel*, but it has not been mentioned in later cultural studies (Whiten et al. 1999). Heel tapping was performed by three of four sexually mature males (TA, FF, and PO, but not YL) and a juvenile male, JJ. Adult females were never observed performing the heel tap during this period, although one young female (Vv) in estrus was observed heel tapping in 2004 to attract an adult male for mating (Humle, personal communication). In 53 of 56 observed cases, the heel tap was directed toward a nearby estrous female, especially when soliciting her for consorting, and the behavior was often used jointly with branch shaking, stamping, or leaf clipping. The heel tap was performed in a non-sexual courtship context in only three cases (once by a juvenile male, JJ, and twice by an adolescent male, PO), and as suggested by Sugiyama (1989a), this behavioral pattern was also used for inviting play. All three mature males who heel tapped lateralized it to one or the other of their feet. Two older males (FF and TA) always used their left foot, and the other young male (PO) used his right foot when heel tapping ($n = 7$, 32, and 15, respectively). In the 23 cases in which the behavior was clearly videotaped, the males tapped an average of 5.5 times (SD = 3.9, range = 2–20) per bout.

Another difference between the Mahale and Bossou populations was the occurrence of mother–offspring mating. Despite a much longer observation time at Mahale,

Fig. 26.5 Heel tapping by a Bossou male (drawn from video footage by Michio Nakamura)

I have never witnessed mating between a mother and her sexually mature son. Mature sons do not express interest in their mothers even when she is the only sexually receptive female to whom many males are attracted. At Bossou, three of four sexually mature males had living mothers at the time of study. Two of these sons, FF (23 years old) and YL (11 years old), the alpha and beta males, respectively, mated with their mothers three times each during my study period. The copulations of FF were nearly forced, as his mother, Fn, refused and screamed the entire time. In contrast, YL's mother, Yo, presented and copulated just as she did with other males.

26.4.5 Grooming

Among Bossou chimpanzees, I observed two grooming-related behaviors that are absent from Mahale. The first behavior was *index to palm*, which always occurred during social or self-grooming. The groomer suddenly stopped grooming, often turning his back on the groomee. He moved his lips continuously, as if holding something between them. He then opened his palm and placed a small particle on it (presumably an ectoparasite) from his lower lip, and then put his index finger on the spot, poking, pushing, and dragging the item (Fig. 26.6). Finally, he again put his mouth to his palm, presumably to eat the particle. Other individuals, especially infants and juveniles, sometimes peered into the palm. This pattern was observed for three mature males (TA, FF, and PO) and two mature females (Pm and Ft). The behavior may be comparable to leaf grooming at Mahale (Zamma 2002), as the latter is not observed at Bossou and the contexts are similar. At Taï National Park in the Côte d'Ivoire, chimpanzees instead employ the index hit technique (Boesch 1996b), which is similar to index to palm at Bossou but differs in that a chimpanzee hits his forearm with the index finger, instead of the palm.

The second variation observed during social grooming was the *sputter*: chimpanzees emit this sound as if forcing air through their lips. This behavior was first reported for chimpanzees of Ngogo, Uganda (Nishida et al. 2004), and is completely absent at Mahale. In Ngogo, 27 individuals were confirmed to utter this sound (Nishida et al. 2004), whereas only 3 individuals (JJ, Ft, and PO) sputtered at Bossou. The juvenile male JJ was responsible for 20 of the 24 observed cases of sputtering.

Other well-documented social grooming customs that are frequently observed at Mahale, such as the grooming hand-clasp (McGrew and Tutin 1978; Nakamura 2002) and social scratch (Nakamura et al. 2000), were not observed during my stay at Bossou.

Fig. 26.6 Index to palm by a Bossou male (drawn from video footage by Michio Nakamura)

26.4.6   *Frequency and Number of Performers*

I compared the frequency and number of performers of several of these behavioral patterns to those of tool-use behaviors (Table 26.1). Heel tapping, index to palm, mutual genital touching, and sputtering were observed more often than any foraging tool-use. Because of the brief observation time, there were too few performers of these behaviors (with the exception of leaf clipping) to evaluate whether these patterns were shared by a majority of the group members. Some of the behavioral patterns may be shared within specific age- or sex-classes. For example, heel tapping was performed by three of four mature males. Similarly, mutual genital touching, presumably a pattern of mature females, was performed by seven of nine mature females. Although index to palm was performed by only five individuals in different age- or sex-classes, this small number of performers was not less than the number observed for any single foraging tool-use during my stay. Observations of index to palm require the observer to be at a relatively close distance and to have proper angles of observation; for example, if the chimpanzees are grooming high up in a tree (often the case for some females) or if the groomer is showing only his or her back, then the observer may be unable to see this behavior. This latter reason may be why the pattern had not been recorded until now. Thus, the frequency and number of performers of index to palm were likely to be underestimated, and follow-up observations may indicate that more individuals perform this behavior.

26.4.6.1   Significance of Subtle Social Behaviors

Complex and sometimes turbulent male–male social relationships have been a central topic of studies of chimpanzee social behavior (Mitani et al. 2000; Newton-Fisher 1999; Nishida and Hosaka 1996; Takahata 1990), and, as a result, social behaviors have been studied relatively less often at Bossou because this group includes only a few adult males. Rank reversal among males or patrolling by males as described here may occur infrequently at Bossou; thus, quantitative investigations of such behaviors may be difficult. However, even anecdotal accounts may provide interesting insights into the sociality of chimpanzees. The case of the rank reversal of YL indicates that chimpanzee males have the potential to rise in rank earlier than usual under certain conditions, such as small numbers of males and/or earlier physical growth (Nakamura and Ohashi 2003). The case of possible boundary patrolling at Bossou may cast doubt on current discussion of patrolling, which often emphasizes the territoriality of chimpanzee groups and possible intergroup aggressions (Watts and Mitani 2001). Because the Bossou group is thought to be nearly isolated, without apparent neighboring groups (Matsuzawa 2006a; Sugiyama 1999), the behavioral pattern known as patrolling may serve a function other than investigation of a neighboring group and may not be directly related to intergroup aggression. This behavior may not be so much for territorial purposes in the context of Bossou as it is frequently linked to times when males sought another adult male with an adult cycling female often ranging at the border of

Table 26.1 Frequency of several observed behavioral patterns and tool-use at Bossou

Behavior[a]	Frequency of events[b]	Number of performers	Performers[c]
Heel tapping	56 (0.23)	4	TA, FF, PO, JJ
Index to palm	13 (0.05)	5	TA, FF, PO, Pm, Ft
Mutual genital touching	17[d] (0.07)	7[e]	Ka, Nn, Fn, Jr, Pm, Vv, Yo
Sputtering	24 (0.10)	3	JJ, Ft, PO
Leaf clipping	77 (0.32)	11	TA, FF, YL, PO, JJ, PE,Ka, Fn, Pm, Vl, Fl
Dipping for driver ants	7 (0.03)	4	TA, PO, Yo, Vv
Using leaves to drink	7 (0.03)	5	TA, Ka, Fn, Pm, Fl
Nut-cracking	2 (0.01)	1[f]	PO
Pestle pounding	2 (0.01)	3	PO, Ka, Ft
Using sticks to get honey	2 (0.01)	1	PO
Other tool-use[g]	10 (0.04)	3	YL, PO, Ft

[a]Includes behavior of focal and nonfocal individuals
[b]Numbers in parentheses indicate frequency/h during 241 h of observation
[c]Males are abbreviated with two capital letters and females with a capital and lowercase letter
[d]Includes cases when only one party touched the other's genitals (see text)
[e]Includes both participants
[f]Performer was not identified in one case
[g]For example, clubbing, throwing objects, playing with objects

the group's home range (Ohashi and Humle, personal communication). However, the fact that the Bossou chimpanzees ranged farther than previously assumed (Ohashi 2006b) highlights the possibility that they may sometimes encounter different groups at the periphery of their range, where observers usually do not venture.

In contrast to the studies of social relationships among males, there has been much less targeted research on females, even in other chimpanzee populations (Nishida 1989; Wrangham et al. 1992). Although females have been less studied, they have often been generalized as "less social" than males. However, this generalization may be premature after considering the large amount of grooming among females at Bossou (Sugiyama 1988) as well as the subtle but frequent social interactions observed among Bossou females (e.g., mutual genital touching). Female behaviors may not be as conspicuous as those of males, but such interactions among adult females may play a role in maintaining the gregarious and peaceful nature of the Bossou group (e.g., Yamakoshi 2004b). Moreover, Bossou females have experienced stable membership because of the lack of new immigrant females. Such unique conditions may provide opportunities to reevaluate the social potential of female chimpanzees.

It is particularly noteworthy that only 2 months of observation revealed several subtle behavioral variations between Bossou and Mahale. Table 26.2 summarizes the findings described in this chapter, together with published behavioral variation between Bossou and Mahale. Many types of foraging tool-use have been reported for Bossou, whereas the only consistently observed foraging tool-use at Mahale was fishing for arboreal ants. In contrast, before my visit to Bossou, virtually no unique social behaviors had been reported for this population, which suggests that there is greater behavioral variation in wild chimpanzees than we currently believe, particularly in the context of social behaviors.

Although I have emphasized the importance of studying variation in social behaviors, there are several reasons why such variation has not been recorded. Tool-use has attracted the most attention in studies of chimpanzee culture or behavioral variation. At Bossou, tool-use has been the main topic of research, and the same types of tool-use have been repeatedly investigated by a number of researchers from different perspectives (see chapters in Parts 3 and 4 of this volume). Hidden in such conspicuous and complex cultural behaviors are more subtle and simpler patterns that differ among wild chimpanzee populations. The low frequency of occurrence may not adequately explain the rare mention of these patterns in previous studies, because during my observations, some of the behaviors described here occurred more often than tool-use (see Table 26.1).

An additional reason that such subtle behaviors remain undescribed is the difficulty encountered when attempting to identify their function, or the behavioral patterns appear to be arbitrarily related to the function (Boesch 1995). For example, we still do not understand why three different methods (i.e., leaf groom, index hit, and index to palm) are employed in different chimpanzee populations for the same function of squashing parasites. Within the current adaptive significance framework of behavioral biology, certain behaviors are sometimes difficult to describe when their benefits are at best ambiguous (Nakamura 2003b).

The everyday social lives of chimpanzees are not composed entirely of conspicuous behaviors relating to male dominance strategies. The key to a well-rounded understanding of the social world of chimpanzees may lie in seemingly simple and subtle behaviors.

Acknowledgments I thank COSTECH, TAWIRI, TANAPA, MMNP, and MMWRC of Tanzania and DNRST of Guinea for permission to conduct field research. I also thank G. Ohashi, G. Yamakoshi, and the entire staff of IREB for cooperation at Bossou, Y. Sugiyama and T. Matsuzawa for offering me the chance to visit Bossou, and T. Nishida for continuous guidance and support of my research. The field study was financially supported by grants from the Japanese MEXT (#12375003, #16255007, #19255008 to T. Nishida).

Table 26.2 Behavioral variation between chimpanzees at Bossou and Mahale

Behavior	Context	Bossou	Mahale	Source[a]
Nut-cracking	Foraging/tool-use	○	—	Sugiyama and Koman (1979b)
Pestle pounding	Foraging/tool-use	○	—	Yamakoshi and Sugiyama (1995)
Using tools for drinking	Foraging/tool-use/play?	○	○	Sugiyama and Koman (1979b), *Matsusaka et al. (2006)*
Dipping for driver ants	Foraging/tool-use	○	—	Sugiyama et al. (1988)
Algae scooping	Foraging/tool-use	○	—	Matsuzawa et al. (1996)
Fishing for termites	Foraging/tool-use	+	—[b]	Humle (1999)
Fishing for arboreal ants	Foraging/tool-use	—	○	*Nishida (1973)*
Leaf clipping	Social (courtship)/frustration / tool-use	○	○	Sugiyama (1981a), *Nishida (1980)*
Throwing a stone to make a splash	Social (display)/tool-use	—	○	*Nishida (1994)*
Bending shrubs	Social (courtship)	+[c]	○	Whiten et al. (1999), *Nishida (1997)*
Heel tapping	Social (courtship)/invitation to play?	○	—	Nakamura and Nishida (2006)
Mutual genital touching	Social (greeting)	○	—	Nakamura and Nishida (2006)
Grooming hand-clasp	Social (groom)	—	○	*McGrew and Tutin (1978)*
Social scratch	Social (groom)	—	○	*Nakamura et al. (2000)*
Sputtering	Grooming sound	+	—	Nakamura and Nishida (2006)
Leaf grooming	Ectoparasite handling/tool-use	—	○	*Zamma (2002)*
Index to palm	Ectoparasite handling	+	—	Nakamura and Nishida (2006)
Leaf-pile pulling	Solo-play	—	○	*Nishida and Wallauer (2003)*

Note: "○," observed in at least the majority of the members of an age- or sex- class; "+," observed in only a few individuals; "—," absent or not reported

[a] References for Bossou are in normal font and those for Mahale are in *italics*

[b] No reports for Mahale M group, but reports for Mahale B (Nishida and Uehara 1980; McGrew and Collins 1985) and K (Uehara 1982) groups

[c] I saw only one adolescent male perform this pattern

Part VI
Adjacent Communities

Chapter 27
The Chimpanzees of Yealé, Nimba

Tatyana Humle

27.1   Historical Perspective on Chimpanzee Research in Yealé

In recognition of its biological diversity, unique ecology, and exceptional scenic beauty, the Nimba Mountains were established as a Nature Reserve, the "Réserve Naturelle Intégrale du Mont Nimba," in 1943 in Côte d'Ivoire (5,000 ha) by the French colonial administration. This reserve, alongside the Guinea portion of the massif, forms a World Heritage Site, gazetted in 1981 for Guinea and in 1982 for Côte d'Ivoire (Fig. 27.1; see Chap. 39).

The first reported chimpanzee (*Pan troglodytes verus*) survey in the Yealé region of the Nimba Mountains in Côte d'Ivoire was conducted by Joulian, as a member of a survey team investigating the distribution of nut-cracking behavior by chimpanzees in Côte d'Ivoire (Boesch et al. 1994). Although Joulian found two *Coula edulis* cracking sites beside the Nuon River at the border with Liberia (see Fig. 27.1), Boesch et al. (1994) were cautious in attributing these findings to chimpanzees because humans in the locality also crack this species of nut to consume the kernel within. During his brief survey, Joulian also confirmed the presence of other nut-bearing species such as *Panda oleosa*, *Parinari excelsa*, and *Detarium senegalensis* trees on this side of the massif (Joulian 1994).

In December 1993 and January 1994, Matsuzawa and Yamakoshi (1996) additionally carried out two brief surveys of the Nimba chimpanzees in the area of the Nuon River beside Yealé, including areas of higher altitude. Humle visited the Yealé site several times between 1999 and 2001 (Humle and Matsuzawa 2001, 2004). Unfortunately, research in Yealé was discontinued in 2002 because of political unrest in the country, although the trained local assistants continued to intermittently habituate the chimpanzees. These studies and surveys have nevertheless provided important preliminary insights into the ecology, the behavior, and the unique cultural repertoire of the chimpanzee communities inhabiting the Yealé area of the Nimba range (Humle and Matsuzawa 2001, 2004; Humle 2003b).

T. Humle (✉)
School of Anthropology and Conservation, The Marlowe Building,
University of Kent, Canterbury CT2 7NR, UK
e-mail: t.humle@kent.ac.uk

T. Matsuzawa et al. (eds.), *The Chimpanzees of Bossou and Nimba*,
DOI 10.1007/978-4-431-53921-6_28, © Springer 2011

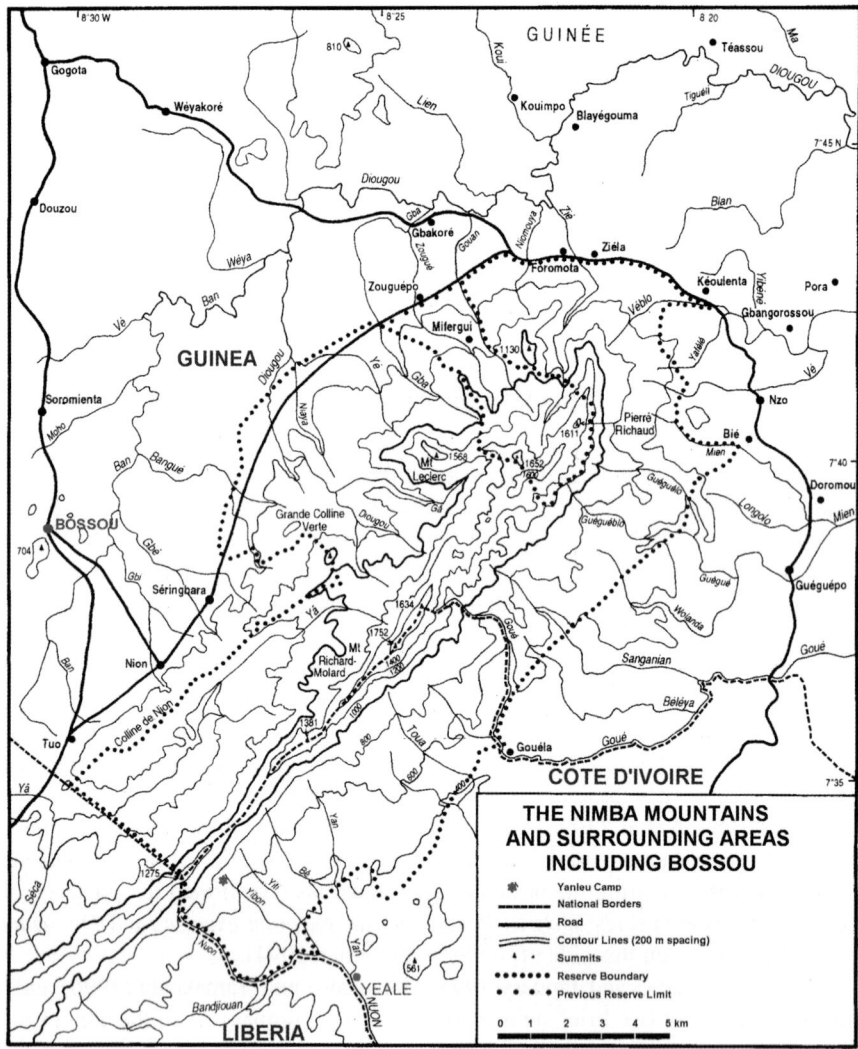

Fig. 27.1 The Nimba mountains and surrounding areas including Bossou, Seringbara, and Yealé, indicating the location of the main temporary camp used (Yanleu Camp) in Yealé, Côte d'Ivoire

27.2 Location and Climate

27.2.1 The Village of Yealé

The village of Yealé (latitude 7°31'21.8" N and longitude 8°25'29.1" W) is located 12 km southeast of Bossou, on the Côte d'Ivoire side of the Nimba Mountains (see Fig. 27.1). This small village, located at the foothills of the Nimba Mountains, practically sits on the border with Liberia. The closest large commercial town in proximity is Danané. The village is home to predominantly Yakuba and Manon people, who, for traditional reasons, typically do not hunt or eat chimpanzees.

27.2.2   The Study Area

A temporary camp site (Yanleu camp: latitude 7°32′50.09″ N and longitude 8°28′03.01″ W) served as a base for most surveys and studies conducted since 1994 (see Fig. 27.1). This temporary camp is located approximately 5.2 km from the village of Yealé within the reserve toward the upper slopes of the mountain. The topography of the Nimba Mountains on the Côte d'Ivoire side of the massif is quite different from that on the Guinean side (see Chap. 28). Indeed, it slopes gradually at first and then exponentially to the summit and is uniformly covered by forest. It therefore lacks the deep valleys, the marked band of high-altitude savanna vegetation, and the more dramatic rock formations characteristic of the Guinea side of the massif. Similarly, however, the Côte d'Ivoire side is distinguished by its evergreen forest of medium altitude (Guillaumet and Adjanohoun 1971). Patches of Marantaceae and Zingiberaceae growth prevail throughout the forest. Some of these areas do succumb to natural fires in the dry season (the peak of the dry season is from December to February), although such events are rarer than on the Guinea side. Between 1,000 and 1,600 m, the habitat is dominated by *Parinari excelsa*, *Uapaca* sp., and *Afrosersalisia cerasifera*, important fruit tree species for the chimpanzees. There is also an abundance of epiphytes in some areas. The flora of the Nimba mountains is extremely diverse. Indeed more than 2,000 plant species have been described, and about 16 are thought to be endemic (Adam 1971–1983). Based on vegetation transects set up in 2000, the dominant species in the Yealé forest are *Rinorea* sp., *Chidlowia sanguinea*, *Gilbertiodendron limba*, *Carapa procera*, and *Ituridendron bequaertii*.

27.2.3   The Climate

Similarly to Bossou and Seringbara, the climate at the foothills of the Nimba Mountains on the Côte d'Ivoire side, as measured in the village of Yealé, is also characterized by a long rainy season (March–November) and a brief dry season (December–February). Rainfall between July 2000 and June 2001 was greater than that recorded at either Bossou or Seringbara with a total annual rainfall of 3,027.6 mm (see Chap. 2). This period was characterized by an average temperature of 26.1°C (range, 13.0–43.3°C). Although temperatures are quite similar to what is recorded on the Guinean side of the massif, rainfall appears to be more important in Yealé, as it is located on the southeastern flank of the range.

27.3 Current Understanding of the Yealé Chimpanzees

27.3.1 Tool-Use and Shelter

As the chimpanzees in the region of Yealé are still relatively unhabituated to human observers, we lack any detailed knowledge of the true extent of their elementary technology (Matsuzawa and Yamakoshi 1996; Humle and Matsuzawa 2001). However, in terms of probe-using tools, we have confirmed ant-dipping (as well as brood extraction) at this site (Humle 2003b). We have thus far been able to confirm that the chimpanzees target *Dorylus arcens* at nests (Humle and Schöning, unpublished data). Because we have not been able to systematically sample army ants throughout the chimpanzees' habitat, we still lack knowledge of the diversity and availability of army ant species at this site. Based on ant-dipping tools recovered between 1999 and 2000, mean tool length at Yealé was 66.5 cm ($n=9$; SD$=19.52$; range, 38–96 cm) (Humle 2003a; see Chap. 9 for details about ant-dipping at Bossou). Compared to Bossou and Seringbara, all ant-dipping tools were woody. The majority of tools recovered were also characteristically stripped of bark.

Coula edulis trees in Yealé are located in the lower-altitude region of the massif (Humle, personal observation). Humle and Matsuzawa (2001) confirmed chimpanzee usage of the cracking sites earlier uncovered by Joulian based on traces of chimpanzee knuckle prints and footprints beside hammer stone and tree anvil assemblages, alongside fresh remains of shells and traces of wear on tree root anvils. Humle and Matsuzawa (2001) found an additional *Coula* cracking site in the area. These three *Coula* nut-cracking sites were utilized by chimpanzees between 1999 and 2000 and continue to be used to this date (Droh et al., personal communication).

Although *Parinari excelsa*, *Panda oleosa*, and *Detarium senegalensis* trees are available to chimpanzees on this side of the massif, chimpanzees have as yet never been confirmed to crack the nuts of these species (Joulian 1994; Humle and Matsuzawa 2001). The latter two species appear, however, to occur at relatively low densities (Humle, personal observation). Humle and Matsuzawa (2001, 2004) confirmed chimpanzee oil-palm (*Elaeis guineensis*) nut-cracking at Yealé with a hammer and anvil stone. Among 127 oil-palm trees surveyed in the area between 2000 and 2001, 2.4% (3/127) provided evidence of nut-cracking. This number is few compared to Bossou, where 22.8% (29/127) of all surveyed oil palms presented nut-cracking ateliers with hammer and anvil stones. In addition, based on indirect evidence, Matsuzawa and Yamakoshi (1996) also proposed that Yealé chimpanzees pound open the fruit of *Strychnos* sp. on embedded rocks, and crack *Carapa procera* nuts and the shell of *Achantina* sp. snails. These behaviors, nevertheless, still await confirmation to this day via further indirect evidence or direct observation.

Matsuzawa and Yamakoshi (1996) found that Yealé chimpanzees have a high propensity for building nests on the ground. Indeed, 35.4% of all nests they encountered during their surveys were constructed on the ground. They suggested that some of these elaborate ground nests might actually serve as night nests. Although ground nesting has also been reported at other chimpanzee sites (Koops et al. 2007; see Chap. 28), ground nesting at night would represent an atypical behavioral pattern for chimpanzees, if eventually confirmed via direct observation. Humle (2003a) found that only 3.7% (14/378) nests were built on the ground during her surveys in the area. However, compared to Matsuzawa and Yamakoshi (1996), these surveys were conducted during the rainy season, suggesting that ground nesting at Yealé may exhibit a seasonal pattern, although no such effect emerged from the detailed study by Koops et al. (2007) of factors influencing terrestrial nesting at Seringbara on the Guinean side of the massif (see Chap. 28). Finally, regardless of the factors dictating ground-nesting behavior in chimpanzees; this behavior appears to be a distinguishing characteristic of Nimba chimpanzees throughout their range.

Based on measures gathered between 1999 and 2001, overall mean nesting height at Yealé was 16.1 ± 0.5 m (range, 0–42.5 m) (Humle 2003a). Night nests also tended to be built slightly higher than day nests. The top ten preferred tree species (controlling for availability) used by Yealé chimpanzees for nesting during this period were in order of preference: *Chidlowia sanguinea*, *Gambeya perpulchrum*, *Rinorea* sp., *Dacryodes* sp., *Plagiostyles africana*, *Carapa procera*, *Heritiera utilis*, *Musanga cecropioides*, *Pseudospondias microcarpa*, and *Pentaclethra macrophylla* (Humle 2003a). The habitats of choice for nesting by Yealé chimpanzees were primary and riparian forest, accounting together for 74.9% of all nest groups encountered. Oil palms were rarely used for nesting at Yealé, although the oil palm is a highly preferred nesting species at Bossou (Humle 2003a; Humle and Matsuzawa 2004; Droh et al., personal observation).

27.3.2   Feeding Ecology and Medicinal Use of Plants

In the Nimba Mountains region of Yealé, between 1997 and 2001, whether through direct or indirect observation (food or fecal remains), we confirmed the consumption of 61 species, including 77 plant parts (Humle, unpublished data; see Table 27.1). *Nauclea diderrichii*, which produces a succulent fleshy fruit, appears to play an important role for the chimpanzees at this site, especially because of its availability in times of fruit scarcity. At Yealé, as at Bossou and Seringbara, fruit availability is lower during rainy season months (Humle, personal observation; see Chaps. 2 and 28). Terrestrial herbaceous vegetation, such as plant species belonging to the Marantaceae and Zingerberaceae families, appears to constitute an important fallback food for Yealé chimpanzees during periods of fruit scarcity.

Because of the extensive reliance on the oil palm for food by the chimpanzees of Bossou (see Chap. 2), we decided to explore further oil-palm use among the chimpanzees of Nimba, including those at Seringbara and Yealé (see Chap. 28). Humle and Matsuzawa (2004) thus surveyed on a monthly basis 127 oil palms in the Yealé region, which was estimated via vegetation transects to hold 4.15 oil palms/ha (Humle 2003a). In addition, to indirectly examine patterns of oil-palm use, these surveys served to track and quantify any differences in phenology of the oil-palm tree between Yealé, Seringbara, and Bossou, including tool availability, nut quantity, quality, fruit and petiole availability, and competition for oil-palm resources by other animal species. It finally emerged that Bossou chimpanzees

Table 27.1 Preliminary food list of chimpanzees at Yealé located between the Nuon and the Yan River

Taxonomic name	Family	Part eaten
Aframomum citratum	Zingiberaceae	Fruit, pith
Aframomum excapum	Zingiberaceae	Fruit, pith
Aframomum longiscapum	Zingiberaceae	Fruit, pith
Afrosersalisia cerasifera	Sapotaceae	Fruit
Albizia zygia	Mimosaceae	Gum
Ananas comosus	Bromeliaceae	Fruit, petiole
Aningeria altissima	Sapotaceae	Fruit
Antiaris africana	Moraceae	Fruit
Bosquea angolensis	Moraceae	Fruit
Canarium schweinfurthii	Burseraceae	Fruit
Canthium horizontale	Rubiaceae	Fruit
Carica papaya	Caricaceae	Fruit
Celtis adolfi-frederici	Ulmaceae	Fruit
Cola caricaefolia	Sterculiaceae	Fruit
Cola cordifolia	Sterculiaceae	Fruit
Coula edulis	Olacaceae	Nut
Detarium senegalense	Caesalpiniaceae	Fruit
Dialium dinklagei	Caesalpiniaceae	Fruit
Elaeis guineensis	Palmae	Flower, fruit, nut, petiole
Ficus barteri	Moraceae	Fruit
Ficus exasperata	Moraceae	Fruit, leaf
Ficus mucuso	Moraceae	Fruit
Ficus sur	Moraceae	Fruit
Gambeya perpulchrum	Sapotaceae	Fruit
Grewia barombiensis	Tiliaceae	Fruit
Hannoa klaineana	Siimaroubaceae	Fruit
Harungana madagascariensis	Hypericaceae	Fruit
Hyselodelphis violaceae	Marantaceae	Pith
Ituridendron bequaertii	Sapotaceae	Fruit
Landolphia dulcis	Apocynaceae	Fruit
Landolphia owariensis	Apocynaceae	Fruit
Lasiodiscus fasciculiflorus	Rhamnaceae	Fruit
Mangifera indica	Anacardiaceae	Fruit
Maranthochloa sp.	Marantaceae	Fruit, pith
Megaphrynium macrostachyum	Marantaceae	Fruit, pith, young leaf
Musa sinensis	Musaceae	Fruit
Musanga cecropioides	Moraceae	Flower, fruit
Myrianthus arboreus	Moraceae	Fruit
Myrianthus libericus	Moraceae	Fruit
Nauclea diderrichii	Rubiaceae	Fruit
Nauclea latifolia	Rubiaceae	Fruit
Octoknema borealis	Octoknemataceae	Fruit
Pachypodanthium staudtii	Annonaceae	Fruit
Parinari excelsa	Rosaceae	Fruit

(continued)

Table 27.1 (continued)

Taxonomic name	Family	Part eaten
Parkia bicolor	Mimosaceae	Fruit
Polycephalium capitatum	Icacinaceae	Fruit, leaf
Pseudospondias microcarpa	Anacardiaceae	Fruit
Salacia cornifolia	Celastraceae	Fruit
Salacia togoica	Celastraceae	Fruit
Sarcophrynium sp.	Zingiberaceae	Fruit, pith
Sherbournia bignoniflora	Rubiaceae	Fruit
Tetrorchidium didymostemon	Euphorbiaceae	Leaf
Thaumatococus daniellii	Marantaceae	Fruit, young leaf, pith
Theobroma cacao	Sterculiaceae	Fruit
Trichilia heudelotii	Meliaceae	Fruit
Uapaca sp.	Euphorbiaceae	Fruit
Uvariopsis guineensis	Annonaceae	Fruit
Vitex sp.	Verbenaceae	Fruit
Vitex doniana	Verbenaceae	Fruit
Vitex ferruginea	Verbenaceae	Fruit

demonstrated the greatest frequency of oil-palm use, while Seringbara chimpanzees exhibited none (see Chap. 28), and Yealé chimpanzees showed all uses observed at Bossou excepting pestle pounding and mature leaf pith-feeding. Because we found no clear difference in proximate environmental variables underlying observed variations in use among the three sites, we concluded that these differences in oil-palm use were cultural. Assuming individual interchange between these communities, at least in the recent past, and the involvement of social learning in the intracommunity transmission and maintenance of these uses, these results raise interesting questions regarding diffusion of behavior between neighboring chimpanzee communities.

Finally, Matsuzawa and Yamakoshi (1996) confirmed the presence of leaf swallowing of *Polycephalium capitatum* leaves among Yealé chimpanzees. Humle (unpublished data) later reconfirmed the prevalence of this behavior upon recovering once again whole leaves of *Polycephalium capitatum* in the feces. This species is also swallowed whole at Bossou and Seringbara (Matsuzawa and Yamakoshi 1996; Huffman et al. 1998; see Chap. 28). Huffman and Wrangham (1994) proposed that leaf swallowing of hairy species, such as *P. capitatum*, helps chimpanzees expel intestinal parasites and gain some relief from strongyle nematode infections (Huffman and Wrangham 1994).

27.3.3   Threats

Chimpanzees are at threat from hunters and poachers who come from outside the region and who, therefore, do not necessarily hold similar taboos to the local villagers against killing or eating chimpanzees. Indeed, Liberian hunters are known to hunt indiscriminately, including chimpanzees, in the Nimba forest of Côte d'Ivoire.

Because the boundary between the two countries within the forest is poorly demarcated, hunters and poachers freely circulate in the forest. It is, therefore, urgent that official patrols be organized to ensure proper law enforcement and to more effectively discourage hunters' and poachers' incursions into strictly protected areas of the massif.

As erosion resulting from the iron-ore mining activities that were conducted by the Liberian American Swedish Minerals Company (LAMCO) in Liberia from 1963 until 1992 (see Chap. 39) is impacting rice yield at the foothills of the mountain in both Liberia and Côte d'Ivoire, including in Yealé, there is an increasing risk that villagers will seek to abandon rice paddy cultivation for hillside cultivation, thus encroaching into chimpanzee habitat and having a negative impact on their survival. Increased threat of open-air iron-ore mining in the whole transboundary region also poses an undeniable threat to Yealé chimpanzees (see Chaps. 39 and 40).

27.4 Future Perspectives and Conclusion

Preliminary surveys and information from local people suggest that three groups of chimpanzees may reside in the region of Yealé, each one adjacent to one of three major rivers found in the Reserve: the Nuon, the Yan, and the Toua (see Fig. 27.1). In the region beside the Nuon River, Yealé, Côte d'Ivoire, the chimpanzee population was estimated at about 50 individuals, with a density of 0.5 chimpanzees/km^2 (Boesch et al. 1994; Hoppe-Dominik 1991; Humle, unpublished data). Clearly, little is known about (1) the precise number of chimpanzees in the region, (2) the number of communities prevailing on this side of the massif, and (3) their precise ranging patterns. Continuous research presence in the region is urgently needed if these questions are to be answered with more certainty and if we are to more comprehensively understand the behavioral ecology and material culture of the Nimba chimpanzees throughout their range (see Chaps. 28 and 29). Fortunately; research resumed in the Yealé region in 2008 and should contribute to our current knowledge of the status and behavior of the Nimba chimpanzees (see Chaps. 29 and 39).

Finally, the different chimpanzee communities of the Bossou–Nimba region are invaluable as they embody one of the few studied remaining viable populations of the Western subspecies of chimpanzee (*Pan troglodytes verus*). They embody the remarkable ability of chimpanzees to adapt to a range of environmental, topographic, and climatological conditions whose diversity is so unique to this region of West Africa. Clearly, our duty is to conserve this population and to preserve this invaluable World Heritage site; however, our challenge is great in the face of growing incompatibilities between conservation and development with the growing threat of iron-ore mining in the region (see Chaps. 39 and 40).

Acknowledgments I wish to thank Professor N'Guessan Yoa Thomas, Director of Research of the "Ministère de l'Enseignement Supérieur et de la Recherche Scientifique," and Cdt. Sombo, Director of National Parks, and Capitaine Cisse, Director of the Nimba Reserve, from the "Ministère de l'Environnement et des Forêts" of Côte d'Ivoire, for granting me permission to work in Yealé in the Nimba Mountains between 1999 and 2001. I am especially grateful to Tetsuro Matsuzawa and Gen Yamakoshi for their initial efforts and their encouragement to pursue work in Yealé. We are all especially grateful to all the Yealé guides – David Droh, Anatole Gogo, Philibert Pahon, Anthony Gopou, and Alexis Wonseu – for their hospitality and their invaluable assistance in the field.

Chapter 28
Chimpanzees in the Seringbara Region of the Nimba Mountains

Kathelijne Koops

28.1 Seringbara Study Site

28.1.1 Location and Status

The Seringbara study site is located on the western side of the Nimba massif in the foothills adjacent to the small village of Seringbara in southeastern Guinea (7°37′50.0″ N, 8°27′44.7″ W). The study area covers about 25 km² and is located 6 km to the southeast of the Bossou research site and 10 km from the Yealé study site on the other side of the Nimba Mountains in Côte d'Ivoire (see Fig. 27.1 in Chap. 27). The Seringbara region of the Nimba Mountains is separated from the Bossou hills by a 4-km-long stretch of savanna. However, in recent years the Green Corridor project (see Chap. 37) has made significant progress in reforesting this "green passage," thus increasing connectivity between Bossou and the Nimba Mountains (Hirata et al. 1998a; Matsuzawa 2006a, 2007; Matsuzawa and Kourouma 2008). The Nimba Mountains form a natural boundary between Guinea, Côte d'Ivoire, and Liberia and are a protected area, the Réserve Naturelle Intégrale du Mont Nimba, in Guinea and Côte d'Ivoire (see Chap. 39). On the Guinean side, the reserve extends over 12,700 ha and in Côte d'Ivoire over 5,000 ha. Both reserves form a World Heritage Site (Lamotte 1998b) covering a surface area of approximately 22,000 ha. In addition, the Guinean part of the Nimba massif has been classified as a Biosphere Reserve since 1980 and includes the Bossou and Déré ecosystems.

K. Koops (✉)
Department of Biological Anthropology, Leverhulme Centre for Human Evolutionary Studies,
University of Cambridge, Fitzwilliam Street, Cambridge CB2 1QH, UK
e-mail: kk370@cam.ac.uk

T. Matsuzawa et al. (eds.), *The Chimpanzees of Bossou and Nimba*,
DOI 10.1007/978-4-431-53921-6_29, © Springer 2011

28.1.2 Habitat and Climate

The Seringbara region of the Nimba Mountains is characterized by evergreen forest of medium altitude (Guillaumet and Adjanohoun 1971) and includes the foothills of Mont Leclerc, 1,586 m; Signal Sempéré, 1,652 m; Grands Rochers, 1,604 m; and Mont Richard Molard, the highest peak at 1,752 m. The Nimba Mountains show immense topographic diversity, ranging from rocky peaks and high-altitude plateaus to deep valleys and rounded hilltops. The hills are covered by dense primary tropical forest interspersed with richly forested valleys, in places dominated by terrestrial herbaceous vegetation of the Zingiberaceae and Marantaceae families. Numerous fast-flowing rivers cut through the forest year round, and the Nimba Mountains constitute a vast water catchment (WCMC 1992). Above 800–900 m the slopes of the mountains become steeper and the vegetation changes into montane forest with patches of terrestrial herbaceous vegetation. On the upper slopes, above about 1,000 m, high-altitude grasslands dominate the landscape, here and there broken up by gallery forests reaching up to 1,600 m (Fig. 28.1). The high-altitude savannas are maintained by annual natural fires in the peak of the dry season, which lasts from December to February. At this time of the year the dry and dusty Harmattan winds blow from the Sahel along the northwest coast of Africa, and wind speeds in Nimba can reach gale force. The rainy season (i.e., mean monthly rainfall >50 mm) covers 9 months of the year, from March through to November (Koops et al. 2007).

Fig. 28.1 The Seringbara region of the Nimba Mountains (photograph by Kathelijne Koops)

28.2   Chimpanzee Research at Seringbara

28.2.1   *Chimpanzee Surveys*

Since 1976, chimpanzee surveys in the Seringbara region have been intermittently ongoing. Sugiyama visited the area several times and conducted interviews with the local people. He did two preliminary surveys of the forested hills around the village, seeking to establish the presence of chimpanzees (*Pan troglodytes verus*). Based on these observations, he thought chimpanzees were only seasonally transient in the area (Sugiyama, personal communication). In 1999, two short surveys revealed six nests and feeding remains in the forest around Seringbara, and chimpanzee vocalizations were regularly heard (Shimada 2000). These findings strongly suggested the presence of at least one permanent chimpanzee community in the area. More systematic research in the Seringbara region conducted in 2000 (January–February, June–September) and 2001 (June–September) provided further evidence of chimpanzee presence, such as nests, feeding remains, and discarded tools (Humle 2003b; Humle and Matsuzawa 2001, 2004). In 2003, a permanent research site was established in the Seringbara forest, and several field camps were set up within the chimpanzees' home range (see Fig. 28.2). Habituation efforts were intensified, and there appeared to be at least two chimpanzee communities present in the Seringbara area based on direct observations and nest counts along transects (Koops 2005). The home range of the most intensively studied Tongbongbon community covers approximately 20 km^2, and the overall chimpanzee density is calculated at 1.72 chimpanzees/km^2 with Distance 5.0 software (Koops, unpublished data).

Fig. 28.2 The research camp of the Seringbara study site, Nimba Mountains (photograph by Kathelijne Koops)

28.2.2   Tool-Use

Tool-use by the Nimba chimpanzees, as well as comparisons between Bossou and Nimba material culture, were the initial foci of research in Seringbara (Humle 2003b; Humle and Matsuzawa 2001, 2004). Both similarities and differences in tool-use between the Seringbara chimpanzees and the nearby Bossou community emerged. Chimpanzees around Seringbara were found to use sticks as ant-dipping wants to eat army ants (*Dorylus* spp.) similar to those used by the Bossou chimpanzees. However, possible differences in tool length and the species of *Dorylus* targeted remain to be investigated in depth. Interestingly, Seringbara chimpanzees also may use digging sticks to dig up the underground nests of these ants (Humle 2003b). Such use of a digging stick in ant feeding has been seen only once at Bossou (Sugiyama 1995b). Another surprising difference in tool-use was that Seringbara chimpanzees do not use a hammer and anvil to crack open nuts of the oil palm (*Elaeis guineensis*), even though the chimpanzees in Bossou rely extensively on nut-cracking for food, especially at times of fruit scarcity (Yamakoshi 1998; Humle and Matsuzawa 2004). Nor do the chimpanzees in the Seringbara region exploit any other nut-bearing species reported as being cracked at other sites, such as *Detarium senegalensis*, *Parinari excelsa*, or *Parinari glabra*, all of which are available within the home range of these chimpanzees (Humle and Matsuzawa 2001; see Chap. 6). In addition, chimpanzees in Seringbara do not engage in pestle pounding, that is, using a detached palm leaf for pounding and softening the palm heart before eating it, although Bossou chimpanzees do (Humle and Matsuzawa 2004). Surprisingly, a new type of percussive technology has recently been described in the Seringbara population, which involves the fracturing of large and fibrous fruits of *Treculia africana* using stone and wooden 'cleavers' as tools and stone anvils as substrate (Koops et al. 2010). Detailed comparison of the variation in oil-palm use for feeding purposes between Bossou and Nimba showed no clear differences in underlying proximate environmental variables. These results therefore suggest that the differences seen in use are most likely culturally determined (Humle and Matsuzawa 2004). Considering the close spatial proximity between Bossou and Seringbara and therefore the great likelihood of individual interchange between the communities, at least in the recent past, these findings suggest cultural discontinuity even among neighboring chimpanzee communities.

28.2.3   Ground Nesting

Nesting patterns and characteristics of the chimpanzees in the Seringbara region were investigated in depth (Koops 2005; Koops et al. 2007). Between August 2003–May 2004 and April–August 2006, this study tested specifically a range of ecological and social hypotheses that might account for the occurrence and distribution of ground nests in Nimba (Fig. 28.3). Ground nesting, a common behavior

Fig. 28.3   Elaborate ground nest at Seringbara, Nimba Mountains (photograph by Kathelijne Koops)

in the Nimba chimpanzees, yet extremely rare in Bossou, was first described at Yealé, Côte d'Ivoire (Matsuzawa and Yamakoshi 1996), and later at Seringbara (Humle 2003b; Humle and Matsuzawa 2001). Generally, chimpanzees prefer to nest in the tree canopy, so although ground nesting is occasionally observed at a few other study sites, the frequent occurrence is unique to the Nimba chimpanzees. Chimpanzees of the Nimba Mountains commonly make both elaborate ("night") and simple ("day") nests on the ground, as many as 35% being reported at Yealé (Matsuzawa and Yamakoshi 1996; see Chap. 27) and 6–20% reported at Seringbara (Koops et al. 2007; Koops, unpublished data).

Koops et al. (2007) tested two ecological hypotheses for ground nesting at the Seringbara site: (1) climatic conditions, such as high wind speeds at high altitudes, might deter the chimpanzees from nesting in trees; and (2) lack of appropriate arboreal nesting opportunities might drive the chimpanzees to nest on the ground. To address possible social correlates of ground nesting, I investigated whether ground nesting is a sex-biased behavior. Hair samples were collected from ground nests and DNA analyses (i.e., Amelogenin and ZFX-ZFY sexing methods) were used to identify the sex of ground-nesting individuals. In addition, the spatial association between tree and ground nests was analyzed. Vegetation plots were marked around ground nests to assess tree availability, and two weather stations at low and high altitudes provided data on rainfall and wind speed. Based on 994 nests in 2003–2004 and 293 nests in 2006, I concluded that the occurrence and distribution of ground nests were affected neither by climatic conditions nor by a lack of appropriate nesting trees. However, ground nesting in the Seringbara chimpanzees appeared to be sex biased, as males built most of the ground nests found (Koops et al. 2007). Moreover, ground nests were made closer to tree nests than tree nests were to other tree nests within the same nest group (Koops, unpublished data). Also, elaborate ground nests were associated with tree nests significantly more often than simple ground nests. These results support the hypothesis that ground nesting in Nimba is socially, rather than ecologically, determined. They also suggest that elaborate and simple ground nests may be functionally different. Simple nests on the ground may provide a more comfortable rest in the daytime, whereas elaborate night nests on the ground may function as a social strategy. Males may nest on the ground to guard an estrous female in the trees above. However, the underlying assumptions – that elaborate ground nests reflect nighttime use and simple ground nests are used during the day – have yet to be tested. In sum, the question as to *why* ground nesting is sex biased, as well as the function of this behavioral pattern, remain to be explored further.

28.2.4   *Feeding Ecology*

At Seringbara, 86 species of plants, including 66 species of trees, 6 species of vines, and 14 species of terrestrial herbaceous vegetation, have been confirmed through direct observation, feeding remains, or fecal analysis to be consumed by the chimpanzees (Koops and Humle, unpublished data). Plant parts consumed include fruit, leaf, bark, gum, and pith. Nine of those species are not available at Bossou, but among the species in common, Bossou chimpanzees consume all species consumed by Nimba chimpanzees. During 2003–2004, seeds from tree fruit occurred in as much as 99% (75/76) of the fecal samples collected, confirming a heavy reliance on fruit in the diet (Koops 2005). Monthly phenology transects showed that the availability of ripe fruit eaten by chimpanzees was lowest in the middle of the rainy season and peaked during the dry season (Koops 2005). As suggested for Yealé (see Chap. 27), terrestrial herbaceous vegetation, such as plant species of the Marantaceae and Zingiberaceae families, was available year round and appeared to constitute an important fallback food for the Nimba chimpanzees.

28.2.5   *Habituation*

From mid-2003, habituation efforts in the Seringbara region increased from occasional surveys and short visits to an almost continuous research presence in the forest (average of 20 days per month). Research efforts were divided between phenological surveys, nest counts, and tracking of the chimpanzees. On tracking days, researchers listened for chimpanzee vocalizations and searched for feeding remains, fresh nests, and traces on chimpanzee trails. After finding the chimpanzees, the researchers moved slowly toward the apes and tried to behave in the least threatening manner (e.g., sit down, talk softly).

In 2003–2004 (8 months), the encounter rate by the principal researcher (K. Koops) was 0.14 observations per day in the forest. The mean observation duration for this study period was 13 min. In 2006 (5 months), the encounter rate was 0.15 observations per day in the forest, but the observation duration increased to an average of 65 min. Furthermore, the approach distance to the chimpanzees greatly decreased. Although in 2003 the chimpanzees usually fled at a distance of at least 50 m, some individuals now approach the researchers to a distance of less than 10 m. Least afraid of the researchers were adolescent males and females (Fig. 28.4). Although adult males and females often left after several minutes, adolescents tended to approach and to stay in the vicinity for as long as 2 h (Koops, personal observation). Behavioral responses toward the researchers varied greatly across

Fig. 28.4 Chimpanzees of the Tongbongbon community in the Seringbara region of the Nimba Mountains: (**a**) Poni (adolescent male), named after a Bossou chimpanzee, and (**b**) Lilé (adolescent female) (photographs by Kathelijne Koops)

years and across encounters (see next section). The chimpanzees either fled without vocalizing, or they stayed in the area for a while and resorted to branch shaking, waa barking, and branch throwing. Recently, some chimpanzees of the Tongbongbon community have started to (partly) ignore the observers, after the usual initial excitement upon discovering the researchers' presence (Koops, personal observation).

28.2.6   Behavioral Variants

28.2.6.1   Hand Clapping

Hand clapping, a previously unknown communication gesture in wild chimpanzees, was observed once in May 2004 in an adult female chimpanzee in the Seringbara region (Koops and Matsuzawa 2006). It seemed to be directed at, or at least provoked by, the presence of the researchers. Upon seeing the observers, the chimpanzee barked and screamed and shook branches. These behaviors were subsequently combined with hand-to-foot and hand-to-hand clapping. This clapping appeared to be a threat or display, similar to branch shaking. The combination of waa barking and clapping may alert other chimpanzees in the area to potential danger (e.g., humans). Also, the behavior may reflect a combination of fear and frustration in response to the presence of human observers. During the total observation period of about 2 h, the adult female showed two hand-clapping bouts and three hand-to-foot clapping bouts before leaving the area. As the chimpanzees are still only partially habituated, it remains to be seen if hand clapping is idiosyncratic, habitual, or customary among the Seringbara chimpanzees.

28.2.6.2   Leaf Biting

On three occasions (May 2006, November 2007, May 2008), three different chimpanzees, that is, an adolescent female, an adolescent male, and a juvenile male, were seen to exhibit "leaf biting." Leaf biting can be defined as systematically biting off leaves at the stem from a detached twig without consuming the leaves. In all cases, the chimpanzees broke off a twig containing many leaves, bit off the leaves one by one, and let them drop to the forest floor. This behavior differs from the previously described "leaf clipping" display (Nishida 1980) because leaf blades are not ripped along the midrib but rather bitten off as a whole. Also, leaf biting is not clearly audible, whereas leaf clipping produces a conspicuous ripping sound and is considered an auditory rather than a visual signal (Nishida 1980). The context of leaf biting appeared similar to self-grooming and seemed to reflect ambivalence toward the researchers. The individuals were partly habituated and did not flee from the observers, yet were not fully relaxed. Leaf biting alternated with feeding, self-grooming, and occasional screaming at the observers. It remains to be seen whether leaf biting is a community-wide behavioral pattern and if it may be used in different contexts.

28.2.6.3 Leaf Swallowing

The use of leaves for medicinal purposes, that is, leaf swallowing, was not described previously in the Seringbara chimpanzees (Humle and Matsuzawa 2001). Bossou chimpanzees swallow the leaves of *Ficus mucuso* and *Polycephalium capitatum* (Matsuzawa and Yamakoshi 1996). However, whole, unchewed leaves of *P. capitatum* have now been found in chimpanzee feces at Seringbara (Koops, unpublished data), as reported earlier for the chimpanzees at Yealé (Matsuzawa and Yamakoshi 1996; see Chap. 27).

28.2.6.4 Leaf Cushions

In the Seringbara region, both simple and elaborate ground nests were commonly built (see Sect. 28.2.3). Besides ground nests, usually made of saplings and terrestrial herbaceous vegetation, a few leaf cushions, or vegetation seats, have been found. These leaf cushions resembled those at Bossou (Hirata et al. 1998b), both in their simplicity and in their apparent function as a protective barrier from the wet ground. They differed from ground nests, as no nest structure and no bending, breaking, or hooking of vegetation was present. Leaf cushions in Seringbara typically contained just a few loose leaves and twigs, placed together on the ground, and apparently used as a cushion upon which to sit. As these observations were based on indirect evidence, confirmation of the exact use of leaf cushions awaits direct observations.

28.3 Future Directions

Current research on the Seringbara chimpanzees addresses the question of *why* these chimpanzees show some types of elementary technology and not others. The aim is to investigate the effect of ecological conditions on the use of elementary technology in both foraging (i.e., ant dipping, nut cracking, termite fishing), and shelter construction (i.e., nest building).

In foraging, the prevalence of elementary technology in a group of chimpanzees may be explained by two, not mutually exclusive, ecological hypotheses (based on McGrew et al. 1997). First, the availability of target species (i.e., ants, nuts, termites) as well as tool materials may affect the use of elementary technology. Second, the availability of alternative food sources, such as fruit and terrestrial herbaceous vegetation, may influence the dependence on elementary technology. In addition to considering these ecological hypotheses, the possible influences of social and cultural factors must be considered. The use of stone tools to crack open oil-palm nuts is ideal for an exploration of the influence of knowledge on traditions among the chimpanzees of Seringbara. Suitable nuts and tool materials for oil-palm nut-cracking are available at low densities in the home range of the Seringbara chimpanzees (Humle and Matsuzawa 2004), suggesting that the chimpanzees lack only the knowledge to crack nuts. To investigate this, an experimental approach is adopted and the chimpanzees are provided with suitable palm nuts and stone tools in an outdoor laboratory (sensu Matsuzawa 1994; see Chaps. 7, 16, and 17). To avoid human disturbance, motion-triggered video cameras are set up to obtain a detailed record of all chimpanzee activity (sensu Sanz et al. 2004). This new approach may shed light on the knowledge these chimpanzees have of nut-cracking, which in turn may help us understand some of the processes involved in cultural transmission between geographically adjacent chimpanzee communities.

The construction of a shelter, or nest, each night in which to sleep is a universal form of elementary technology in chimpanzee (Fruth and Hohmann 1994, 1996), but the function of nest building awaits systematic investigation. At Seringbara, it is possible to examine the function of both *arboreal* and *terrestrial* nest building. Several ecological hypotheses for tree nesting are considered, such as safety from predators, thermoregulation, and antivector/parasite strategy (sensu McGrew 2004). As already noted, ground nesting at Seringbara is not explained by basic environmental factors (Koops et al. 2007). To further investigate a social hypothesis of male mate guarding it is necessary to determine the sex of both ground nesters and of others who make tree nests above associated ground nests. Also, it remains unclear whether only some members of the community make ground nests or if it is a community-wide behavioral pattern. Individual genotyping could clarify the identity of individuals who habitually nest on the ground (McGrew et al. 2004).

DNA analyses may also help increase our knowledge of the population size, population structure, and genetic diversity of the Seringbara chimpanzees (see also Shimada et al. 2004). In addition to DNA-based sex identification, such a genetic approach requires the use of microsatellite genotyping techniques and mitochondrial DNA sequencing. This information is vital and urgent for several reasons. First, mining in the Nimba Mountains is beginning at present. It is essential to have premining baseline data on population genetic structure, so that the impact of mining on the chimpanzee population in the area can be monitored and evaluated. Furthermore, DNA analyses in combination with increased habituation may reveal the extent of migration between the Bossou and Nimba communities (see Chap. 34) and the effects of the Green Corridor project (see Chap. 37). The future of chimpanzees in Bossou and Nimba, along with many chimpanzee populations across Africa, depends on such crucial efforts to increase connectivity between isolated forest blocks. By acting now, we may be able to conserve the genetic diversity of wild chimpanzees and ensure their long-term survival.

Acknowledgments In Guinea, I thank the Ministère de l'Enseignement Supérieur et de la Recherche Scientifique, the Direction Nationale de la Recherche Scientifique, and the Institut de Recherche Environnementale de Bossou for the permission to do this research. Special thanks are due to the Seringbara team – Kassié Doré, Fromo Doré, Fokayé Zogbila, and Paquilé Chérif – for their dedication and hard work in the field. I thank Tetsuro Matsuzawa, William McGrew, and Tatyana Humle for support and advice. Also, I thank Tatyana Humle, William McGrew, and Sonja Koski for critical comments and for logistical support during preparation of this manuscript in 2008 while in the field in Guinea. The research was financially supported by grants from the Ministry of Education, Culture, Sports, Science and Technology (nos. 12002009 and 16002001), the Japanese Society for the Promotion of Science core-to-core program HOPE and 21COE (A14) to T. Matsuzawa, a Leakey Foundation Grant to T. Humle (2003–2004), and by grants from Lucie Burgers Foundation for Comparative Behavioural Research (the Netherlands), Schure-Beijerinck-Popping Foundation/KNAW (the Netherlands), Gates Cambridge Trust and St. John's College, Cambridge (United Kingdom), and International Primatological Society and IUCN/SSC Primate Specialist Group, Great Ape Conservation Action Fund to K. Koops.

Chapter 29
Chimpanzees in the Eastern Part of the Nimba Mountains Biosphere Reserve: Gouéla II and Déré Forest

Nicolas Granier

29.1 The Eastern Part of the Nimba Mountains Biosphere Reserve

The eastern part of the Nimba Mountains Biosphere Reserve (NMBR), called by local people the "Vépo region," refers to the Guinean territory localized "behind Vé river" (i.e., "south of Vé"). It contains two core areas of the Biosphere Reserve, which are the uneven southern slope and foothills of the Guinean Nimba range (hereafter Gouéla II) and the hilly Déré forest (Fig. 29.1). These two strictly protected areas are separated by 10 km of buffer zone consisting of lowlands covered by fields and fallow lands with small residual patches of damaged forest. Drained by the upper part of the Cavally River, the Vépo region presents fertile and arable soil, attracting cultivators. As a consequence, it is nowadays one of the first rice production areas of the Lola Prefecture, and this development is to the detriment of the forest.

A road linking Guinea to Côte d'Ivoire crosses the eastern part of the Reserve's buffer zone. Several villages and settlements occur along this road, with two major poles of human concentration: N'Zo, the county town, and Gouéla, the border village. Historically, the first humans to settle in the Vépo region were the Kono people (Germain 1984). Nowadays, it is also populated by other ethnic groups: forest people, in majority Guerzé, Manon, and Yakuba, as well as exogenous populations found mostly in N'Zo and Gouéla, such as Mandingos and Fula. The recent soaring population growth and the inherent subsistence activities of local people generate important anthropic pressures upon surrounding ecosystems.

N. Granier (✉)
Behavioral Biology Unit, Department of Environmental Sciences, University of Liège,
Quai Van Beneden, 22, 4020 Liège, Belgium
e-mail: nicogranier@yahoo.fr

T. Matsuzawa et al. (eds.), *The Chimpanzees of Bossou and Nimba*,
DOI 10.1007/978-4-431-53921-6_30, © Springer 2011

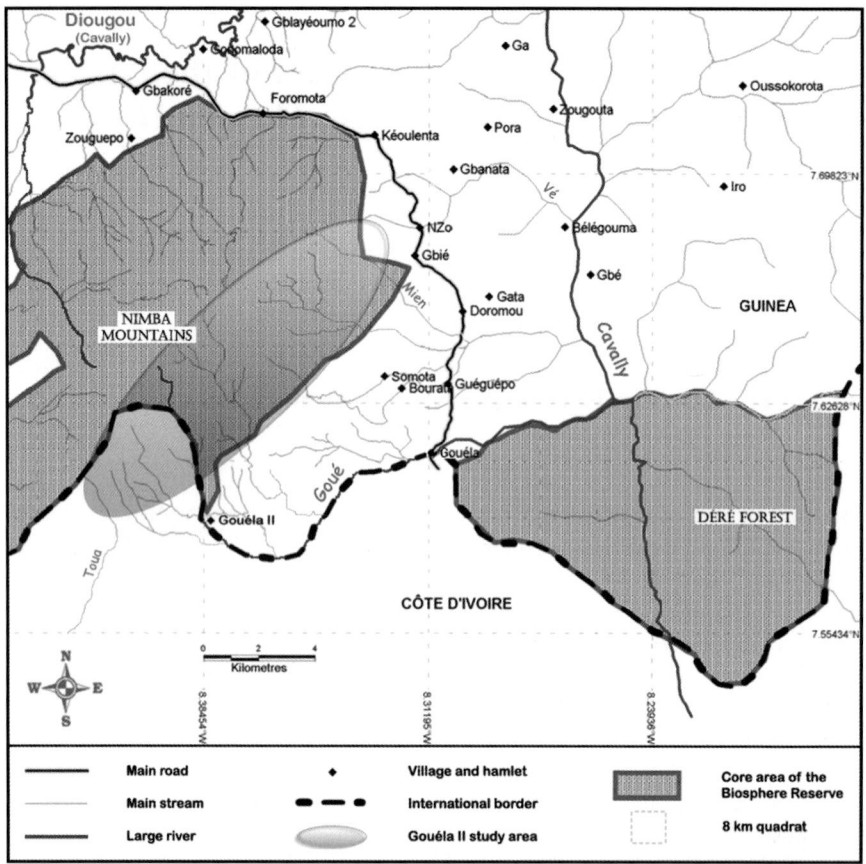

Fig. 29.1 The eastern part of the Nimba Mountains Biosphere Reserve (NMBR). This map presents the eastern part of the NMBR at the border between Guinea and Côte d'Ivoire. The two study areas, Gouéla II (*gray shadowed oval*) and the Déré Forest are represented, as well as the surrounding human settlements and hydrographic network, including the names of important rivers

29.1.1 Gouéla II

Gouéla II is the name of a century-old Guinean encampment for cultivation localized near the Goué River (which marks the border with Côte d'Ivoire) just at the limit of the Reserve's core area (07°35′49.8″ N; 08°22′48.3″ W; see Fig. 29.1). It is accessible only after more than 3 h walking, departing either from N'Zo or Gouéla. By extension, the region under study, which corresponds to the transboundary part of the Nimba southern slope stretching between the Mien and the Toua Rivers, was named Gouéla II. This forest has been strictly protected since 1944, as are the entire Nimba Mountains (Lamotte 1998b; see Chap. 39).

In Guinea, the crest and slopes of Nimba above 800 m elevation are covered by altitude grasslands. The foothills of the Nimba relief present intermittent stretches of savanna, which correspond to iron-bearing plains with thin soil and scattered shrubby species. Between these lowland and altitude herbaceous ecosystems, Guineo-equatorial rainforest covers the slopes of the massif (Fig. 29.2). A fifth habitat type described by Schnell (1998) is the secondary vegetation, composed of heliotropic ground plants with a relatively low density of wooded species. This habitat occupies a growing space at the limit between altitude grassland and altitude forest as a result of the negative impact of uncontrolled and repetitive bushfires.

A footpath links N'Zo (Guinea) to Yealé (Côte d'Ivoire) through Gouéla II, following the Reserve's limit. Along this path stands a network of human settlements interconnected by secondary trails and populated by remote communities living by farming, hunting, and natural resources exploitation. Despite a good preservation status of the Gouéla II core area of the NMBR, the adjacent forested buffer zone suffers from annual clearance for cultivation, which increases the isolation of Nimba ecosystems. A relatively significant poaching pressure using both guns and snares is exerted on the fauna inhabiting this core area. Among some 29 mammal species surveyed, Granier et al. (2007b) reported the presence of the emblematic African buffalo, the protected Jentink's duiker (*Cephalophus jentinki*), and 10 primate species, 3 of them being listed in the IUCN Red List of Endangered Species: the chimpanzee (*Pan troglodytes verus*), the Diana monkey (*Cercopithecus diana*), and the western pied colobus monkey (*Colobus polykomos*) (IUCN 2009).

Fig. 29.2 View of the Nimba Mountains from Gouéla II encampment. This picture illustrates the vegetation types present in the study area: altitude grassland, Guineo-equatorial rainforest, and plain savanna (photograph by Nicola Granier)

29.1.2 Déré Forest

This triangular-shaped forest at the border between Guinea and Côte d'Ivoire is contiguous to the Ivorian Tiapleu and the "massif des Dans" forests (see Fig. 29.1). Jealously protected by local villagers, it was called 20 years ago the "black forest" because of its close canopy, as well as the traditional ceremonies and initiation rites that took place within. To address the small-scale agricultural colonization that started in the early 1990s, the Man and Biosphere (MAB) program of UNESCO established the Déré Forest as a core area of the NMBR in 1993. However, no Guinean legal text has yet enforced this decision, and a logging company, named Valauris S.A., took advantage of the situation by building roads and cutting timber inside the forest between 1999 and 2001. The resulting land settlement has favored the immigration of hundreds of farmers seeking arable land to cultivate rain-fed rice using slash-and-burn techniques. Thus, an important part of this Reserve's core area is nowadays made up of farmbush, while the remaining patches of forest, which includes large and diverse tree species, constantly become smaller. Since 2002, the settlement of Ivorian rebels arrogating that the Déré forest belongs to Côte d'Ivoire has worsened this environmental problem.

The last mammal survey carried out in the Déré Forest by Granier et al. (2007b) reported the presence of 17 species including 1 primate species, the greater spot-nosed monkey (*Cercopithecus nictitans*), and 2 artiodactyl species quoted in the IUCN Red List of Endangered Species: the Jentink's duiker and the pygmy hippopotamus (*Hexaprotodon liberiensis*) (IUCN 2009).

29.2 Chimpanzee Research

29.2.1 History

The presence of chimpanzees on the southern slope of the Nimba Mountains was reported more than 65 years ago by Lamotte (1942). However, particular interest for the species in this region truly began in 1993 with the establishment by Matsuzawa of the Yealé research site in Côte d'Ivoire (Matsuzawa and Yamakoshi 1996; Humle and Matsuzawa 2001; see Chap. 27). In Gouéla II forest, only two short-term surveys were conducted to estimate the status of chimpanzees (Sugiyama 1995b; Shimada 2000), and both confirmed the presence of the species. In the Déré Forest, the only primatological data came from a Rapid Assessment Program conducted by Conservation International on several taxonomic groups of wildlife (McCullough et al. 2006). The presence of chimpanzees was not reported during their 3-day survey.

29.2.2   Present Study

29.2.2.1   Objectives

A behavioral and ecological study of nonhabituated chimpanzees' habitat use was initiated in the eastern part of the NMBR in 2006. The general aim of this still ongoing work is to describe the ranging and grouping patterns of the species in this region. Specifically, it aims to understand how seasonality in food availability and anthropic pressures leading to habitat modification influence the spatiotemporal distribution of chimpanzees (Sugiyama and Koman 1992; Chapman and Peres 2001; Basabose 2005). It is nowadays commonly admitted that different chimpanzee communities have different cultural traditions, which express through behavioral variations in material skills, knowhow, habitat use, or resource exploitation (Boesch 1996a; Matsuzawa and Yamakoshi 1996; Whiten et al. 1999; Humle and Matsuzawa 2001). Consequently, the study of an unknown community always presents important issues, which brought me to investigate the chimpanzees of the eastern part of the NMBR. This study also has conservation-oriented objectives, consisting of acquiring accurate data on this endangered species in its exceptional and peculiar environment to help effectively protect it (Kormos et al. 2003a; IUCN 2009).

29.2.2.2   Methods

To reach these objectives and obtain a preliminary overview of chimpanzee presence, I first conducted interviews of local communities in March 2006. Thirty-two were performed in 18 human settlements located between the two focal core areas of the NMBR. Interviewees were all hunters and/or cultivators, contacted either individually or in a group, in a random and opportunistic manner. However, a systematic questioning procedure was repeatedly used, with questions addressing chimpanzee presence, abundance, distribution, and relationships with humans.

Then, during three fieldwork periods (March–April 2006, December 2006–May 2007, and February–June 2008), field studies of chimpanzee habitat use were carried out in the Gouéla II and Déré forests. Evidence of chimpanzee presence was recorded from three types of survey (transect, recce, and scouting survey). These itineraries were systematically positioned across the study areas and covered by walking (Table 29.1). All chimpanzee presence indicators, including nests, feces, trails, footprints, traces, feeding remains, tool-use sites, and vocalizations, were thoroughly recorded and geo-referenced together with changes in habitat types encountered during the walks.

Sampling effort was lower in Déré than in Gouéla II, with 163 and 350 km walked, respectively (see Table 29.1). Twelve parallel transects 300 m long were established in Gouéla II. Their origins were systematically located 10 ft. apart (309 m), starting in plain savanna and heading upward to the north, going through

Table 29.1 Sampling effort from the three survey itinerary types

	Gouéla II			Déré forest		
Survey type and number	T ($n=12$)	R ($n=4$)	Ss ($n=51$)	T ($n=0$)	R ($n=3$)	Ss ($n=18$)
Number of passage	8	4	1	0	3	1
Walk distance (km)	28.8	96	225	0	79	84
Total distance (km)	349.8			163		
Total number of day	165			29		

This table shows the number of each survey itinerary type in Gouéla II and in the Déré Forest. The number of passages and the walked distance per site on each type of survey itinerary and in total are also figured, as well as the number of days it required
T transect; *R* recce; *Ss* scouting survey

all vegetation types. Each transect was walked eight times. Four recces set in Gouéla II consisted of loops with a mean length of 5.9 km (± 0.7 SD), stretching between the plain and altitude herbaceous ecosystems; each was walked four times. Both transects and recces were periodically walked to record data, with a passing frequency of 3 weeks for transects and of 1 month for recces. In Déré, three recces were set as loops rising perpendicularly from the Cavally River to the limit of the classified forest (mean length = 8.7 km ± 1.3 SD). The recces were walked three times, without any temporal regularity. Scouting surveys consisted of opportunistic walks covered just once following chimpanzee tracks; 51 surveys were walked in Gouéla II and 18 in Déré.

29.3 Preliminary Results in Gouéla II

29.3.1 Selective Use of Altitude Forest and Irregular Presence

Numerous animal trails roughly following the contour lines of the Nimba range were observed at the limit between altitude forest and grassland, between 700 and 800 m in elevation. These trails crossed large expanses of secondary vegetation and, surprisingly, revealed numerous indicators of chimpanzee presence, i.e., 1.14 per kilometer ($n=87$ km walked on such trails). Almost two thirds of all the recorded evidence was found in altitude forest and adjoining secondary vegetation (61% of all presence indicators; $n=393$), whereas the sampling effort in these two habitat types represented 43% of total. In addition, vocalizations heard ($n=18$) always came from the higher parts of the forest. Chimpanzee tracks recorded in the lower parts of the forest (above 700 m; $n=97$) constituted a quarter (24.7%) of all the recorded indicators of presence. Interestingly, the presence of ripe edible fruits was recorded in the vicinity of 87% of these low-altitude records. Finally, collected chimpanzee feces contained seeds of tree species such as *Harungana madagascarensis*,

Musanga cecropioïdes, Trema guineensis, Aframomum sp., and *Ficus* sp., plant species widely represented in the secondary vegetation of Nimba, most of which fruit during the period of fruit scarcity (March–June). Further investigations are ongoing to provide more insights into the potential relationship between chimpanzees' use of secondary vegetation and food availability within this habitat type.

29.3.2 Nesting Behavior

A total of 337 nests was seen and recorded. The total number of nesting groups could not be identified because several nesting sites are periodically reused by chimpanzees, and discrimination of different-age nests is not reliable enough to determine group composition. However, a mean size of 5.4 nests per group was calculated from the fresh and unequivocally identified nesting groups ($n=23$). Groups of 2 or 3 nests were frequently observed, and the largest one was composed of 22 fresh nests in the higher part of Sakona River (Fig. 29.3). This area seems to be an important nesting site because large groups of fresh and recent nests were twice observed and new nests seen during each visit. Nesting sites were located between 649 and 843 m altitude, mostly in galleries (36% of total nest number) and altitude forest (34% of total). Nests were often built in very steep places with a ground declivity superior to 22.5% (63% of total). The mean height of nests from the ground was 7.8 m ± 4.6 SD. Both ground declivity and nest height were measured using a clinometer and calculated post hoc.

These preliminary results together strengthen the assumption that plant food availability may be a determining factor of the chimpanzee's spatiotemporal distribution in Gouéla II. The higher part of the Gouéla II transboundary area may be part of the home range of at least one community of chimpanzees. It may constitute a peripheral zone of its (their) habitat(s), punctually visited by small parties or solitary individuals, although no periodicity has yet been revealed.

29.3.3 Tool-Using Behaviors

Eight ant-dipping sites were observed on the ground in the altitude forest (mean altitude, 784 m). The wands used by chimpanzees were systematically identified and measured. *Aframomum* sp., *Dacryodes* sp., *Microdesmis keayana*, and *Mareya micrantha* were the species most commonly used to catch ants, seen in at least two collection sites. Tool length ranged between 18 and 73 cm ($n=44$; mean=46.8 ± 15.5 SD), with a diameter ranging from 0.2 to 0.8 cm (mean=0.4 ± 0.1 SD). Both Sugiyama (1995b) and Shimada (2000) also reported evidence for ant dipping during their short-term surveys in Gouéla II.

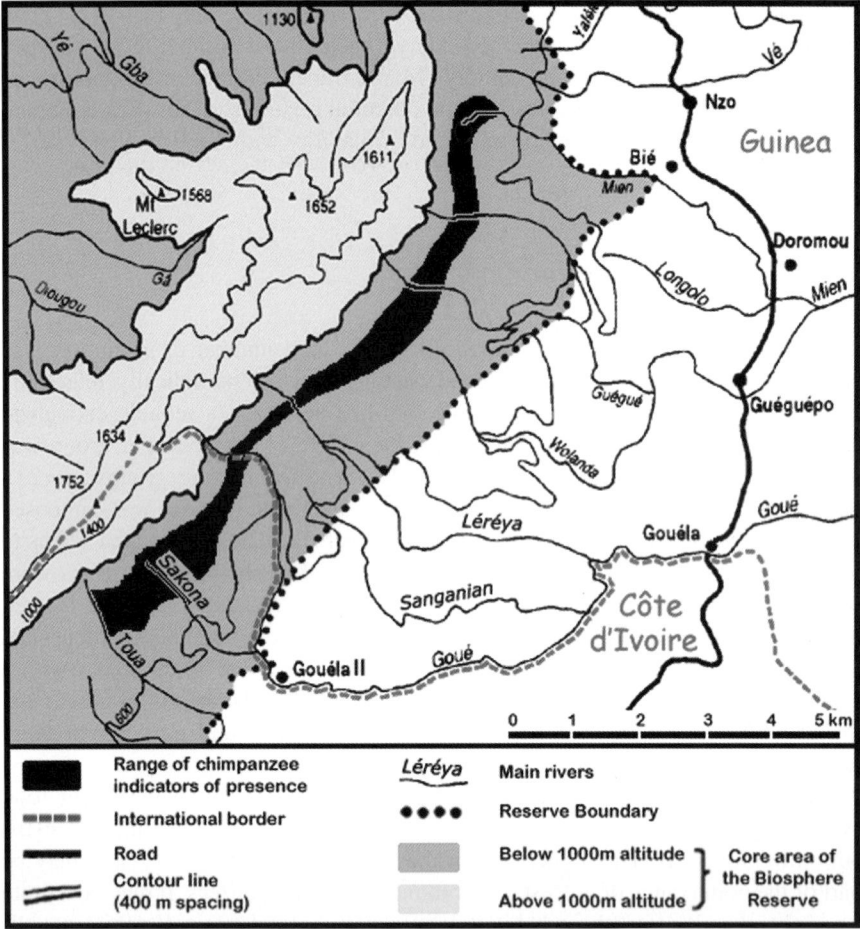

Fig. 29.3 Spatial distribution of chimpanzee presence indicators in Gouéla II. This map shows details of the Gouéla II study area at the border between Guinea and Côte d'Ivoire, including the names of all rivers and the nearest human settlements. The range in which all chimpanzee indicators of presence were observed appears in *black*

It is assumed that Gouéla II chimpanzees may crack open nuts using stones, as nut trees (*Elais guineensis* and *Coula edulis*) occur in the Mien River catchment area (northern end of Nimba), and this behavior was reported from interviews. However, I have not yet uncovered any evidence of nut-cracking in the area.

Finally, four interviewees have reported chimpanzee fishing in little streams, behavior that was corroborated by an observation made by Michel Zogbélémou, my guide from Gouéla II. Chimpanzees would build a dam with trees and leafy branches in a narrow part of the stream and use their hands to catch fish from the upstream water. Mention of an individual from the Bossou chimpanzee community eating fish was also made by Sugiyama (1987).

29.4   Preliminary Results in Déré Forest

It emerges from interviews with local communities that before the logging exploitation chimpanzees were present in the black forest. However, only three reports of a present chimpanzee occurrence in the Déré forest were made from the same interviews ($n = 32$). These affirmations were nonetheless corroborated in March 2006 by observation of a very old nest built in an *Aningueria altissima* tree near the Ivorian border, in the southeastern part of the forest (Granier et al. 2007a). A farmer settled in this area of the classified forest reported that chimpanzees would annually visit the place at the end of the rainy season, coming from the "Massif des Dans" just across the border. Although the age of the nest seemed to correspond to the end of the rainy season, I could not confirm this assumption. Still, it is highly plausible that the presence of chimpanzees and of other large fauna in the Déré forest has largely decreased during the past decade, mainly caused by the negative influence of human activities and habitat destruction. Given the actual high rate of forest clearance, prospects for chimpanzee research in Déré are extremely poor, unless enormous long-term efforts of environmental conservation are urgently undertaken.

29.5   Perspectives on Future Chimpanzee Research and Conservation

29.5.1   Research Perspectives

Further data collection is still necessary to provide a detailed picture of factors influencing habitat use of chimpanzees in the eastern part of the NMBR. Up to now, data collection occurred exclusively during the dry season; consequently, the rainy season in 2009 will be spent collecting missing data. Then, botanical and fruit phenology information will be added to the spatial analysis of chimpanzee indicators of presence to obtain a year-round image of seasonality. Further investigations have also been initiated to understand the influence of food availability and human/predator presence on chimpanzees' choice of nesting sites in difficult-to-reach places.

It is well known that chimpanzees from the Bossou community regularly travel to the Nimba Mountains and to Liberia (Sugiyama 1999; Ohashi 2006b). Yet, very few data are available regarding intercommunity exchanges and encounters between the Bossou and Nimba communities. Because the chimpanzee is a highly mobile species that exploits a very large habitat regardless of national borders and shows individual migrations, it would be of great interest to implement a large-scale study of the variables driving the chimpanzee's ranging and grouping patterns. The objective is to put forward tri-national perspectives on the Nimba chimpanzee's habitat use, intercommunity exchanges, and material cultures, by working from an overall point of view.

After analysis of the preliminary results presented here, it appeared essential to amend the study area to obtain a more cohesive area, in keeping with research objectives. Thus, I decided to exclude Déré Forest and to focus exclusively on an enlarged part of the Nimba range by including Yealé (Côte d'Ivoire, 12 km southeast from Gouéla II), where previous chimpanzee research has been carried out (see Chap. 27). However, aware of the huge conservation issues of Déré Forest, I decided to maintain contact with its resident populations.

29.5.2 Conservation Perspectives

Chimpanzee conservation is tightly connected to the problems of habitat and the sustainable preservation of biodiversity. In addition, there is growing evidence suggesting that the alleviation of threats facing Nimba chimpanzees requires an integrated collaboration between research and conservation. Consequently, the challenge facing chimpanzee protection nowadays is to provide coherent conservation policy and measures that would integrate the different political, socio economic, and protective statuses of the Nimba Mountains in each of the three countries, plus the known elements of chimpanzee life history and ecology. Such integrative perspectives are consistent with the proposed approach of research just described here.

The current very poor conservation status of the Déré Forest is the result of an intricate mix of almost all the critical threats facing biodiversity conservation nowadays: industrial activities (logging in this case) and human-related activities such as human politics (soaring population growth, migrations, wars, difficulties in cross-border management, etc.) or community subsistence activities (slash-and-burn agriculture with perpetual lack of arable land, nonrational hunting or poaching, fishing, gathering, uncontrolled use of fire, etc.) (see Chaps. 39 and 40). It emerges from this situation that the eastern part of NMBR does not constitute a cohesive unit, mainly because of the different conservation status, problems, and needs of its two core areas. It would be rather consistent to apprehend it in two distinct parts: Gouéla II, linked to the problems of research and conservation in the NMBR core area, and the Déré Forest, which has concretely became more related to the buffer zone management issue. Such a differential management would emphasize the dissimilarities of these two focal areas, while improving efficiency of conservation measures in the single but diversified entity of the Nimba Mountains Biosphere Reserve.

Acknowledgments I wish to thank all staff of the Direction Nationale de la Recherche Scientifique et Technologique (DNRST) and the Institut de Recherche Environnementale de Bossou (IREB) for their help in conducting this research, especially their respective directors, Tamba Tagbino and Makan Kourouma. I am particularly grateful to the guides and local authorities of N'Zo subprefecture: Michel Zogbélémou, Jacques Bamba, Leonard Gamaleu, Mamadou Zogbélémou, Mamadou Sylla, and Korfou Diallo, for their indispensable help and dedication.

Special thanks are due to Tetsuro Matsuzawa and Marie-Claude Huynen for their support and advice, as well as Marie-Claude and Laura Martinez for their comments on the present manuscript. The research was financially supported by grants from the Japanese Ministry of Education, Culture, Sports, Science and Technology, the Japanese Ministry of Environment, and the Japanese Society for the Promotion of Science to T. Matsuzawa, and by a grant from IUCN/SSC Primate Specialist Group, Great Ape Emergency Conservation Action Fund, to N. Granier.

Chapter 30
Diécké Forest, Guinea: Delving into Chimpanzee Behavior Using Stone Tool Surveys

Susana Carvalho

30.1 Study Site

30.1.1 Location and Conservation Management

The Forest of Diécké is the second largest protected reserve in the administrative region of *Guinée Forestiére*. It is located in southeastern Guinea, where the border appears to blend in with Liberia. Between latitudes 7°39′ N and 7°21′N and longitudes 9°06′W and 8°47′ W, the forest of Diécké is situated by road about 100 km west of Bossou (Bossou-Nzerekoré-Diécké road), and 950 km of the capital, Conakry. This moist dense forest is one of the main biodiversity hotspots in West Africa (World Wildlife Fund (2007); Fig. 30.1; see Chap. 40). This area has more than 700 km² of almost undisturbed evergreen or semi-deciduous forest, ranging about 30 km from west to east, and 35 km from north to south, and is presently surrounded by human settlements strategically exploiting its resources.

The Forest of Diécké was classified as a protected reserve during colonial times, between 1945 and 1955 (Delorne 1998). Today, to pursue research in the Reserve, permits are required: these are granted by the *Centre Forestier de N'Zérékoré*. The region has seen a rapid increase in human immigration that began in 1989 with the influx of more than 500,000 refugees from Liberia and Sierra Leone.

This influx led to enormous increases in local demand for resources and to years of deforestation and conversion of areas of primary to secondary forest (Black and Sessay 1997). In recent years, however, the area has been managed by the *Projet de Gestion des Ressources Forestières, Progerfor*, in cooperation with German partners, based in the Centre Forestier de N'Zérékoré. This project is focused on conservation/reforestation, damage minimization, and regional development and monitoring of hunting activities. For this purpose, the forest was divided into 30 plots with three

S. Carvalho (✉)
Leverhulme Centre for Human Evolutionary Studies, Department of Biological Anthropology,
University of Cambridge, Henry Wellcome Building, Fitzwilliam Street,
Cambridge CB2 1QH, UK
e-mail: scr50@cam.ac.uk

T. Matsuzawa et al. (eds.), *The Chimpanzees of Bossou and Nimba*,
DOI 10.1007/978-4-431-53921-6_31, © Springer 2011

Fig. 30.1 Broad view of the Diécké Forest in the Korohouan area (photograph by Susana Carvalho)

levels of protection: The *série d'amelioration* (improvement area) level is a peripheral area, where forest fragments of slash-and-burn agriculture, mostly for the cultivation of rice or cassava and the planting of oil palms (*Elaeis guineensis*), are found. The *série d'utilisation durable* (area of sustainable utilization) consists predominantly of altered primary forest and secondary forest. Finally, the *série de protection intégral* (integrally protected area) is the core of the Diécké Forest, shaped mainly by extensive patches of closed-canopy rainforest (Fig. 30.2). This area is one of the last remaining areas of Western Guinean Lowland forests and maintains a great biodiversity of flora and fauna.

30.1.2   Portrait: Ecology, Geomorphology, and Human Harvest of Resources

The core area of Diécké can mostly be described as evergreen lowland forest and swamp and riparian forest. The average altitude is 350 m, with the lowest point at 50 m above sea level and the highest, Mont Jna, at 800 m. However, recent updates concerning the classification of these forests suggest that many of the Western Guinean Lowland Forests should be considered as "late secondary stands" because, throughout history, some of these areas have been cyclically under human impact, and, as a result, may not be considered as forests in their primary state (World Wildlife Fund 2007). Therefore, Diécké should be considered a mosaic forest, as the diversity of flora inside the protected area ranges from characteristic tree species,

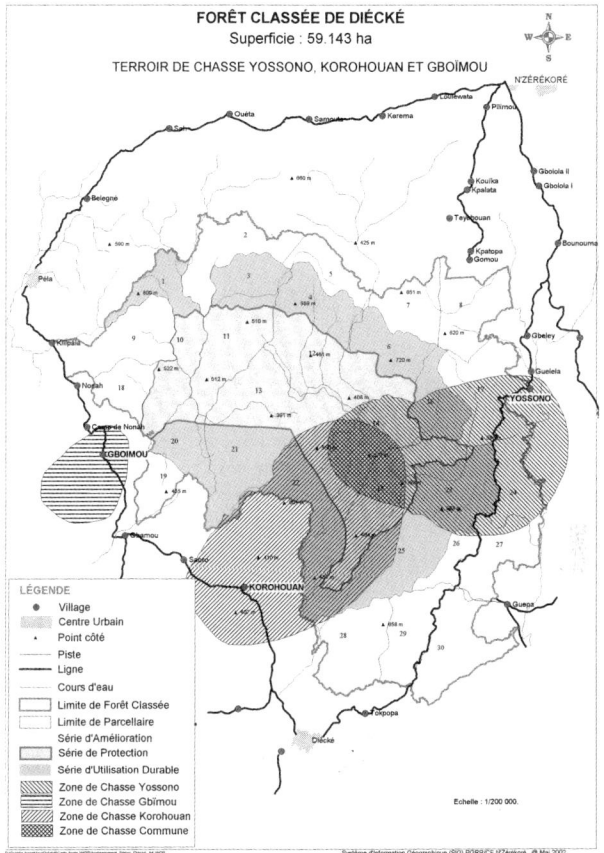

Fig. 30.2 Geographic information system (GIS): The division of the Diécké forest into 30 parcels and three levels of protection. The zones where hunting has been identified are also controlled by the project of conservation and regional development (© GIS; PGRR/CF, N'Zérékoré, Guinea, 2002)

with canopies reaching over 30 m in height – similar to the humid forests of Liberia and Côte d'Ivoire – to vegetation found in disturbed canopies, as well as in swamps, and riverine and secondary forest areas. The climate presents average temperatures of 32°C during the dry season and 23°C during the wet season, and is characterized by a short dry season lasting from January to March (Black and Sessay 1997; Carvalho et al. 2007). The forest is shaped by three main rivers carving the land to the south: on the western side the Nyé River, and on the eastern side the Gbin and Gbin-bé Rivers. The many floodable streams often generate swampy areas (Carvalho et al. 2007, 2008).

Along with the Ziama Forest, Guinea, this reserve is currently an important refuge for endemic species of fauna and flora, and is one of the 12 major sites for biodiversity conservation in West Africa (World Wildlife Fund 2007). The speciation

304   S. Carvalho

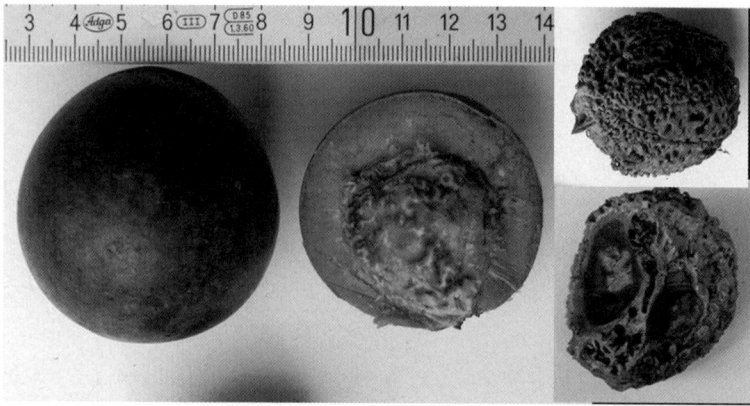

Fig. 30.3 The *Panda oleosa* nut is the hardest to crack open. In the Diécké Forest, chimpanzees crack open this hard nut with stone hammers and stone or root anvils. The new fruit presents a pulp with orange-colored endoderm (*left*). When the fruit dries up, the hard-shelled nut is cracked to extract three nut seeds lodged inside (*right*). The edible kernels are white (photographs by Susana Carvalho)

process that led to the present biodiversity occurred during the Pleistocene, between 250,000 and 15,000 years BP with repeated cycles of retraction and expansion of these forests during drier (ice ages) and hotter periods (Caldecott and Kapos 2005).

Some of the floral species that inhabit the mixed lowland tropical forests such as Diécké are typically *Pycnanthus angolensis*, *Piptadenia africana*, *Alstonia congolensis*, *Antiaris* spp., *Khaya grandifolia*, *Entandophragma utilis*, *Heriteria utilis*, *Mansoniea altissima*, *Dialium* spp., *Lophira alata*, *Uapaca guineensis*, etc. (White and Abernethy 1996; Bah et al. 1997). Of these, several species were found during the ongoing research and confirm the description of Diécké as a mosaic forest. In addition, we corroborated the presence of *Panda oleosa* (Fig. 30.3), *Cola cordifolia*, *Coula edulis*, *Raffia hookeri*, *Parinari excelsa*, *Landolphia owarensis*, *Strychnos spinosa*, *Acridocarpus australocaledonicus*, and *Aframomum* spp. The presence of important fauna has also been recorded, through traces or direct observations: it includes the western chimpanzee, *Pan troglodytes verus*, the pygmy hippopotamus, *Hexaprotodon liberiensis*, the dwarf crocodile, *Osteolamus tetraspis*, the dwarf forest buffalo, *Syncerus nanus caffer*, and the blue duiker, *Cephalophus monticola*.

Geologically, Diécké belongs to an ancient West African formation, belonging to the Precambrian granitic shield (Clark 1967). Diécké's geological formations date from the Neo-Proterozoic (530–628 Ma) and are composed mainly of granitic and granodioritic complexes, as well as gneisses and ferruginous quartzite complexes. However, it is also possible to find young alluvial deposits along the riverine valleys inside the forest. Diécké belongs to the geomorphologic unit called Piémont de N'Zérékoré (Dux et al. 2002; Gradstein et al. 2004) and the soils are generally poor, lateric, and therefore easily exhausted by agricultural activities.

Nevertheless, because it is one of the most productive regions of Guinea, the area has been occupied by human populations for centuries. Migrations of different ethnic groups coming from the western savannas have occurred since at least 1,350–1,500 AD (Black and Sessay 1997). At present, the western side of Diécké is primarily Guerzé, with the exception of Korouhan (Manon village), while the eastern side is predominantly Manon. Main villages around the protected reserve, from west to east, include Nonah (7°31'50.9" N; 9°04'27.7" W), Gboimou (7°30'52.3" N; 9°04'23.3" W), Gbamou (7°27'55.7" N; 9°03'41.4" W), Saoro (7°27'06.2" N; 9°01'22.2" W), Korouhan (7°26'09.6" N; 8°59'22.2" W), Diécké (7°20'23" N; 8°57'15.8" W), and Yossono (7°33'10.5" N; 8°48'55.8" W).

Deforestation in this area has been a problem since the beginning of the colonial era (1905–1907). However, the unexpected influxes of migrants during the late 1980s have contributed to accelerated rates of deforestation (Black and Sessay 1997). The immense demand for resources for fuel, construction, and supplies is exacerbating the exhaustion of the land and water sources, timber harvesting, and bushmeat hunting. Hunting pressure in this area is very high and hard to monitor, in part because of the lack of forestry agents (Bourque and Wilson 1990; Kormos et al. 2003b). Extensive plantations of oil palm (SOGHIPAH loggings) and rubber developed rapidly in recent years, converting areas of primary forest into secondary forest. Furthermore, plantations of crops, such as rice, cassava, coffee, kola, cacao, and maize expand every year (personal observation). Problematically, the Diécké Forest in particular, and all the entire prefecture of Yomou in general, has been one of the most isolated areas of Guinea. Thus, conservation awareness-raising programs or wildlife research activities have been rare and have had little influence in this area.

30.2   Chimpanzee Research in Diécké Forest: State of the Art

The aforementioned scenario, coupled with the large size of the area, contributed to an almost complete lack of systematic data or information concerning the status of the western chimpanzee in Diécké. Sugiyama and Soumah (1988) estimated the existence of 50 individuals in this area, based on questionnaires that were collected nationally across Guinea. Surveys carried out by Ham in 1997 (Kormos et al. 2003b) combined questionnaires and a 5-km transect in Diécké (where 14 nests were found), and estimated 209–307 individuals in this area. The presence of chimpanzees was also confirmed directly by Ham in 1997, with two observations, one of them in the Yossono area (Kormos et al. 2003b).

In 1999, the Kyoto University Primate Research Institute (KUPRI) research team, led by Tetsuro Matsuzawa, carried out the first more extensive survey in the forest of Diécké. The main goal was to locate a new study site for gathering comparative data with Bossou. After 3 days during their preliminary survey, they confirmed chimpanzees in the forest in the Nonah area. Furthermore, they confirmed the use of stone and wooden tools to crack panda and coula nuts (Matsuzawa et al. 1999). This discovery suggested that additional research in this forest could provide an important contribution, not only to the conservation of chimpanzees but also to our understanding of chimpanzee material culture. Additional surveys carried out by Takemoto and Humle in 2000 expanded the area surveys around Nonah and Yossono and confirmed several stone tool-use sites in the forest (Humle and Matsuzawa 2001). They recorded one coula nut-cracking site and three panda sites in the Yossono area in 1999–2000 (Humle and Matsuzawa, personal communication).

In 2006, KUPRI initiated regular research in this area, combining archaeological and primatological methods, with the main aim of recording nut-cracking sites and stone tool-use behavior, as well as to collect information on the ecology of the forest and on the current status of chimpanzee within (see Chap. 15; and Carvalho et al. 2007, 2008; Biro et al. 2010; for details on the surveys). Between January 2006 and May 2008, six survey trips were carried out in the Nonah and Korouhan areas (during 35 days, mainly in the dry season) and four temporary camps were set up inside the forest, while local teams were organized and trained to work as local guides (see below for survey results). Research has developed since with the establishment of the first encampment (*Lehtout* in the Korouhan area) in the core of the reserve in May 2008 (Fig. 30.4). By the end of 2009, we discovered new nut-cracking sites and set up transect lines to measure raw material availability (nut trees and stones). In addition, traces of chimpanzee presence were recorded inside and outside the protected area and, more importantly, the chimpanzees of Diecké were observed and recorded directly for the first time (Carvalho, unpublished data; Fig. 30.5).

Fig. 30.4 Construction of the first encampment for chimpanzee research in the Diécké forest: Lehtout camp (in May 2008) (photograph by Susana Carvalho)

Fig. 30.5 First photographic evidence of a Diécké chimpanzee: a juvenile male observes us from the top of a *Landolphia owarensis* tree (photograph by Susana Carvalho)

30.2.1 The 2006–2008 Surveys: Goals, Methods, and Results

The main goal of these surveys was to detect indirectly or directly the practice of nut cracking by wild chimpanzees and to compare possible differences with Bossou. Diécké harbors different nut species (e.g., panda, coula) than the species found at Bossou (i.e., oil palm). The presence of nut-cracking technology in Diécké allows recording of the tool-use process, which provides comparative and complementary data to those obtained in Bossou. Importantly, it also provides an opportunity to carry out analyses of regional and typological variations (Carvalho et al. 2008). Other objectives were to enlarge the surveyed areas and to establish a new research site for comparative studies across wild chimpanzee communities. To accomplish the purpose of observing the tool-use process, finding new nut-cracking sites in the Diécké forest was an essential goal (see Chap. 15).

Sousa and Carvalho carried out their first excursion to the Diécké area in January 2006. This excursion was a pilot trip, necessary to make social contact with local people from different villages around the reserve, to establish which sites would be more suitable and in need for future regular research, and to evaluate the validity of local information about the presence of chimpanzees in the forest near each village. Face-to-face interviews were done in the main villages. Although this method may be inaccurate, it is recognized as a useful technique to gather important information before the beginning of research (Kühl et al. 2008). It proved to be valuable, because it was possible to discard erroneous locations based on nonplausible information (e.g., that chimpanzees had been directly observed in the reserve around village Y, while vocalizing and beating their chests with their hands). Nonah and Korouhan were chosen as focal villages to start training research teams and to survey their respective forest areas. Nonah had already hosted previous researchers, and Korouhan had the potential of providing access to some important biodiversity spots inside the forest. Moreover, high hunting pressure was recorded at both sites, and research presence and conservation education initiatives were viewed as potentially highly beneficial to alleviate anthropogenic pressures acting on these areas.

Five surveys were done in Diécké up to 2008, departing from Nonah or Korouhan with the same local teams, using temporary camps. The surveys were carried out by the KUPRI-International team, accompanied by two teams of local guides. Diécké is a vast area, and the presence of only a handful of nut-cracking sites had previously been reported before this study (Matsuzawa 1999; Humle and Matsuzawa 2001). The survey procedure relied, mainly, on the local guides' knowledge of the forest, walking along existing tracks, similarly to reconnaissance surveys (Kühl et al. 2008). During a selected survey to search for nut-bearing trees, it was therefore often necessary to open new routes with machetes because of the high density of the forest in some areas. Visibility was often low. However, once it was apparent that nut-cracking places were located near watercourses, the survey was focused on these locations, as proposed by archaeological surveying methods (Renfrew and Bahn 1998; Banning 2002). Tropical rainforest is an uncommon place to perform archeological surveys for reasons of reduced ground visibility, bioturbation processes, and dense vegetation, all of which create obstacles to the application of the methods for surface surveys (Renfrew and Bahn 1998; Mercader et al. 2002; Mercader 2003; Carvalho 2007; and see Chap. 15 for details on this problem). When a nut-cracking site was found, a centered survey was done, within a 1-km radius of the site, to record the occurrence of raw material sources, water, and chimpanzee shelters (nests or beds), feeding signs, or footprints. All nut-cracking sites, as well as chimpanzee traces, were introduced into a GIS database (see Chap. 15 for methods) (Fig. 30.6).

Seven chimpanzee nut-cracking sites were found, six of which were composed of stone hammers and anvils and one of which comprised an anvil root and a stone hammer. We recorded in total 37 stone hammers (plus 31 fragments) and 28 nonmovable anvils. After logging each of the nut-cracking sites, these were monitored for studying the operational sequence of nut-cracking behavior. Operational sequences were established, with evidence of selection, transport, reutilization, and the discarding of stone tools (Carvalho et al. 2008). Three recorded instances of tool transport within two nut-cracking sites (Fig. 30.7: SB2 in the Nonah area had three new tools transported to the site) allowed us to verify the reutilization of hammer stones while, at the same time, it confirmed the presence of chimpanzees in the area.

For the recorded tools, there was a significant positive correlation between the *function* of the tool and some of its physical characteristics. Tools with an anvil function were significantly larger ($r=0.864$; $P<0.01$) and longer ($r=0.814$; $P<0.01$) than tools with a hammer function. *Macro use-wear traces* were present in the observed tools indicating, as at Bossou, the *reutilization* of these tools over

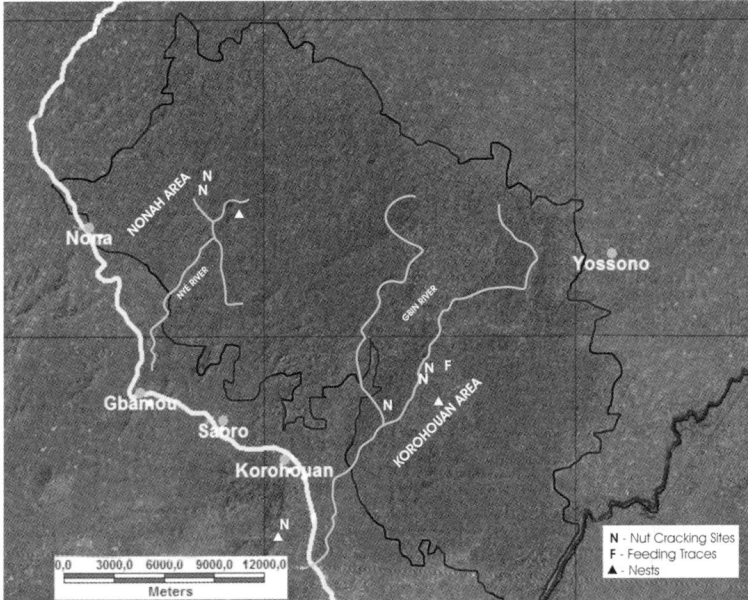

Fig. 30.6 GIS view of the Diécké Forest with the surveyed areas. All the data collected were included to provide the spatial analysis

Fig. 30.7 One nut-cracking site of *Panda oleosa* nuts (SB2) in the Nonah area was visited by the chimpanzees. Note the large size of the stone hammer (*right-hand side*) in situ after being used, left on the top of the outcrop anvil, with the nut shell remains (photograph by Susana Carvalho)

time. All the nut-cracking sites were situated close to the nut-bearing trees and relatively close to watercourses, the latter being the most likely source of eroded raw material to transport and use as hammers consequent to the lack of movable stones in the forest area. The chimpanzee stone-tool technology employed at Diécké presented clear differences compared with that recorded at Bossou (e.g., raw materials; tool mobility or complexity). Our main results suggest that chimpanzees select and adjust their tools according to particular stone attributes, maintaining patterned dimensions that differ between the regions of Diécké and Bossou (see Carvalho et al. 2008; Biro et al. 2010; for an extensive description).

These findings indicate regional variation in this technology, which may reflect the local ecological conditions experienced by different chimpanzee communities. For example, chimpanzees in Diécké use significantly larger hammers and anvils to crack the very hard panda nut (the hardest nuts among all the species consumed by chimpanzees), compared to those used by the Bossou chimpanzees to crack open the softer oil-palm nut. Some of these hammers are so large that individuals may need to use both hands when pounding the nuts. Yet, the transport of new hammers to monitored nut-cracking sites, recorded in December 2006 (Carvalho, unpublished data), confirms that transport is also part of the operational sequence of the nut-cracking behavior. It occurs even when it appears costly and unnecessary, indicating possible preferences for particular tools and the capacity to recognize tool efficiency, similarly to what has been observed at Bossou and among captive chimpanzees (see Chap. 20). In addition, Diécké anvils are all nonmovable, with flat and stable surfaces, giving an efficient option to crack such hard nuts.

During our pilot reconnaissance surveys, we had no direct sightings of chimpanzees. However, we recorded 28 arboreal nests in three distinct areas ($n=25$, Mont Medou; $n=2$, Mont Jna; $n=1$, Taffa), two areas with feeding remains (*Landolphia owarensis*), and two trails of footprints in Nonah (Mont Jna) and Korohouan (Mont Medou).

Regarding the different species of nut-bearing trees, both coula and panda trees were observed in the integrally protected area, whereas no oil palms were recorded in this area during 2006 (Carvalho et al. 2007). However seven oil palms were recorded in the Korouhan area, inside the reserve, in May 2008. It is also possible to find (feral or human-planted) oil-palm trees in the peripheral area. Local people around Diécké Forest, as in Bossou, crack oil-palm nuts using stone tools (Fig. 30.8).

30.3   Future Directions

Chimpanzee research in the Diécké Forest has just begun. In the near future, we aim to expand our knowledge of the status of the western chimpanzee in this reserve, and to gather more information on the technological abilities of these great apes. Future surveys will determine the size and the home range of the possible different groups following the latest practice guidelines for surveys and monitoring of great ape populations (Kühl et al. 2008). Additional monitoring of stone-tool use will be

Fig. 30.8 *Left:* Woman nut-cracking in Bossou (Guinea) with quartz hammer and anvil. Note the fractured hammers discarded around the anvil. *Right:* Woman nut-cracking in Nonah village (Diécké, Guinea) with quartz hammer, anvil, and wedge (photographs by Susana Carvalho)

performed using motion-triggered video cameras (Sanz et al. 2004), while density and abundance of individuals and/or of raw materials will be estimated through distance sampling with line or point transects. In addition, an archaeological survey (Renfrew and Bahn 1998; Banning 2002) and excavation with three-dimensional analysis of a nonhuman primate nut-cracking site was carried out in the Diécké Forest in 2009 (sensu Harris 1989; Mercader 2003; Mercader et al. 2002, 2007).

Ideally, we seek to combine direct and indirect observations of the Diécké chimpanzees. Unfortunately, we expect sightings to be infrequent because chimpanzees likely associate human presence with hunting, given the strong hunting pressure in the area. Thus, it is necessary to develop this new study site, with continuous research presence, and, simultaneously, to carry on promoting environmental education and cooperation with the villagers around Diécké. We must increase our efforts to try to eradicate the practice of hunting for bushmeat in the so-called integrally protected areas of the reserve, to minimize the ecological degradation, and to give hope to the continued survival of the almost unknown chimpanzees of the Diécké Forest.

Acknowledgments I wish to thank the Direction National de la Recherche Scientifique et Technologique, République de Guinée, and Dr. Kourouma Makan, Director of the Institut de Recherche Environnementale de Bossou, for permission to conduct fieldwork at Diécké. I would also like to thank Dr. Papa Cécé Condé from the Centre Forestier de N'Zérékoré, Dr. Werner Grimmelman (Progerfor) and Prof. Tetsuro Matsuzawa for their support and collaboration. We are grateful to the British Embassy in Conakry for funding the construction of the first latrines in the village of Korouhan. The research was supported by Grants-in-Aid for scientific research from the

Ministry of Education, Science, Sports, and Culture of Japan: MEXT-16002001, JSPS-HOPE, JSPS-21COE-Kyoto-Biodiversity, and F-06-61 of the Ministry of Environment, Japan, to Tetsuro Matsuzawa. S.C. was supported by The Municipality of Leiria; Cambridge European Trust (RIB 00107), FCT-Portugal (SFRH/BD/36169/2007), The Wenner-Gren Foundation, Queens College Cambridge, and the Leakey Trust (U.K.). I finally wish to thank B. Zogbila, H. Gbéregbé, J. Doré, P. Goumy, J.M. Kolié, J. Malamu, L. Tokpa, A. Kbokmo, C. Koti, O. Mamy, C. Clement, Justin, C. Kanou, Jean, and the villages of Nonah and Korohouan for the field support and J. Morgadinho for the design work. Last, I am particularly grateful to C. Sousa for guidance during my 2006 research and to P. Kelmendi, S. Koski, T. Humle, and W.C. McGrew for comments on the manuscript.

Chapter 31
From Bossou to the Forests of Liberia

Gaku Ohashi

31.1   Chimpanzees at Bossou

The village of Bossou is located in the southeastern corner of Guinea, West Africa. The village is surrounded by several hills, which chimpanzees (*Pan troglodytes verus*) use as the core area of their home range (see Chap. 2). The villagers of Bossou think of chimpanzees as the reincarnations of their ancestors, and this is why Bossou chimpanzees have survived and can survive in such close proximity to human presence and habitation. Bossou chimpanzees have been studied since 1976, more than 30 years, by teams of researchers and students either from or collaborating with Kyoto University, Japan (Matsuzawa 2006a; Sugiyama and Koman 1979a, b; see Chap. 1). Long-term research at Bossou has revealed various aspects of chimpanzee behavior, such as unique tool-use behaviors, their feeding ecology, social behavior, and population dynamics (see relevant chapters in this volume). However, the Bossou group itself is extremely endangered. Since the beginning of our study more than 30 years ago, we have never recorded any instances of female immigration into the Bossou group, and the great majority of female chimpanzees natal to Bossou disappear around the time of their sexual maturation (see Chap. 3). As a result, the percentage of aged individuals is increasing in the group (see Chap. 3). To make matters worse, the number of Bossou chimpanzees suddenly decreased to 12 as a result of an epidemic of respiratory disease in 2003 and currently numbers 13 chimpanzees (Matsuzawa 2006a; see Chaps. 25 and 32). For the conservation of the Bossou chimpanzees, we have to care not only about one chimpanzee group, but also surrounding groups, to ensure the genetic well-being of the population through individual interchange.

G. Ohashi (✉)
Japan Monkey Centre, Inuyama, Aichi 484-0081, Japan
e-mail: gaku.ohashi@gmail.com

T. Matsuzawa et al. (eds.), *The Chimpanzees of Bossou and Nimba*,
DOI 10.1007/978-4-431-53921-6_32, © Springer 2011

31.2   Bossou Chimpanzees Go Across National Border

Recently, we discovered that Bossou chimpanzees cross the national border between
Guinea and Liberia (Ohashi 2006b; Fig. 31.1). The village of Bossou is actually situated only approximately 4 km away from the Liberian border. On February 14, 2006,
11 chimpanzees, including my focal individual (PE, an adolescent male – see
Appendix A for abbreviations of the chimpanzee names), were found on the hill of
Gban, in the core area of the group. At 8:29 A.M., this party started to travel on the
ground. Because of the thick vegetation, we lost the chimpanzees at 8:39 A.M. At
10:47 A.M., we relocated my focal chimpanzee. The party consisted then of six chimpanzees: three adult males (TA, FF, and YL), two adult females (Pm and Yo), and one
adolescent male (PE). They were resting on the ground, but they soon started to move
off again. At 10:58 A.M., the six chimpanzees entered a swamp and fed on *Nephrolepis
biserrata* leaves for about 5 min. They moved again and arrived at a pineapple field.
The field is located on the western slope of a hill named Zono, which Bossou chimpanzees often visit during consortship periods. For about 10 min, they ate pineapple
fruit before entering the forest. At 11:46 A.M., three chimpanzees (YL, Yo, and Pm)
started to eat *Pseudospondias microcarpa* fruit, and the other chimpanzees rested

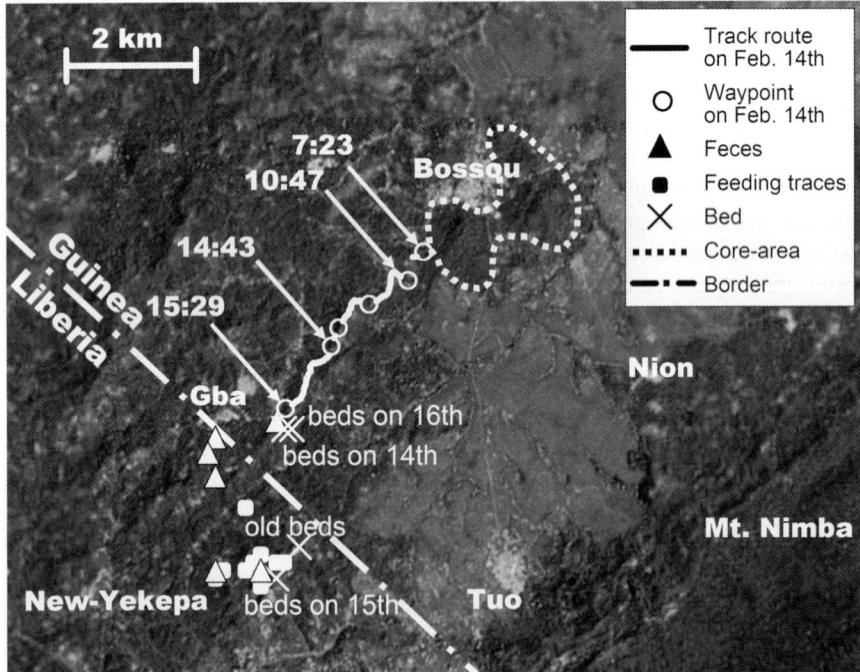

Fig. 31.1 Route employed by Bossou chimpanzees when visiting forests in Liberia. During this
visit, they spent a night across the national border (modified from Google Earth image)

on the ground. At 12:23 P.M., the six chimpanzees climbed the hill, and arrived at another *P. microcarpa* tree at 12:52 P.M. They ate fruit and rested in the tree. At 2:14 P.M., they stared to travel on the ground again. On the way, at 2:43 P.M., Yo started to eat *Pennisetum purpureum* pith, but the other five chimpanzees continued to travel. The five chimpanzees entered the primary forest surrounding the village of Gba, Guinea, on the border with Liberia. At 3:27 P.M., three adult males (TA, FF, and YL) uttered pant-hoots, climbed trees, and crossed a valley. Because the forest on the other side of the valley is a sacred area for the village, we had to cease following the chimpanzees at 3:29 P.M.

On February 15, we looked for the chimpanzees around the village of Gba. We found six fresh beds (nests) in the vicinity. Judging from the fresh feces beneath the beds, we presumed that the six chimpanzees had slept there on the night of the 14th (see Fig. 31.1). When we visited the hill near the Liberian border, we found feces at the top, but we could not find any signs of the chimpanzees around the village of Gba.

On February 16, six new beds (Fig. 31.1), fresh traces, and feces were found in the forest beside the village of New-Yekepa in Liberia. Judging from the traces, the chimpanzees appeared to have eaten sugar cane pith, *Parkia* fruit, *Landolphia* fruit, *Myrianthus* fruit, pineapple fruit, *Aframomum* pith and fruit, and palm petioles.

On February 17, four new beds were confirmed in the forest of Gba (see Fig. 31.1). Judging from the beds, the six chimpanzees had returned to the Guinean side on the evening of the 16th. In the afternoon, all six chimpanzees were confirmed in the core area of Bossou.

We also recorded several old beds in the forest around the village of New-Yekepa, Liberia, suggesting that Bossou chimpanzees had visited this forest on previous occasions. In the situation just described, the chimpanzees only stayed in Liberia for 2 days. Because of their short stay and the difficulty of observing chimpanzees ranging in the peripheral areas of their home range, researchers in the past probably missed traveling patterns of Bossou chimpanzees extending into Liberia.

According to the local people, chimpanzees exist in the northern part of Liberia. It may therefore be possible that many of the chimpanzees who disappeared from the Bossou group currently live in forests in Liberia. However, few data are available on chimpanzee behavior and presence near the border because research in Liberia ceased completely during the civil war (Kortlandt and Holzhaus 1987; see Chap. 39). To clarify the actual status of chimpanzees in this northern region of Liberia, we began surveys in the area in 2006.

31.3   Extensive Research in Nimba County, Liberia

After the above-described event, we visited the area again and conducted interviews around Nimba County in Liberia. Based on the feedback from the local people, we were able to confirm three locations potentially harboring chimpanzees in the region: Bonla, the Nimba Mountains, and Kpayee-Lepula (Fig. 31.2). We then proceeded to survey all three of these forest areas.

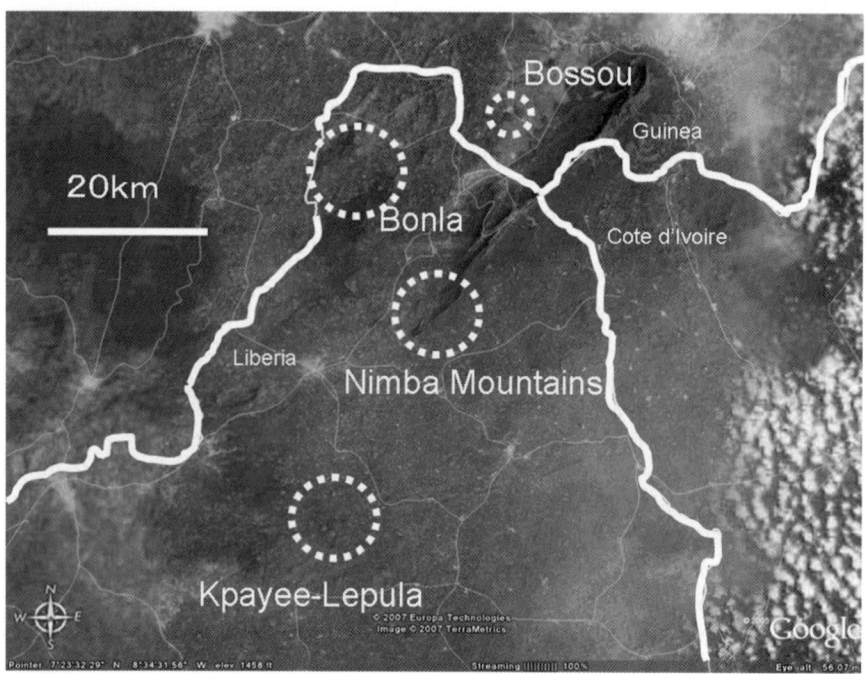

Fig. 31.2 Survey area in Nimba County in Liberia. We visited three areas: Bonla, Nimba Mountains, and Kpayee-Lepula (modified from Google Earth image)

31.3.1 Bonla

Behind New-Yekepa, where Bossou chimpanzees visited, the forest is vast. However, before my visit, no information was available on chimpanzees around this village. We continued to interview local villagers from east to west of the area, until we reached the village of Bonla where the presence of chimpanzees in the surrounding forest was confirmed by villagers. The village of Bonla is located 20 km from Bossou. In March 2006, we stayed in Bonla for 4 days, and entered the forest. Although we failed to observe chimpanzees, we heard pant-hoots several times. In February 2009, we visited Bonla again. The mountains where we had previously heard many pant-hoots were almost all burnt as a result of slash-and-burn agricultural activities. We could not confirm any traces of chimpanzees in the area. It seems that extensive slash-and-burn agricultural practices in this area may be robbing chimpanzees of their habitat. According to the local people, hunters also sometimes shoot chimpanzees around the village. From the viewpoint of Bossou, Bonla is probably the nearest area where chimpanzees can be found to the west of Bossou. We therefore need to clarify their home range and ensure sufficient habitat protection to promote interchange of individuals between groups in the region. From the point of view of tool use, Bonla chimpanzees were confirmed to dip for army ants (*Dorylus* spp.) and crack open *Coula edulis* nuts (Fig. 31.3).

Fig. 31.3 Nut-cracking trace in Bonla forest. Chimpanzees appear to crack open coula nuts in this area (photograph by Gaku Ohashi)

31.3.2   Nimba Mountains

Although chimpanzees have been studied in the Nimba Mountains on both the Guinean and Ivorian sides, little is known of chimpanzees on the Liberian side (Granier et al. 2007a; Humle and Matsuzawa 2001, 2004; Koops et al. 2007; Koops and Matsuzawa 2006; Matsuzawa and Yamakoshi 1996; Shimada 2000; see also Chaps. 27–29). The interviews we carried out had confirmed the presence of chimpanzees on the Liberian side. To corroborate these assertions, we visited the Nimba Mountains in March 2006 for 2 days. The forest itself consisted of primary forest. During our brief visit, we confirmed very old beds but could not find any new traces. However, many spent cartridges were scattered on the forest floor and numerous trails were observed, probably reflecting regular incursions by hunters into the forest. Many snares were also found in the forest (Fig. 31.4). Although Liberian authorities try to protect the area as a National Park, hunting pressure seems to be very high. Before the civil war in Liberia, the Nimba Mountains were mined for iron ore near the border with Guinea and Côte d'Ivoire. It may now be very difficult for the chimpanzees in this area to have any contact with other groups located further north (see Chaps. 27–29). The mining activities may have led to the isolation of the chimpanzees located in this southern region of the massif.

31.3.3   Kpayee-Lepula

When we conducted interviews in Liberia, we received the information that there is an area where villagers "protect" chimpanzees as in Bossou. The name of village

Fig. 31.4 Snare hunting activities in the Nimba Mountains (photograph by Gaku Ohashi)

is Kpayee-Lepula. The village is located about 50 km away from Bossou. Since March 2006, we have visited this village five times. During our brief surveys, we observed chimpanzees twice (Fig. 31.5). During both sightings, the chimpanzees were observed in a tree, and did not appear very afraid of us. Once the chimpanzees saw us, they first uttered alarm calls, but then after a short time, an adolescent male started to leaf-clip, while the others continued to feed on fruit. A juvenile male even approached toward us on the ground. Indirect evidence indicated the existence of tool-use behaviors. Interestingly, chimpanzees at Kpayee-Lepula seemed to crack open oil-palm nuts with a pair of stones, as Bossou chimpanzees do (Fig. 31.6). We also confirmed coula nut-cracking sites. During further surveys, however, chimpanzees appeared not to use this forest during the rainy season. We therefore believe that these chimpanzees may use forest more to the south as their core area. With a large river running through the forest from east to west, chimpanzees may cross the river during times when water levels are low as during the dry season. Further surveys should be conducted from other villages to assess habitat usage of chimpanzees in this area.

31.4   Threats to Chimpanzee Habitats

Not only Guinea, but also Liberia, are mineral-rich countries. The Nimba Mountains are, so to speak, a mass of iron ore (see Chap. 39 for further details). The mining operation in Liberia has now been suspended, but the old mining site of the Liberian American Swedish Minerals Company (LAMCO) has divided the forest into two.

Fig. 31.5 A male chimpanzee at Kpayee-Lepula. When he found us, he uttered alarm calls and performed leaf-clipping (photograph by Gaku Ohashi)

Fig. 31.6 We found oil-palm nut-cracking traces at Kpayee-Lepula. The chimpanzees use a pair of stones to crack open nuts, as Bossou chimpanzees do (photograph by Gaku Ohashi)

Fig. 31.7 One of the many diamond mining sites around Kpayee-Lepula (photograph by Gaku Ohashi)

It may be difficult for chimpanzees to move in a meridional direction. Around Kpayee-Lepula village, many villagers are also digging and mining for diamonds (Fig. 31.7), a highly polluting activity that results in erosion and habitat destruction.

Mining activities change not only the ecological environment but also the social environment. Once mining operations start, many migrant workers come into the area in search of jobs or the "lucky strike." This human influx can dramatically change the local economy and indirectly threaten chimpanzees in the area.

Concerning local activities, slash-and-burn agriculture is still dominant. People burn the forest, and plant rice, cassava, and vegetables. These fields are only used for a year. After harvesting the crops, farmers normally abandon the field. Although the villagers also possess livestock, they usually obtain protein from bush meat and fish. They will kill livestock for consumption primarily only during important ceremonies. Although the Liberian Authorities try to manage wildlife, it is another matter to restrict human activities. Consent of the local residents is especially needed.

31.5   Role for the Manon People in Chimpanzee Conservation

The national border divides not only the chimpanzees' habitat but also the geographic distribution of the local Manon people. Manon people live in a tri-national area including Guinea, Côte d'Ivoire, and Liberia. They communicate daily with each other beyond the national border. Liberians often visit Bossou to attend the weekly market. When Bossou villagers become sick and require medical care, they often cross the border and visit hospitals in Liberia. Normally, border agents do not require local Manon people to show a passport or a national ID card for these temporary visits. Liberia is different from Guinea in many respects; they differ in their administrative system and official language. However, at the local level, the way of life of the people is quite similar. Our experience at Bossou can probably help promote conservation efforts in Liberia. By understanding the customs, beliefs, and traditions of the local Manon people, we may be able to catalyze a peaceful coexistence between human and wildlife in the region.

Acknowledgments I wish to thank the Direction Nationale de la Recherche Scientifique et Technologique (DNRST), and the Institut de Recherche Environnementale de Bossou (IREB), for permission to conduct research in Guinea. I am also grateful for the support of the Superintendent of Nimba County for his warm welcome and kind support of my preliminary work in Liberia. I also thank Prof. Tetsuro Matsuzawa for his supervision. Special thanks are due to the local assistants – Boniface Zogbila, Pascal Goumy, Jiles Doré, Paquilé Chérif, and Henry Gbéregbé – for their hard work in the field. The research was financially supported by grants from the Ministry of Education, Culture, Sports, Science and Technology (nos. 12002009 and 16002001) to Prof. T. Matsuzawa, a grant of JSPS AA Science Platform Program to Prof. T. Furuichi, and the Global Environment Research Fund (F-061) of the Ministry of the Environment, Japan, to Dr. T. Nishida.

Part VII
Conservation

Chapter 32
The 2003 Epidemic of a Flu-Like Respiratory Disease at Bossou

Tatyana Humle

32.1 Disease: One of the Major Threats to Wild Chimpanzees

The development of field sites and long-term surveys of wild chimpanzees across Africa has resulted in an increasing number of reports of infectious and parasitic diseases affecting wild chimpanzees (*Pan troglodytes*) and other species of great apes (Formenty et al. 2003). Because of our remarkable physiological and genetic similarities, chimpanzees and humans share a multitude of potentially virulent viruses, bacteria, parasites, and fungal pathogens. Indeed, Wolfe et al. (1998) numbered at least 140 diseases of variable virulence shared by both humans and great apes. The encounter rate between chimpanzees and humans or their waste has considerably increased in recent years, primarily caused by ever-escalating human encroachment into chimpanzee habitat. This situation has given rise to resource competition between the two species, which may also exacerbate the risk of disease transmission. Research and ecotourism programs have also seriously aggravated the threat of disease transmission and may continue to put wild chimpanzees at risk unless strict guidelines and health standards are urgently put in place, respected, and maintained (Köndgen et al. 2008; Macfie and Williamson 2010).

All long term studied communities of chimpanzees across Africa have witnessed at one time or another in their history an epidemic outbreak of disease resulting in the loss of at least one member. For example, in 1966, an outbreak of a polio-like virus affected 12 chimpanzees in Gombe, Tanzania, killing six of them and resulting in lifetime paralysis in the other six (Goodall 1986; Wallis and Lee 1999). An even more serious disease, Ebola hemorrhagic fever, was first reported in wild chimpanzees in November 1994. This Ebola epidemic killed 25% of the members of a community that comprised 43 individuals in the Taï forest, Côte d'Ivoire (Formenty et al. 1999). However, the demographic patterns of this chimpanzee community suggest that epidemics resulting in a sudden population decline also likely occurred in the past (Boesch and Boesch-Achermann 2000). What is

T. Humle (✉)
School of Anthropology and Conservation, The Marlowe Building,
University of Kent, Canterbury CT2 7NR, UK
e-mail: t.humle@kent.ac.uk

T. Matsuzawa et al. (eds.), *The Chimpanzees of Bossou and Nimba*,
DOI 10.1007/978-4-431-53921-6_33, © Springer 2011

certain is that during the past 15 years repeated epidemics of Ebola fever have caused dramatic declines among both wild chimpanzees and gorillas, most particularly in Gabon and the Republic of Congo (Huijbregts et al. 2003; Walsh et al. 2003, 2005).

Another serious concern, in recent years, has been the increased occurrence of respiratory disease outbreaks (Köndgen et al. 2008). These outbreaks have been reported at all long-term study sites of chimpanzees in Africa. Bossou is no exception. Matsuzawa (1997b) reported the death from a respiratory disease of an infant chimpanzee of the Bossou community in January 1992. Until 2003, only isolated mortality events had been noted in this community (Sugiyama 1984; Matsuzawa et al. 1990). However, in November 2003, a virulent epidemic of a flu-like respiratory disease struck the community and resulted in the death of five individuals (Matsuzawa 2006a). I describe here some of the current knowledge that we have concerning respiratory outbreaks in chimpanzees. I then further describe the course of events during the 2003 epidemic at Bossou, while highlighting the urgency in implementing and complying to concrete measures aimed at minimizing the event of such epidemics in the future.

32.2 History of Respiratory Syndrome Outbreaks Among Great Apes

Respiratory disease is the most common cause of morbidity and mortality among wild great apes habituated to human presence, especially for research or touristic purposes (Woodford et al. 2002; Goodall 1986; Homsy 1999; Nishida et al. 2003; Hanamura et al. 2008). Several outbreaks of respiratory disease have been reported during the past 30 years (Wallis and Lee 1999). Indeed, in 1988, an outbreak among mountain gorillas (*Gorilla beringei beringei*) in Rwanda resulted in the death of six individuals and the sickening of 27 others. Although the source of the disease was never identified, Sholley and Hasting (1989) proposed that the disease was most certainly of human origin, as these gorillas were the subject of tourism. In 1996, during an outbreak among the chimpanzees at the Gombe National Park in Tanzania, 11 chimpanzees succumbed to respiratory infections (Wallis and Lee 1999; Lonsdorf et al. 2006). The members of this community had apparently been provisioned for purposes of habituation and therefore had experienced close proximity with local staff, thus raising the speculation that they had been infected by a human pathogen.

Although anecdotal accounts of respiratory outbreaks abound, until recently no outbreak event had successfully identified the pathogen or pathogens responsible and linked its origin to humans. Based on the compilation of data across seven reported respiratory outbreaks that occurred at three different chimpanzee field sites (Table 32.1), mean percentage morbidity of respiratory outbreaks was extremely high, with a minimum of $81.8 \pm 12.2\%$ (range, 35.0–100.0%), and mean percent mortality was $14.4 \pm 3.1\%$ (range, 5.6–26.3%).

Table 32.1 Summary of morbidity and mortality of reported outbreaks of respiratory epidemics at three field sites of wild chimpanzees across Africa

Site	Month	Year	References	Percent			Proportion deceased			
				Morbidity (minimum)	Morbidity (maximum)	Mortality	All	More than 10 years old	5–10 years old	Less than 5 years old
Bossou	January	1992	Matsuzawa (1997b)	NA	NA	5.6	1/18	0/9	0/5	1/4
	November–December	2003	Matsuzawa et al. (2004)	81.8	100	26.3	5/19	3/13	0/3	2/3
Taï	May	1999	Köndgen et al. (2008)	100	100	18.8	6/32	5/15	1/7	0/10
	March	2004	Köndgen et al. (2008)	100	100	18.2	8/44	0/22	3/10	5/12
	February	2006	Köndgen et al. (2008)	92	92	2.9	1/34	0/19	0/5	1/10
Mahale	June–July	2006	Hanamura et al. (2008)	35	49	18.5	12/65	4/?	1/?	7/?
	September	1993	Hosaka (1995)	NA	NA	10.9	11/101	5/?	3/?	3/?

NA not available

Note: Although other outbreaks are reported in the literature, morbidity or population size and age-class structure are not always available. Therefore, only seven epidemics are reported here

Based on necropsy samples collected from Taï chimpanzees during the course of three separate respiratory epidemics between 1999 and 2006, Köndgen et al. (2008) found that, similarly to human cases, a mix of bacterial and viral respiratory pathogens occurred in the lungs. The most common bacterium was *Streptococcus pneumoniae*, which was found in all respiratory outbreaks. *S. pneumoniae* may induce fatal pneumococcal pneumonia or even meningitis in great apes (Ott-Joslin 1993). In addition, all samples tested positive for one of two paramyxoviruses: human respiratory syncytial virus (HRSV) or human metapneumovirus (HMPV) (Köndgen et al. 2008). Both viruses are shed in respiratory secretions but also have been detected in feces or sweat (von Linstow et al. 2006). Köndgen et al. (2008) suggest that these paramyxoviruses may predispose chimpanzees to secondary bacterial infection caused by *S. pneumoniae* or *Pasteurella multocida*, whose presence was also confirmed during one outbreak. This suggestion is supported by the fact that, based on the data compiled in Table 32.1, the age classes most affected by respiratory epidemics are the weakest and most vulnerable individuals. Indeed, mean percent infant (0–5 years old) mortality was $28.7 \pm 11.8\%$ (range, 0–66.7%), as opposed to $11.3 \pm 7.1\%$ (range, 0–33.3%) for individuals more than 10 years old and $8.9 \pm 6.0\%$ for individuals between 5 and 10 years old (range, 0–30.0%) (see Table 32.1).

Finally, respiratory disease outbreaks are widespread among chimpanzees and other great ape species and may cause significant individual losses within communities. Although the pathogens involved may not necessarily be the same during each outbreak and across all sites, it is apparent that wild chimpanzees are highly susceptible to respiratory diseases and human viruses and bacteria. Whether all outbreaks thus far reported are necessarily of human origin or are part of a natural disease cycle remains unclear. However, it is clear that respiratory diseases are a serious threat to wild chimpanzees and great apes in general throughout Africa.

32.3 The 2003 Epidemic at Bossou, Guinea

32.3.1 Notable Precursor Events

Before describing the course of events that took place during the respiratory epidemic that struck the Bossou community in November–December 2003 (Matsuzawa et al. 2004; Matsuzawa 2006a), I first would like to point out a few important events that occurred prior to the outbreak. These might or might not be of any significance, but are mentioned here since they may be pertinent and relevant to the origin of the epidemic and the high mortality that ensued. In August 2003, the village of Bossou witnessed an epidemic of cholera that resulted in the death of 12 people (I am noting this event, as I noticed that before the respiratory epidemic that struck wild chimpanzees at Mahale, Tanzania, in 1993, a cholera epidemic had also been reported a few months earlier in the closest village to the study group; cf.

Hosaka 1995). Three months later, on November 13, 2003, tourists visited Bossou and observed two adult male chimpanzees of the community for no more than 20 min. Although an observation distance of no less than 20 m was initially respected, both males (Yolo and Poni), who were the most habituated members of the community at the time, approached the group of tourists to within 5–7 m, thus not allowing for a safe distance to be respected. At the time, tourists were not required to wear masks during their outings with the chimpanzees.

32.3.2   Symptoms and Course of Events

Although all chimpanzees, including the two males, observed during the 5 days following this touristic event were healthy, our local research guides failed to locate any member of the community between the 19th and 23rd of November. On November 24, all eight chimpanzees [two adult males (Foaf, Yolo), three adult females (Yo, Jire, Fana), two juveniles (Fanle, Jeje), and one infant (Jimato)] observed that day showed symptomatic signs of respiratory disease, including nasal discharge and coughing. Such symptomatic features had been recorded in the past because almost every year, especially in the dry season, when minimal temperatures are the lowest at night, the chimpanzees often suffer from colds. However, in this case, the morbidity was greater than usual and the coughing appeared more acute.

Both Nina and Kai, two old females presumed to be more than 50 years old at the time, were last seen on November 17. Kai's decomposed body was found in the Bossou chimpanzees' core area on December 6. Nina was never observed again, and her body was never found. We therefore presumed that she had also been affected by the epidemic and succumbed to the epidemic. Poni, a male aged 10 years old in 2003, was last sighted on November 18, 5 days after the visit of the tourists. His dead body was found on December 7 in an advanced state of decomposition.

Jire, an adult female, and her 13-month-old male infant, Jimato, were last sighted together on November 26. Jire was then seen again on December 3 carrying Jimato's dead body on her back clamping his arm between her neck and shoulder (see Chap. 25). Jire persisted in carrying Jimato's body until February 9, 2004, when she finally dropped it accidentally and failed to retrieve it for more than 24 h (see Chap. 25). The mummified body of Jimato was then recovered by T.H. on February 10 (see Fig. 32.1 for example of mummified body). Vuavua, a 12-year-old female, and her 2.5-year-old female offspring, Veve, were last sighted together on November 18. No sightings of Vuavua and Veve were made between November 19 and December 1. Vuavua was then sighted without Veve on December 2 (see Sect. 32.3.3).

Finally, because the recovered corpses of Kai and Poni were already in an advanced state of decomposition, and neither showed symptoms of illness on the day they were last seen, we could unfortunately not determine when they started showing symptoms and thus the time interval between the first symptomatic signs and death. In addition, we were unable to gather necropsy samples to identify the agent(s) of death. Among the 14 individuals observed between November 24 and

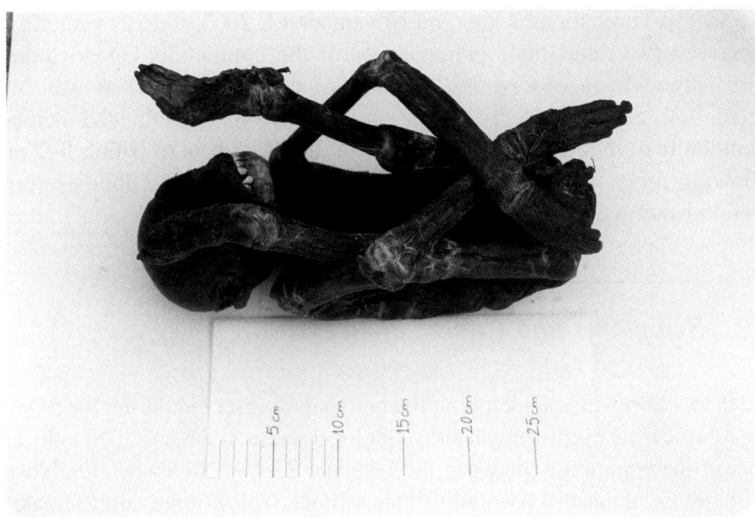

Fig. 32.1 Mummified body of Veve after retrieval of the body by researchers in 2004 (photograph by Tatyana Humle)

December 6, we were able to confirm signs of morbidity in 11. However, we were not able to observe all 14 daily throughout this period. We cannot therefore exclude the possibility that morbidity was not 100%. All morbid individuals showed a decrease in food intake and physical activity, resting for unusually high proportionate periods of time. In addition, between November 19 and December 6, the chimpanzees were unusually quiet and therefore extremely difficult to locate by local assistants and researchers on site.

The percentage mortality based on the total community size at the time was high with the loss of more than a quarter of the community's members: the two oldest females (Kai and Nina), one young adolescent male (Poni), and the two youngest members (Jimato and Veve) of the community. All five individuals were thus potentially weaker than the others either because of their age or their status. Poni was the only adolescent male in the community at the time and may therefore have experienced particularly high levels of stress. We indeed have good reason to believe that the immune system of these five individuals was weaker or more stressed compared to that of other community members, yielding the possibility that they succumbed to a secondary infection as proposed by Köndgen et al. (2008) for similar respiratory outbreaks recorded at Taï.

32.3.3 Veve's Story

Because Veve's mother, Vuavua, was seen for the first time without her on December 2 and was sighted alone during the following 10 days, we had assumed (1) that Veve, a 2.5-year-old infant, had succumbed to the epidemic and (2) that, unlike Jire, Vuavua, as a first-time mother, had actually abandoned the dead body of her infant. However, we found Veve alive on December 11, alone in a coffee plantation located between the two main hills within the core area of the Bossou chimpanzees. Because we had observed Vuavua ever since December 2, we can safely assume that Veve had survived without her mother for at least 10 days. Veve was obviously weakened. She was heavily infested with ectoparasites and showed all the symptoms of respiratory disease, with coughing and a runny nose.

We conducted a 24-h watch over Veve, during which time we provided her with clean watery fruit (bananas, papaya, pineapple, oranges), as well as marantaceae pith and fruit. We wore masks and gloves at all times. Veve constructed her own arboreal night nest the first night, as well as the second. During the day, she would eat and rest on the ground. On the morning of December 13, all the chimpanzees of the community visited the coffee plantation, which they typically traverse when traveling between the two hills of Gban and Guein within their core area. Upon sighting Veve, Vuavua retrieved her and continued her passage through the coffee plantation accompanied by all the other members of the community. Veve became the center of attention of the whole community for the rest of the day. She was intensely groomed by her mother and other members of the community. During the following days, Veve then seemingly started regaining some strength, feeding independently most of the time, while also suckling on occasion, albeit for short durations, suggesting that Vuavua no longer produced milk. Finally, despite this successful reunion, Vuavua was seen carrying Veve's dead body on December 30 after the two had last been sighted together on December 28. Vuavua carried Veve's corpse until she abandoned it on January 17, 2004 (Biro et al. 2010; see Chap. 25) (Figs. 32.1 and 32.2).

32.4   Conclusion and Implemented Measures

Clearly, as already reported by Burbridge in 1928, great apes, including chimpanzees, are highly susceptible to respiratory diseases. Since the 2003 epidemic at Bossou, we have set in place a policy for a minimum observation distance between human and chimpanzees (7 m for researchers, students, and local staff; 20 m for tourists), and for the compulsory wear of a mask when in proximity to chimpanzees by tourists, local staff, and volunteers, as well as all researchers or students, on site (Fig. 32.3). In addition, we reinforced the policy that no local assistant, student, or researcher should track chimpanzees and/or enter the forest if showing signs of illness. Strict guidelines for human waste or debris disposal within chimpanzee habitat were also reinforced. All non-organic wastes or debris within chimpanzee habitat are to be removed; organic waste is either gathered for subsequent hygienic disposal or, depending on circumstance, buried deep underground away from possible contact by chimpanzees. Local assistants were instructed on how to dispose of their waste, and local people living in the vicinity of chimpanzee habitat are encouraged

Fig. 32.2 Wearing masks during observation of the chimpanzees at Bossou: (**a**) Kathelijne Koops, Ph.D. student, taking notes while observing the chimpanzees (photograph by Jiles Doré); (**b**) Jiles Doré, one of our local field assistants, taking a photograph of Bossou chimpanzees (photograph by Kathelijne Koops)

Fig. 32.3 Vuavua (young female, aged 13 years) carries the mummified corpse of her daughter, Veve, at Bossou (photograph by Tatyana Humle)

to better manage disposal of organic waste and to increase their hygiene standards and minimize risk of disease transmission. In 2003, we thus promoted the construction and use of latrines.

Ultimately, all sites should ensure that a maximum of field assistants (including of course all researchers and students) are vaccinated for hepatitis A and B, measles, mumps, rabies, rubella, typhoid, meningococcal meningitis, and polio (Beck et al. 2007). In addition, all researchers, students, and local staff should regularly be screened for tuberculosis before following and spending time in proximity to chimpanzees (Woodford et al. 2002). Moreover, if chimpanzees occur in close proximity to human settlements as at Bossou, a program of vaccination should be encouraged to minimize the prevalence of anthropozoonotic diseases in the area. Appropriate health standards if promoted and implemented can thus benefit both humans and chimpanzees in the locality.

Because very little is known about disease transmission and its impact on wild chimpanzees, upon occurrence of new outbreaks, it is vital that, whenever possible, appropriate necropsy samples are gathered for precise identification of etiological agents. However, in many cases only well-trained and well-equipped staff (preferably with veterinarian background or training) may safely carry out such sampling forensic procedures. As sampling methodologies are being improved (Chi et al. 2007; Goldberg et al. 2007; see Chap. 35), we can hope that with appropriate training and equipment workers in situ may be able, under favorable circumstances, to conduct such procedures. Health monitoring is also essential across all sites with habituated great apes. It is important for researchers, students, and field assistants alike to pay careful attention to the health of the chimpanzees they observe, including incidences of diarrhea, coughing, vomiting, and any other symptomatic signs of ill health (see Chap. 36) (Lonsdorf et al. 2006).

Finally, the 2003 outbreak resulted in considerable loss to the genetic and mitochondrial diversity of the Bossou community (see Chap. 34). The community with its current 13 members is, therefore, highly vulnerable to the event of another outbreak. We therefore have a duty to prevent any future outbreaks by implementing strict standards and health policies for researchers, students, local assistants, and tourists, and by improving the hygiene and sanitary conditions of the local villagers.

Acknowledgments I am particularly grateful to Tetsuro Matsuzawa, Gen Yamakoshi, Kathelijne Koops, Dora Biro, and Misato Hayashi, as well as all our local assistants at Bossou, especially Guano Goumy and Tino Zogbila, for their assistance and support throughout this difficult period that was the 2003 epidemic of respiratory disease at Bossou.

Chapter 33
Microclimate and Moving Pattern

Hiroyuki Takemoto

33.1 Introduction

Takemoto (2004) demonstrated that although seasonal changes in the rate of ground usage were affected to some extent by the vertical or horizontal distribution of food, the main effect seemed to be microclimatic fluctuations in the forest (Fig. 33.1). In tropical forest, various vertical structures can be observed not only in biomass production but also in temperature and relative humidity (RH) (Richards 1996). In such kind of habitat, chimpanzees are expected to use higher places in cold conditions such as during the wet season to reduce their costs of thermoregulation, because higher places exhibit higher temperatures compared with the forest floor. Similarly, chimpanzees will use areas nearer the ground more often to prevent increasing their metabolism when ambient temperatures are extremely high.

At Bossou, however, until recently, no detailed data on vertical differences in seasonal microclimatic fluctuations were available. I present here the difference of air temperature between ground level and tree crown in the forest of Bossou and discuss the seasonal differences in terrestriality of the chimpanzees.

33.2 Materials and Methods

33.2.1 Measurement of Microclimate

Air temperature in the forest was measured with a digital temperature/humidity data logger (HN-CHN; CHINO Corporation) with a determination range (±precision) of −10 to 50°C (±0.5°C) and 0–100% RH (±2%). I established two measuring areas for vertical structure of air temperature. One area was located in the primary forest on

H. Takemoto (✉)
Primate Research Institute, Kyoto University, 41-2 Kanrin, Inuyama, Aichi 484-8506, Japan
e-mail: takemoto@pri.kyoto-u.ac.jp

T. Matsuzawa et al. (eds.), *The Chimpanzees of Bossou and Nimba*,
DOI 10.1007/978-4-431-53921-6_34, © Springer 2011

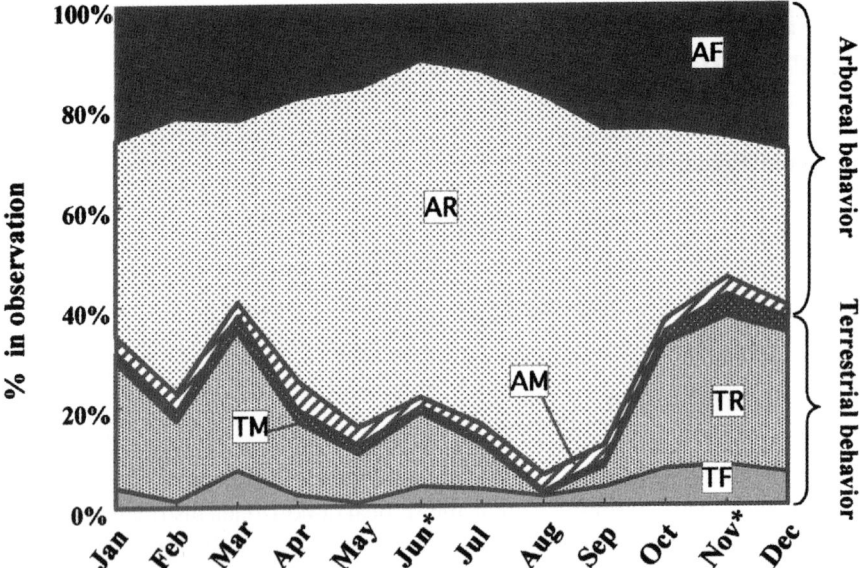

Fig. 33.1 Seasonal change in activity budgets (Takemoto 2004). *AF* arboreal feeding; *AR* arboreal resting; *AM* arboreal moving; *TF* terrestrial feeding; *TR* terrestrial resting; *TM* terrestrial moving; *Asterisk* June and November represent averages of two different years. Seasonal change in time spent for terrestrial behavior (TF+TR+TM) was correlated negatively with the air temperature inside forest

the hill of Gban, where the forest canopy is around 30 m in height; the other was a patch of secondary forest on the hill of Guein, where the canopy ranges from 18 to 22 m in height. Both these areas are located within the home range of the chimpanzees. The data loggers were fixed on a tree trunk with a wooden roof and placed at 1.5 and 16–17 m from the ground, respectively. In the secondary forest, the data logger was fixed just beneath the tree crown, whereas in the primary forest, the data logger was located much lower than the tree crown. The vertical difference of the temperature in the forest was recorded automatically every 10 min.

33.2.2 Observation of Chimpanzees

Four focal chimpanzees (two adult males and two adult females) were chosen. One focal chimpanzee was followed for an entire day or for as long as possible. The activities of the focal individual and all other visible community members were recorded using a scan sampling method every 10 min (sensu Altman 1974). Dependent and younger chimpanzees less than 7 years old were excluded from the data set because of body size effects on thermoregulation.

Activity budgets comprised six behavioral categories: arboreal feeding (AF), arboreal resting (AR), arboreal moving (AM), terrestrial feeding (TF), terrestrial resting (TR), and terrestrial moving (TM). Research periods were from August 4 to

September 16, 2007 in the rainy season, and from January 2 to February 15, 2008, in the dry season. Observation duration was 18 days during the rainy season and 24 days during the dry season.

33.3 Results and Discussion

33.3.1 Air Temperature in the Forest

Average daytime air temperature varied drastically, from 23.4 to 34.6°C, depending on season and height above ground (Table 33.1). In the wet season, the vertical structure in the forest revealed differences of 1.5 and 3.7°C in average daytime temperature on Gban and Guein, respectively. In the dry season, the air temperature difference between the two heights was even greater: 2.1°C on Gban and 6.3°C on Guein. In all cases, ground-level temperature was lower than higher up in the canopy. The vertical difference in air temperature was greater in the dry season compared with in the wet season.

33.3.2 Seasonal Difference in Activity Budgets

The activity budgets of the chimpanzees are shown in Fig. 33.2. Time spent for terrestrial behavior (TF + TR + TM) increased from 14% during the wet season to 60% during the dry season. Resting behavior dramatically varied across seasons; it was greater during the rainy season than the dry season. However, there was no difference in arboreal feeding across seasons. This result is concordant with previous findings (Takemoto 2004).

33.3.3 Vertical Structure of Air Temperature and Terrestriality of Chimpanzees

The range of the thermoneutral zone in ambient temperature for primates is approximately 5°C, although this range may fluctuate across species. Lower critical temperature and upper critical temperature for humans are 25 and 28°C, respectively

Table 33.1 Averages of daily maximum air temperature inside forest

Gban, primary forest (tree crown, 22 m)			Guein, secondary forest (tree crown, 18 m)		
Height	Wet season (°C)	Dry season (°C)	Height	Wet season (°C)	Dry season (°C)
16 m	25.0	30.0	17 m	27.1	34.6
1.5 m	23.5	27.9	1.5 m	23.4	28.3
Difference	1.5	2.1	Difference	3.7	6.3

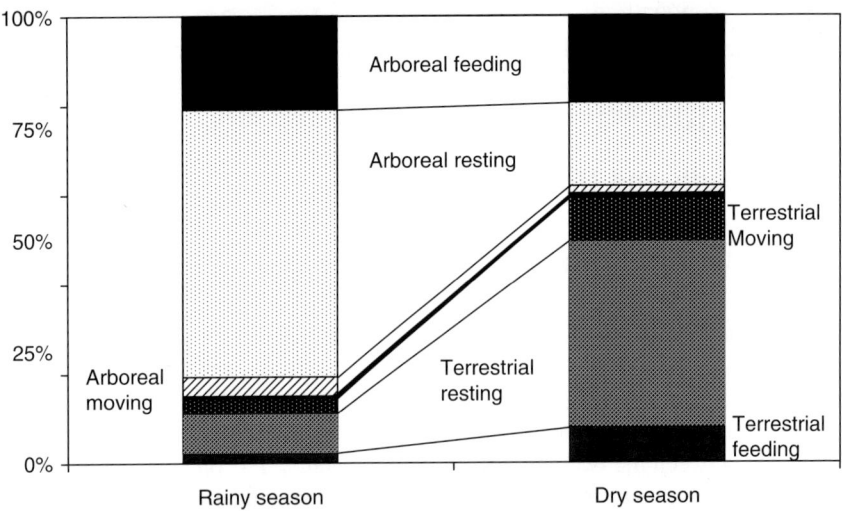

Fig. 33.2 Difference in activity budgets between rainy season and dry season 2007–2008. Time spent for terrestrial behavior was 14.5% in rainy season and 59.3% in dry season

(Wilkerson et al. 1972), and other primates are slightly above or below these points. It appears that ambient temperature around the forest canopy in the dry season (34.6°C) is extremely high or that near the ground (23.5°C) is considerably low for chimpanzees.

No inference can be made on the difference in microclimate between the two forest types, as no data just beneath the forest canopy in the primary forest of Gban were collected. Normally, closed canopies tend to increase the difference in vertical structure of microclimate in tropical forests. If so, the possibility exists that the difference in air temperature between canopy and ground in primary forest exceed 6.3°C.

Trees in Bossou bear fruits abundantly between January and April, although a small peak can also occur in September (Yamakoshi 1998; Takemoto 2004; see Chap. 2). During this study, fruit availability was four times greater in the dry season than in the wet season. Therefore, there is no reason to assume that fruits in trees explain arboreality in chimpanzees.

The reason why terrestrial behavior increased in the dry season is not related to the vertical distribution of food but, rather, is a strategy for reducing thermoregulatory costs by choosing suitable space for resting in relationship to the vertical structure of the tropical forest's microclimate. Microclimate in tropical forest may greatly affect not only spacing patterns but also day range or food choice, influencing party size and other social behavior.

Acknowledgments I thank the Ministère de l'Enseignement Supérieur et de la Recherche Scientifique et Technologique, in particular the Direction Nationale de la Recherche Scientifique et Technologique (DNRST) and the Institut de Recherche Environnementale de Bossou (IREB), for granting me the permission to carry out research at Bossou. This work was supported by KAKENHI (19657074).

Chapter 34
Genetic Variation in the Chimpanzees of Bossou and Nimba

Makoto K. Shimada

34.1 Introduction

34.1.1 Background to Mitochondrial DNA Analysis

Currently, various methods and platforms for genotyping are available; however, only a limited number of methods can be applied to noninvasively collected samples from wild mammals because of sample quality and quantity. Mitochondrial DNA (mtDNA) is present in a much higher number of copies, 100–1,000 times more, in a cell than is nuclear DNA, and sequencing mtDNA is not subject to artifacts such as stutter band or allelic dropout, often problematic with microsatellite typing (Woodruff 2004). mtDNA typing has not only the advantage of experimental ease in obtaining reliable data but also of allowing tracing of genealogy because mtDNA is clonal and free of recombination. Because the mtDNA of males is not passed down to the next generation, mtDNA typing is only suitable for detecting the matrilineal genetic relationship among individuals and populations.

34.1.2 Aims

To reveal matrilineal genetic relationships among the original members of the Bossou community and also between Bossou and Nimba chimpanzee communities (*Pan troglodytes verus*), we conducted mtDNA genotyping using noninvasively collected samples from chimpanzees at Bossou and Nimba, West Africa.

M.K. Shimada (✉)
Primate Research Institute, Kyoto University, 41-2 Kanrin, Inuyama, Aichi 484-8506, Japan
and
National Institute of Genetics, Mishima 411-8540, Japan
and
Institute of Comprehensive Medical Science, Fujita Health University,
1-98 Dengakugakubo, Kutsukake-cho, Toyoake, Aichi 470-1192, Japan
e-mail: mshimada@fujita-hu.ac.jp

T. Matsuzawa et al. (eds.), *The Chimpanzees of Bossou and Nimba*,
DOI 10.1007/978-4-431-53921-6_35, © Springer 2011

The chimpanzee community at Bossou is one of the smallest among the long-term study sites of chimpanzees. It has consisted of about 20 individuals since the start of observations in 1976 (Sugiyama 1999). Chimpanzees in Bossou mostly confine their activity to a core area of about 6 km². Because the core area is surrounded by savanna and gallery forest, the Bossou chimpanzee group has never been observed to encounter neighboring groups. The Nimba Mountains area, located on the border between the Republic of Guinea, Côte d'Ivoire, and Liberia, is separated from Bossou, the closest neighboring population, by savanna (see Chaps. 2, 29, and 39).

34.2   Methods

34.2.1   Sample Collection

Samples, including hair, feces, and wadges (chewed-up fruit remains), were collected noninvasively from wild chimpanzees living at Seringbara, Yealé, and Gouéla in the Nimba Mountains (see Chaps. 27–29). From Seringbara, Yealé, and Gouéla, we collected 26, 23, and 4 samples, respectively. Because the sample size from the Gouéla site was so small, we did not include this population in the population structure analysis. In Bossou, we collected hairs, feces, and urine samples. We collected hairs with sterilized tweezers from recently used night beds. (We use the term bed because, in the ecological literature, "nest" refers to a shelter for bearing and nursing offspring.) We stored hairs collected from each bed in a disposable plastic tube containing absolute ethanol. We collected the feces by wiping the surface of the feces with a cotton swab soaked in a saline/ethylenediaminetetraacetic acid (EDTA) (1 mM) solution and washing the cotton swab in 2 ml saline/EDTA solution. We then added 10 ml absolute ethanol. Urine was collected using sterile, disposable plastic syringes and deposited in a sterile tube with two volumes of 70% ethanol (sensu Hayakawa and Takenaka 1999). We collected wadges using sterilized tweezers and stored them in a disposable plastic tube containing absolute ethanol. These samples were stored in the field at room temperature, away from direct sunlight. Hair samples were transferred to a −20°C freezer, and feces and wadge samples were transferred to a refrigerator in the laboratory. Although the samples gathered in the Nimba Mountains were from unidentified individuals, all the samples collected at Bossou came from known and identified individuals (Sugiyama 1999).

34.2.2   DNA Extraction

The method of DNA extraction was as follows. For hair, we extracted DNA from a single hair using ISOHAIR (Nippon Gene) according to the manufacturer's instructions. For urine, we followed the method described in Hayakawa and Takenaka (1999). From feces, DNA was extracted using the QIAmp DNA Stool Kit (Qiagen) according to the manufacturer's instructions with the following modifications. After centrifugation (800×*g*, 10 min at room temperature), the precipitate was suspended in 1.6 ml ASL buffer and incubated at room temperature for 30–60 min. The final elution of DNA was incubated into AE buffer for 20–30 min. This procedure is essentially the same as that used by Morin et al. (2001). To avoid cross-contamination during DNA extraction, we wore hair caps and gloves at all times and used filtered tips.

34.2.3   Mitochondrial DNA Sequence Determination and Analyses

We determined about 605 bp in the hypervariable region I of the mtDNA control region. We designed primer sets, L15926 (5′-TAC ACT GGT CTT GTA AAC C-3′, corresponding to positions 15326–15344 of the complete chimpanzee mtDNA sequence of Horai et al. (1995), the DDBJ/EMBL/GenBank accession number = D38113), and H16555 (5′-TGA TCC ATC GTG ATG TCT TA-3′, corresponding to positions 15971–15990 of D38113), and as an nested primer set L15933 (5′-GGT CTT GTA AAC CGG AAA CG-3′; 15332–15351 of D38113) and H16538 (5′-TCT TAT TTA AGG GGA ACG TGT G-3′; 15954–15975 of D38113). The polymerase chain reaction (PCR) condition is described in Shimada et al. (2004). Sequencing with automated sequencers (ABI377 or ABI310; Perkin Elmer) was done from both ends using the same primers used in PCR. For the genotyping of identified individuals from the Bossou community, we sequenced at least four samples per recorded matriline (i.e., mother–offspring record), regardless of the individual. We excluded the data if sample quality was low in the chromatogram and/or sequences diverged from the corresponding region of published chimpanzee mtDNA sequences.

34.3   Results

Twenty distinct haplotypes could be identified from those sequences. We submitted these obtained mtDNA haplotype sequences to DDBJ/EMBL/GenBank (AB189231–AB189251) and published typing results and success rate elsewhere (Shimada et al. 2004, 2009; see also Appendix E).

34.4   Discussion

34.4.1   Relationship Among the Bossou–Nimba Populations and Other West African Chimpanzees

Figure 34.1 shows the phylogenetic relationship between *P. t. verus* mtDNA haplotypes with published sequences and the three other subspecies of chimpanzee, using the bonobo (*Pan paniscus*) as the outgroup. The mtDNA diversity observed at Bossou and Nimba is similar to that documented in comparable studies within the

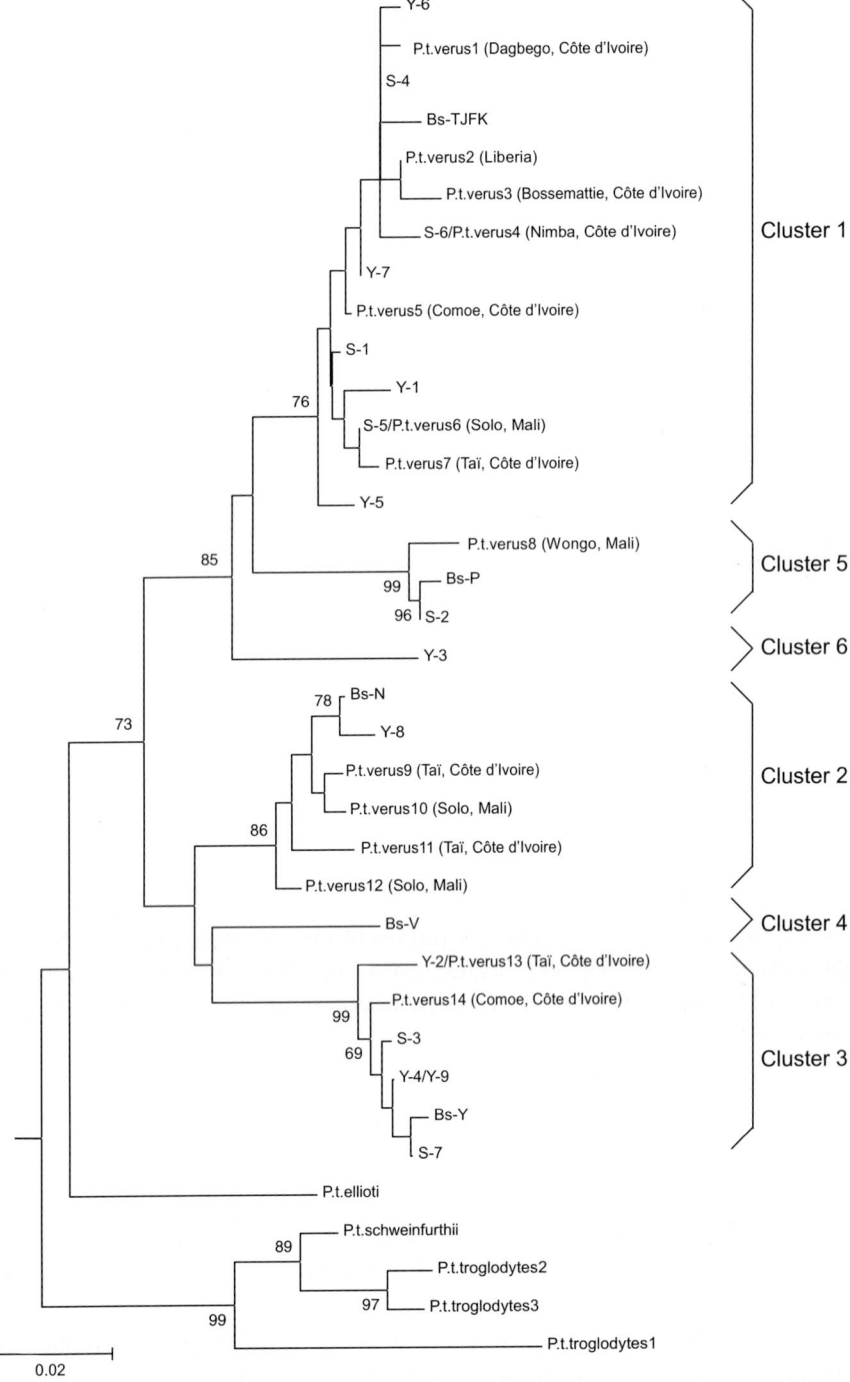

Table 34.1 Analysis of molecular variance (AMOVA) among the Bossou chimpanzee community and two sites in the Nimba Mountains – Seringbara and Yealé

Variances compared	s^2	P	SE
Between Bossou and Nimba groups (σ_a^2)	0.583 (6.27%)	0.341	0.023
Between the two Nimba communities (σ_b^2)	0.251 (2.70%)	0.231	0.015
Within communities (σ_c^2)	8.468 (91.03%)	0.016	0.004
Total (σ_T^2)	9.303	–	–

subspecies. The mtDNA haplotypes do not cluster according to geographic region. In fact, the six clusters found in the Bossou and Nimba region represent all the major mitochondrial clades of the *verus* subspecies. Similar distribution patterns of mtDNA haplotypes have also been reported in other West African chimpanzee populations (Morin et al. 1994a; Goldberg and Ruvolo 1997a, b; Gagneux et al. 1999b, 2001). The absence of a clear population structure of mtDNA variation suggests panmixture of the ancestral population of the *verus* subspecies, and that insufficient time has passed since the separation of the current habitats for geographically distinct population structures to evolve. These results also suggest that the Bossou–Nimba groups derived from the ancestral population of the West African chimpanzee without passing through a serious bottleneck. The AMOVA (analysis of molecular variance) results also indicate that the mtDNA revealed no distinct population structure among the chimpanzees of the Bossou–Nimba region (Table 34.1).

34.4.2   Recent Migration of Chimpanzees Between Bossou and Nimba

Only three mtDNA haplotypes were shared among groups in the Bossou–Nimba region (Shimada et al. 2004). This finding suggests that there is a low level of gene flow via females in this region, although the mtDNA haplotype tree and AMOVA suggest the absence of a clear population structure, which may be associated with high levels of gene flow. There are at least two explanations for this paradox. The first possibility is that we missed shared haplotypes. More extensive sampling in each of these populations may have yielded more shared haplotypes. The other explanation is that haplotypes are well mixed in ancestral populations of West African chimpanzees, which dispersed rapidly, and was followed by restricted gene flow between current fragmented habitats.

Fig. 34.1 A neighbor-joining tree of chimpanzee mtDNA variants including those described in this study and previously published data. The root is located using bonobo (*Pan paniscus*) sequence as an outgroup. Haplotype names denote their main sampling site (i.e., Bs-, S-, Y-, indicates Bossou, Seringbara, and Yealé). Because sequence data previously published were shorter than those in this study, aligned regions were shorter than presented in Table S2 in Appendix E. Consequently, sequences in this study were trimmed and some of them became identical, such as Y-4 and Y-9

34.4.3   Demography and Matrilineage Genetic Information

According to our field observation records, there were eight adults or subadult individuals (later referred to as original adults) at Bossou when field observations began in 1976. Their genetic relationship was unknown at the time. Sequencing of the mtDNA control region showed that four of the eight original adults shared an identical mtDNA haplotype. Furthermore, nearly half of all community members recorded since the beginning of the field study carried this same haplotype (Shimada et al. 2009). This observation raises the question whether this skewed matrilineal genetic composition affects the risk of extinction of matrilineal haplotype and mate choice and emigration patterns in the Bossou chimpanzees. Furthermore, the reduced mating opportunity with non-kin individuals brought on by this skewed matrilineal genetic composition may explain field observations that have confirmed copulations between sexually independent males and maternally related females of reproductive age (Ohashi and Humle, personal observation).

34.4.4   Feature of Sampling Material Type: Caveat for Hair Samples

We collected five types of samples for the mtDNA sequencing: (1) hairs collected from night beds, (2) hairs collected from the ground, (3) urine, (4) feces, and (5) wadges. All four types, except wadges, were gathered from known individuals in Bossou. We compared the experimental difficulty across these four sample types. The comparison shows that hair samples are the most difficult to sequence and most often yield inconsistencies compared to results from other sample types. We successfully obtained mtDNA sequences from 44 of 53 (83%) hair samples, 34 of 45 (76%) urine samples, and 5 of 6 (83%) fecal samples. On average, we could determine mtDNA sequences from 80% of the samples. Although there was no significant differences in final amplification success rate among these three sample types ($c^2=0.7$, $d.f.=2$, $P>0.5$), it is worthy to note that 24% of the samples (all hair samples) yielding mtDNA sequences required two-round or nested primer PCR (Table S3 in Appendix E). Because two-round or nested primer PCR protocol is prone to cross-contamination (Garcia-Quintanilla et al. 2000), this suggests that hair samples are difficult to amplify and that the results obtained with hair samples, especially by two-round PCR, are less reliable than with other types of samples. The use of hair samples collected from the ground is generally unreliable because of the possibility that hairs from other individuals could have dropped in the same area (see Table S3 in Appendix E). Therefore, we concluded that hair samples collected from the ground are the least reliable and excluded them from our results, as well as any results produced by two-round PCR. These exclusions, however, do not change individual haplotype assignment (Table S3 in Appendix E).

Finally, we could distinguish between the inconsistencies that originated from mix-up of hair samples in the field and from technical difficulties in the laboratory by eliminating results obtained by two-round PCR. Our most striking example suggesting the mix-up of hairs was that from a sample of hairs collected from a night bed built and used by one adult female (Kai). This sample contained three different haplotypes, suggesting that other individuals had spent time in that bed. Because adult chimpanzees make their own bed every night, researchers in previous studies have assumed that hair samples obtained from one bed were dropped by the bed maker (Morin et al. 1994a, b; Gagneux et al. 1999a). The fact that three haplotypes were recorded from one night bed indicates that this assumption is not always valid, as young may reuse or play in the bed constructed by other kin or non-kin individuals (Takemoto, Myowa-Yamakoshi, and Humle, personal observations).

34.5 Summary

mtDNA sequencing of Bossou and Nimba chimpanzee communities and subsequent analyses revealed that no clear population structure in the chimpanzee population of the Bossou–Nimba region, as has been found in other West African chimpanzee populations. This pattern suggests that there has been sufficient gene flow in the common ancestral population of West African chimpanzees. In the current generations, no matrilineal gene exchange was found between Bossou and Nimba communities. Moreover, four of eight original adult members, who were adults or subadults during the initiation of field observations in 1976, are likely matrilineal relatives. Although matrilineal genetic diversity of the Bossou community has not shown any significant reduction so far, there is a high risk of mtDNA haplotype extinction because of the skewed distribution of haplotypes and predicted relaxation of incest avoidance. Finally, hair samples showed inferior results to other sample types, such as urine, feces, and wadges in reliability of experimental results and traceability of sample to individual.

Acknowledgments The studies reviewed in this chapter consist of field research organized by the Kyoto University Primate Research Institute (KUPRI) and experimental research conducted in the Division of Population Genetics at the National Institute of Genetics (NIG). Y. Sugiyama provided hypotheses and managed the fieldwork. N. Saitou supervised the experiments and analyses. S. Hayakawa, S. Fujita, and T. Humle conducted field sampling and provided useful discussions of the results. T. Matsuzawa, H. Takemoto, G. Yamakoshi, M. Myowa-Yamakoshi, S. Hirata, and G. Ohashi, provided field observation records and were extremely supportive in the field. Y. Noaki helped with the genetic analysis. J. Koman, G. Goumi, P. Chérif, and P. Goumi provided valuable assistance in the field. This project was financially supported by the Nissan Science Foundation (1999) to M.K.S., the Inamori Foundation (2000) to M.K.S., Grants-in-Aid for Scientific Research from the Ministry of Education, Science, Sports, and Culture, Japan: research on priority area (No. 14011248) to N.S., international scientific program (No. 10041168) to Y.S., and COE Research (No. 10CE2005) to KUPRI. M.K.S. received the COE fellowships to KUPRI and NIG during this project.

Chapter 35
Intestinal Bacteria in Chimpanzees in Bossou: A Preliminary Study of Their Nutritional Implication

Kazunari Ushida

35.1 Need for Bacterial Culturing in Bacteriological Surveys on Wild Apes

Bacteriological surveys based on the 16S rRNA gene can be conducted at all research fields of wild apes if fresh feces are available. As a result of the recent advances in molecular biological techniques, we can now carry out phylogenetic analyses on both cultured and noncultured bacteria. In our study, fresh feces were collected aseptically into ethanol to fix the bacteria for preservation of DNA. After transportation of samples to our laboratory in Kyoto, we successfully analyzed the bacterial 16S rRNA gene in the feces of wild chimpanzees (Uenishi et al. 2007; Fujita and Kageyama 2007; Ushida 2009).

However, molecular-based phylogenetic analyses of bacteria only poorly provide physiological evidence for their presence in the intestine. Live bacterial cultures are required to evaluate their function in the physiology, particularly in the nutrition, of the host (Gibson and Macfarlane 1995). To obtain live bacteria from the feces of chimpanzees, the feces should be promptly treated after defecation. If feces are aseptically transferred in tubes filled with CO_2 (such as Kenki Porter; Clinical Supply, Tokyo, Japan) and cooled to 4°C, a 24-h period is often tolerated before culturing. However, researchers often encounter are major constraints when collecting fecal samples from wild subjects in the field. Collection of feces just after defecation is often problematic because of the challenge of locating a large enough sample of freshly deposited feces, especially among semi- or nonhabituated individuals. Prompt and appropriate transportation of fecal samples to the experimental facility is also usually impossible in field conditions.

In Bossou, researchers and guides can follow the chimpanzees to collect feces just after defecation. Moreover, chimpanzees in Bossou live close to the village

K. Ushida (✉)
Laboratory of Animal Science, Graduate School of Life and Environmental Sciences,
Kyoto Prefectural University, Shimogamo, Kyoto 606-8522, Japan
e-mail: k_ushida@kpu.ac.jp

T. Matsuzawa et al. (eds.), *The Chimpanzees of Bossou and Nimba*,
DOI 10.1007/978-4-431-53921-6_36, © Springer 2011

where the research facility is located. Therefore, freshly collected feces can be promptly transported to the experimental facility; this is the great advantage of Bossou as a site for bacteriological research on wild chimpanzees. The research facility in Bossou is only a simple building and does not provide 24-h electricity and distilled water; however, it is possible to conduct basic bacteriological surveys based on culturing (Ushida et al. 2010). I present here some of our preliminary findings on intestinal bacteria of wild chimpanzees other than their molecular ecology, as the latter is discussed elsewhere (Uenishi et al. 2007; Ushida 2009).

35.2   In Vitro Incubation of Feces with Plant Polymers

Although Bossou chimpanzees ingest a wide variety of foods (see Chaps. 2 and 22), the major portion of their diet is comprised of plant materials consisting of soluble sugars, starch, soluble polymers such as pectic substances and arabinogalactan (gum), partially soluble hemicellulosic polymers such as substituted xylan, and insoluble polymers such as cellulose. These plant materials are largely indigestible, except for soluble sugars such as fructose and sucrose and some starch elements. Indigestible saccharides are fermented in the large intestine to produce short-chain fatty acids (SCFA) such as acetate, propionate, and butyrate. SCFA is absorbed from the large intestine to contribute to the host's nutrition (Cummings et al. 2004). The dependence on intestinal SCFA, hence bacterial fermentation of indigestible saccharides, is obviously large in the case of herbivorous animals (Stevens and Hume 1995). In the case of humans, 6–9% of maintenance energy is supplied by SCFA (McNeal 1984). There is no estimate of the contribution of colonic fermentation to the maintenance of energetic requirements in chimpanzees. Considering chimpanzees' major food resources, the contribution of colonic SCFA to the host's nutrition should be larger than or at least equivalent to that of humans.

To degrade particular plant materials, the host needs particular bacteria in the large intestine: cellulose is fermented by cellulolytic bacteria such as *Ruminococcus* spp. and *Fibrobacter* spp.; xylan is fermented by xylanolytic bacteria such as *Prevotella* spp. and *Bacteroides* spp.; pectin is fermented by pectolytic bacteria such as *Lachnospira* spp., and so on (Hobson 1988). The partial sequences of these bacteria were detected from the feces of Bossou chimpanzees (Uenishi et al. 2007; Ushida 2009). We thus demonstrated that bacterial fermentation of these plant polymers can be demonstrated using in vitro incubation of feces containing plant polymers.

The digestive capacity of chimpanzees in the wild cannot be assessed by in vivo digestion trials as is done with chimpanzees in captivity (Milton and Demment 1988). Therefore, in vitro incubation techniques (Tilley and Terry 1963; Van Soest 1982) are the only way available to estimate the digestive capacity of chimpanzees in the wild.

35.3   Albizia Gum Fermentation

Chimpanzees in Bossou often ingest the gum exudate of *Albizia* spp., which is chemically identified as arabinogalactan protein (Anderson and Morrison 1990), a typically indigestible plant polymer. We estimated its contribution to the host's nutrition because the gum exudate was fermented to SCFA by the intestinal bacteria of the chimpanzees. The details are given in our article (Ushida et al. 2006). In the rainy season of 2003, a preliminary experiment was conducted (Fig. 35.1). A portion

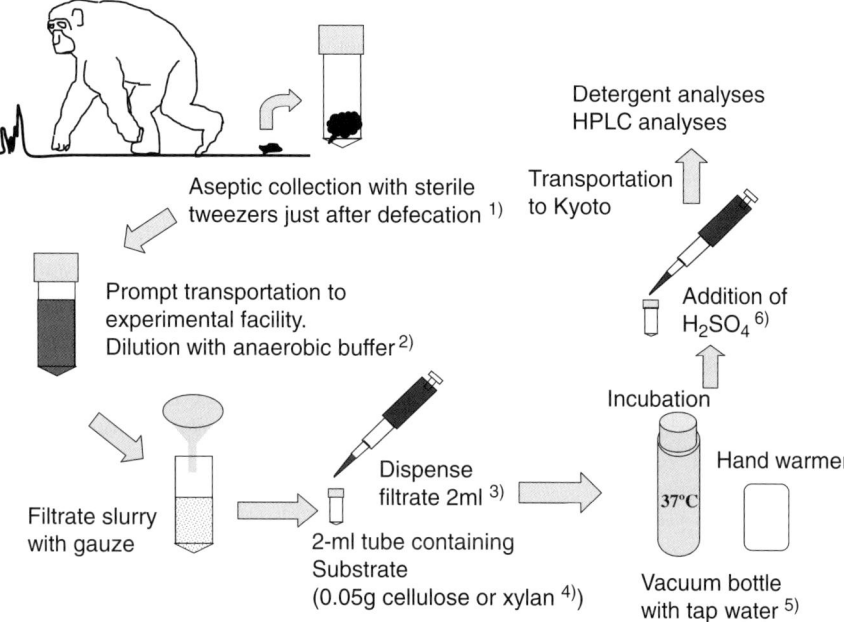

Fig. 35.1 Technique applied to incubate fecal bacteria of chimpanzees in the wild. *(1)* Collection of feces was made just after defecation to avoid the oxygen effect that eventually modifies bacterial composition in eliminating strict anaerobic bacteria. Eventual contamination with soil bacteria should also be avoided. Identification of the individual is important to avoid redundant sampling. *(2)* Anaerobic (pre-reduced) buffer is preferred to avoid eventual elimination of strict anaerobes. *(3)* In bacteriology, tubes typically contain medium and headspace gas. A mixture of nitrogen and carbon dioxide is used as a headspace gas for anaerobic culturing to expel the air within tubes. Such a gas mixture was impossible to obtain under the field conditions available at Bossou. Accordingly, tubes were totally filled with inoculum to expel the air, hence oxygen, from the tube. *(4)* Cellulose and xylan both represent plant fiber indigestible by the chimpanzees' own digestive system. Bacterial cellulose degradation is affected by cellulose crystallinity, and that of xylan is affected by substitution. In this experiment, we used Sigma cellulose (SIGMACELL Type 101), which is widely used, for the crystalline cellulose degradation test. We also used oat spelt xylan (Sigma–Aldrich), which is widely used for xylanase tests. *(5)* Room temperature of 37°C is maintained with the aid of disposable self-heating pad hand warmers. Hand warmers were placed outside of thermoses. Temperature was regularly checked (every 3–6 h) using a thermometer. *(6)* To stop fermentation (to kill bacteria) by reducing pH, we added 20 μl locally available sulfuric acid (H_2SO_4) as available for car batteries

(10 g) of fresh feces of one female chimpanzee (Velu) was combined with 40 ml phosphate-buffered saline and squeezed through nylon tissue (tissue for commercially available tea bags). Filtrant (2 ml) was dispensed into a 2-ml plastic tube containing 0.1 g gum exudate freshly collected from an *Albizia zygia* tree. Anaerobiosis is usually maintained by replacement of headspace gas with a nitrogen-carbon dioxide mixture (Ushida and Sakata 1998). Since such a gas is not available at Bossou, anaerobiosis was obtained by eliminating headspace by filling the inoculum to the top of the tube. For this experiment, the author's body was used to maintain a temperature of around 37°C. At the end of incubation, 200 µl sulfuric acid (~37%; a local product for car batteries) was gently added to stop fermentation. After transportation to the author's laboratory in Kyoto, tubes were centrifuged, and the supernatant was subjected to organic acid analysis. Acetic acid and lactic acid were the major acids produced during the incubation, and total organic acid concentration reached some 40 mmol/l in the culture tubes. In a previous study on arabinogalactan fermentation in the large intestine of pigs, we identified propionate as the major end products of fermentation and *Prevotella ruminicola* as a major degrader of arabinogalactan to produce propionate (Kishimoto et al. 2006). As *Prevotella* spp., such as *P. ruminicola*, were detected as one of the major intestinal bacteria of chimpanzees at Bossou (Uenishi et al. 2007), the small production of propionate from *Albizia* gum by chimpanzee fecal bacteria was somewhat surprising. The small production of propionate may be explained by the production of lactate, which is produced by a wide range of intestinal bacteria as a fermentation end product. However, lactate is seldom detected in the large intestine, because there are bacteria that ferment lactate into acetate, propionate, or butyrate (Tsukahara et al. 2002). The bacteria responsible for *Albizia* gum fermentation need further investigation.

We calculated the contribution of one piece (30 g fresh weight) of *A. zygia* gum to the host's energetic intake to conclude the insignificance of gum exudate as an energy source. However, based on the inductively coupled plasma (ICP) analysis of *A. zygia* gum exudate, findings suggest that it may provide calcium and magnesium at levels fulfilling the chimpanzees' daily requirement (Ushida et al. 2006).

35.4   Cellulose and Xylan Fermentation

As the preliminary experiment of 2003 was successful, we conducted further systematic incubation experiments at Bossou using cellulose and xylan as substrates (see Fig. 35.1). Fresh feces of 11 chimpanzees (Yo, Yolo, Fana, Foaf, Fanle, Velu, Tua, Pama, Peley, Jire, Jeje) were successfully collected in the rainy season of 2004 and rapidly transported (within ~15 min after defecation) to the experimental facility at Bossou. If chimpanzees were located at a distant location from the experimental facility, sampling was abandoned. Although we did not systematically evaluate the time limit between collection and arrival at experimental facility for successful samples, we arbitrarily used 15 min as maximum time allowed to reach the facility. This time limit was sufficient, however, to allow the local field assistants to return to the experimental facility from the summits of surrounding hills.

A portion (10 g) of fresh feces was combined with 40 ml phosphate buffer (0.05 M, pH 6.8) and squeezed through nylon tissue. Filtrant (2 ml) was dispensed into a 2-ml plastic tube containing 0.05 g substrate (cellulose powder or oat spelt xylan; both were purchased from Sigma) to not leave a headspace for the anaerobiosis. After being tightly capped, tubes were incubated in stainless vacuum bottles containing tap water at 37°C. The temperature of the water was kept as constant as possible by checking the temperature every 3–6 h. Bottles were otherwise kept in a polystyrene box gently heated with commercially available hand warmers. At the end of incubation, fermentation was stopped by adding sulfuric acid (see Fig. 35.1). Incubation was repeated twice for each individual, and duplicate tubes were allotted both at the start of incubation and at the end of the 24-h incubation.

After transportation to the author's laboratory in Kyoto, both organic acid and residual substrate were analyzed. Organic acid was analyzed as described above. Residual substrate was determined after extraction by acid detergent solution (cellulose) or neutral detergent solution (xylan) as described by Ushida et al. (1990). Cellulose and xylan degradation for 24-h incubation were $7.5 \pm 2.9\%$ and $12.2 \pm 6.6\%$, respectively ($n = 11$). Increase in total SCFA concentration during the 24-h incubation on cellulose and xylan was, respectively, 12.8 ± 5.0 mmol/l and 37.7 ± 4.6 mmol/l ($n = 11$). Increases in each major SCFA (acetate, propionate, butyrate) were 7.4 ± 2.6 mmol/l, 4.8 ± 2.3 mmol/l, and 0.8 ± 0.4 mmol/l, and 24.1 ± 0.2 mmol/l, 10.8 ± 2.8 mmol/l, and 1.1 ± 0.4 mmol/l, for cellulose and xylan, respectively.

Using the same principle of in vitro incubation, Ehle et al. (1982) provided an in vitro fiber digestibility test of humans on a Western-style diet. They found that crystalline cellulose (Solka floc) was totally indigestible when incubated with human fecal bacteria for 48 h. Accordingly, this means that the cellulolytic activity in the large intestine of chimpanzees is higher than that of humans, at least those on a Western-style diet.

The results suggest that the chimpanzees of Bossou are able to utilize cellulose and hemicellulose (xylan) as a source of SCFA in the large intestine. Milton and Demment (1988) also reported significant digestion of hemicellulose (60–77%) and cellulose (38–70%) of wheat bran with 12–17 h of digesta retention within the large intestine of chimpanzees in captivity.

Chimpanzees are omnivorous animals, but their major food is plant material. The significance of bacterial fermentation in the large intestine for their nutrition is quite evident, because plant materials are composed of many indigestible oligomers and polymers that are fermented by intestinal bacteria to produce SCFA. The nutritional significance of intestinal bacteria is, so far, difficult to assess in wild animals. The in vitro fermentation technique that we have introduced here is applicable to wild chimpanzees, at least in the experimental conditions available at Bossou.

Obviously, the present experiment is still at a level of a preliminary study. However, the results suggested that chimpanzees in Bossou showed higher fiber degradation in the large intestine than humans. Research on the particular fiber-digesting bacteria in the large intestine of chimpanzees is of interest because none of the fiber-degrading bacteria in the large intestine of chimpanzees, which should have a great impact on their nutritional status, have yet been identified. The microbial flora established in the large intestine of chimpanzees may be more complex than that of humans. For example, humans do not have entodiniomorphid ciliate protozoa in their large intestine. We have recently analyzed the great ape-specific ciliate protozoon, *Troglodytella abrassarti*, and found a close phylogenetic relationship with the horse-specific ciliate protozoa, *Cycloposthium* spp. (Irbis et al. 2008). The roles of these ciliate protozoa in fiber degradation in the large intestine of chimpanzees have not yet been evaluated, but it is suggested that these ciliate protozoa have a potential role in host nutrition. When chimpanzees are isolated from other chimpanzees and maintain contact with humans, they lose *T. abrassarti*; this often occurs when chimpanzees are illegally raised as pets in human homes where they are provided few fibrous foods, and sometimes none. The prevalence of diarrhea increases under such circumstances (Garriga, personal communication). In this context, it is important to identify the natural intestinal microbial flora of chimpanzees.

Acknowledgments A part of this study was supported by the HOPE project from the Japan Society for the Promotion of Science. The author thanks Dr. T. Matsuzawa, Kyoto University Primate Research Institute (KUPRI), for the opportunity to work on chimpanzees in Bossou and KUPRI. Sincere gratitude is expressed to Dr. S. Fujita (Department of Veterinary Medicine, Yamaguchi University). Thanks are also due to Dr. T. Humle, Dr. A. Kato, and Mr. G. Ohashi in KUPRI for their help with sample collection. I would also like to thank Dr. Y. Ohashi, Mr. G. Uenishi, and Miss M. Hiraguchi of the author's laboratory for their help with the microbiological analysis. The author is indebted to Mr. P. Goumy, P. Chérif, B. Zogbila, J. Doré, H. Gbéregbé, and M. Doré for their assistance with the collection of chimpanzee feces. Finally, I thank the Institut de Recherche Environnementale de Bossou (IREB) and the Direction Nationale de la Recherche Scientifique et Technologique (DNRST) of the government of the Republic of Guinea for their support.

Chapter 36
Health Monitoring

Shiho Fujita

36.1   Why is Health Monitoring in Wild Great Apes Necessary?

The great apes face many threats to their continued existence in the wild. Of these threats, infectious diseases have the greatest impact because epidemics may result in rapid decline in population size (see also Chap. 32). Moreover, in populations that are habituated for the purposes of ecotourism or research, the great apes are continually faced with a high risk of disease transmission from humans. Because great apes are closely related to humans in phylogeny, pathogens from human sources may be easily transmitted to the great apes. Table 36.1 summarizes published reported outbreaks of disease among wild chimpanzee and gorilla populations that have been habituated to humans. The Bossou community experienced two epidemics: one in 1992 and the other in 2003 (see also Chaps. 25 and 32 for further details). Although the pathogens that underlie these outbreaks could not always be identified, direct and/or indirect contact between humans (including domestic animals) and great apes possibly contributed to the occurrence of the majority of these infectious disease outbreaks (Chapman et al. 2005; Leendertz et al. 2006b). Therefore, periodic health monitoring is necessary in wild populations exposed to human habituation to prevent interspecific transmission of pathogens between humans and great apes. Especially, for Bossou chimpanzees, because of the small population size, epidemics may severely threaten the future of this community.

In a routine clinical investigation, noninvasive specimens such as urine, feces, and saliva are useful because these can be obtained repeatedly non-invasively from habituated animals without capturing them. The collection of blood or tissue samples typically requires the capture and/or containment of animals, which may pose risks of injury to both chimpanzees and investigators, and ethically unacceptable stress to wild individuals, while also interfering with long-term behavioral research

S. Fujita (✉)
Department of Veterinary Medicine, Faculty of Agriculture, Yamaguchi University,
Yoshida 1677-1, Yamaguchi 753-8515, Japan
e-mail: fujita@yamaguchi-u.ac.jp

T. Matsuzawa et al. (eds.), *The Chimpanzees of Bossou and Nimba*,
DOI 10.1007/978-4-431-53921-6_37, © Springer 2011

Table 36.1 Deaths caused by disease (including suspected ones) in wild chimpanzees (C) and gorillas (G)

Site	Year	Disease	No of deaths	References[a]
Tanzania				
Gombe	1966	Polio	6 (C)	1, 2, 3
	1968	Respiratory disease	5 (C)	1, 2, 3
	1975	Respiratory disease	1 (C)	1, 2, 3
	1978	Respiratory disease	1 (C)	1, 2, 3
	1987	Respiratory disease	9–10 (C)	2, 3
	1996	Respiratory disease	8–11 (C)	2, 3
	1997	Scabies	3 (C)	2, 3
	2000	Respiratory disease	2 (C)	3
	2002	Respiratory disease	>2 (C)	3
Mahale	1986	AIDS-like disease	3 (C)	4
	1993–1994	Flu-like disease	11 (C)	4, 5
	2006	Human metapneumovirus (HMPV)	12 (C)	6, 7
Uganda				
Virunga	1996	Scabies	1 (G)	8
Bwindi	1996	Scabies	1 (G)	9
Rwanda				
Volcano National Park	1988	Measles, etc.	6 (G)	8, 10
	1990	Measles	1 (G)	8
Gabon–Congo border	2001–2003	Ebola	15 (C), 50 (G)	11,12
Gabon				
Lossi Sanctuary	2002–2003	Ebola	~5,000 (G)	13
Côte d'Ivoire				
Täi	1987	Monkey pox	1 (C)	14
	1992	Ebola	8 (C)	14,15
	1994	Ebola	12 (C)	14,15
	1999	Human respiratory syncytial virus (HRSV), *Streptococcus pneumoniae*	6 (C)	16
	2001–2002	Anthrax	8 (C)	17
	2004	Human metapneumovirus (HMPV), *Streptococcus pneumoniae*, *Pasteurella multocida*	8 (C)	16
	2006	Human respiratory syncytial virus (HRSV), *Streptococcus pneumoniae*	3 (C)	16
Guinea				
Bossou	1992	Flu-like disease	1 (C)	18
	2003	Respiratory disease	5 (C)	19
Cameroon	2004–2005	Anthrax	3 (C), 1 (G)	20

[a] *1* Goodall (1986), *2* Wallis and Lee (1999), *3* Lonsdorf et al. (2006), *4* Nishida et al. (2003), *5* Hoasaka (1995), *6* Hanamura et al. (2006), *7* Kaur et al. (2008), *8* Homsy (1999), *9* Kalema-Zikusoka et al. (2002), *10* Ferber (2000), *11* Leory et al. (2004), *12* Rouquet et al. (2005), *13* Bermejo et al. (2006), *14* Boesch and Boesch-Achermann (2000), *15* Formenty et al. (1999), *16* Köndgen et al. (2008), *17* Leendertz et al. (2004), *18* Matsuzawa (1992), *19* Matsuzawa et al. (2004), *20* Leendertz et al. (2006b)

and the natural behavior of the animals. This chapter presents some of our preliminary research on health monitoring of wild chimpanzees by using noninvasive
sampling.

36.2 Detection of Clostridium perfringens in Fecal Specimens

C. perfringens is known to be not only a member of the normal intestinal microflora
of domestic animals and humans but also a ubiquitous bacterium in the natural
environment (Allen et al. 2003; Saito 1990). When a disorder of the intestinal
microflora occurs, pathogenic bacteria, including *C. perfringens*, are known to
increase proportionally and cause autogenous infections (Mitsuoka 1982). This
bacterium may be classified into several types according to the toxins it produces
(McDonel 1980; Smedley et al. 2004). It may cause many kinds of diseases in
diverse hosts, including humans, that are sometimes fatal to the host (for review, see
Borriello 1995; Hatheway 1990; Niilo 1980; Songer 1996). Because hyperprolif-
eration of this bacterium in the intestinal tract can be a disease-causing agent even
in a healthy individual, it is advantageous to check the prevalence of this bacterium
in assessing health status.

We carried out the detection of *C. perfringens* in feces from chimpanzees in a
captive group and two wild populations, Bossou (Guinea) and Mahale (Tanzania),
by using the rapid polymerase chain reaction (PCR) method. We used nested PCR
assays, which include a two-step PCR and have a 10^3-fold higher sensitivity than a
single round (Fujita and Kageyama 2007). The bacterium was detected in most fecal
specimens (80%) in captive chimpanzees. In contrast, the detection rate among wild
chimpanzees was lower, with 23% (12/53) of fecal samples from the Bossou group
and 1.2% (1/81) in the Mahale group. These results indicate that the intestinal
microflora differs among chimpanzee populations living under different conditions.

The supply of artificial diets to animals in captivity might be positively correlated with the increase in intestinal *C. perfringens*. That is, the high calorie and low
fiber content of captive diets might favor the proliferation of the bacterium in the
intestine (Fujita and Kageyama 2007). However, even among the wild populations,
detection rate of this bacterium was quite different. Therefore, environmental factors might also influence intestinal microflora. *C. perfringens* is ubiquitous in
human-inhabited areas and is enduring because of sporulation. Therefore, various
environmental factors such as soil, water, and surfaces in and around human living
areas could serve as natural reservoirs for this bacterium. Although at Mahale, local
people, except for rangers, tour guides, tourists, researchers, and field assistants, are
restricted to areas outside the National Park, resulting in less frequent contacts
between humans and chimpanzees, at Bossou the chimpanzees' habitat borders a
village, and the chimpanzees cross roads and paths also used by humans and raid
crops on a regular basis (see Chap. 22). In addition, the local people defecate in the
undergrowth at the forest edge and around their agricultural fields. These indirect
contacts between human and chimpanzees are anticipated to increase the possibility
of *C. perfringens* infection in chimpanzees.

Although all the samples were derived from clinically healthy chimpanzees, the prevalence of *C. perfringens* could reflect a disorder of the intestinal microflora. This simple and noninvasive technique might be useful in the assessment of health status of chimpanzees once baseline levels are established.

36.3   Urinalysis

Urinalysis by using multi-reagent test strips is a rapid, easy, and relatively inexpensive method to assess physiological state and has been applied to great apes in the wild (Kaur and Huffman 2004; Knott 1996; Krief et al. 2005). Although dipstick urinalysis alone is not a reliable method to evaluate health status, these studies have revealed that dipstick urinalysis can provide a reliable assessment of the physiological condition of wild great apes when combined with other types of assessment, such as behavioral observation or fecal examination for parasitic load.

We collected 498 (dry season, 287; wet season, 211) fresh urine samples from 18 clinically healthy chimpanzees in Bossou (Table 36.2). These samples were tested for pH, specific gravity, protein, glucose, ascorbic acid, nitrites, ketones (acetoacetic acid and acetone), erythrocyte, hemoglobin, leucocyte, urobilinogen, and bilirubin by using Pretest 7aII or multi II (Wako Pure Chemical Industries), and for ketones (β-hydroxybutyrate) by using Sanketopaper (Sanwa Kagaku Kenkyusho). Not all samples were analyzed for all parameters because of their limited volume.

The most common test results are presented in Table 36.3. Urine pH of most samples was between 8 and 9. This pH was consistent with previous reports from other wild chimpanzee populations (Kaur and Huffman 2004), and, therefore, is considered to be a normal level in chimpanzees. However, the incidence of alkaline urine was greater during the dry season than the wet season. Incorporation of rain into the samples might have been responsible for the lower pH, although careful attention was paid to avoid contamination when sampling. Urine specific gravity reflects the concentration of particles in the urine, which may fluctuate in response to physiological factors such as intake of fluid. Urine specific gravity ranged between 1.005 and 1.025, and mostly between 1.005 and 1.001, which is a normal level. Ascorbic acid (vitamin C) is normally excreted into the urine and can influence test results for some parameters such as glucose, nitrates, and occult blood. Only 2.5% (7/377) of samples were positive for ascorbic acid.

High levels of protein are excreted into urine by subjects suffering kidney damage and febrile illness. In the present study, protein ($\geq$30 mg/dl) was detected in about 30% of the urine samples. However, it is known that alkaline urine leads to false-positive results in the test for protein. As almost all samples that were positive for protein had a pH between 8 or 9, this may explain the elevated recorded levels of urinary protein. In humans, protein can also be transiently excreted into urine even by healthy individuals after hard exercise. In the present study, protein was detected more frequently in the samples collected during the wet season than those

Table 36.2 Number of urine specimens (number of chimpanzees) for urinalysis by age and sex class

| | Dry season (Dec 2001–Mar 2002) | | Wet season (Jun–Aug 2004) | |
	Males	Females	Males	Females
Adults (>12 years)	22 (2)	158 (7)	51 (3)	110 (6)
Adolescents (8–11 years)	37 (2)	43 (3)	0 (0)	17 (1)
Juveniles (4–7 years)	11 (2)	13 (1)	27 (2)	0 (0)
Infants (0–3 years)	3 (1)	0 (0)	6 (1)	0 (0)
Total	73 (7)	214 (11)	84 (6)	127 (7)

Table 36.3 Most common results of noninvasive urinary test strip analysis conducted among the wild chimpanzees of Bossou

Parameter	Test result	Dry season	Wet season	Total
pH	8–9	262/286 (91.6)	137/204 (67.1)	399/490 (81.4)
Urine specific gravity	1.005–1.010	–	110/112 (98.2)	110/112 (98.2)
Ascorbic acid	0 mg/dl	277/284 (97.5)	93/ 93 (100)	370/377 (97.5)
Protein	Negative or trace[a]	146/286 (51.1)	189/204 (92.6)	335/490 (68.4)
Glucose	Negative[b]	284/284 (100)	202/204 (99.0)	486/488 (99.6)
Ketones				
Acetoacetic acid/acetone	Negative[c]	224/286 (78.3)	176/204 (86.3)	400/490 (81.6)
β-Hydroxybutyrate	0 μmol/l	183/231 (79.2)	119/183 (65.0)	302/414 (72.9)
Erythrocyte	Negative[d]	280/284 (98.6)	198/205 (96.6)	478/489 (97.8)
Hemoglobin	Negative[e]	283/284 (99.6)	205/205 (100)	488/489 (99.8)
Leukocytes	Negative[f]	–	96/112 (85.7)	96/112 (85.7)
Nitrites	Negative[g]	–	88/112 (78.6)	88/112 (78.6)
Urobilinogen	Normal[h]	283/284 (99.6)	204/204 (100)	487/488 (99.8)
Bilirubin	Negative[i]	–	103/112 (92.0)	103/112 (92.0)

Figures indicate the number of samples which showed the most common result/total number of samples tested. In parentheses is the percentage value

[a]<10 ng/dl (negative) or 10–20 ng/dl (trace)

[b]<100 mg/dl

[c]<5 mg/dl as acetoacetic acid and 100 mg/dl as acetone

[d]<20/μl

[e]<0.06 mg/dl

[f]<25/μl

[g]<0.1 mg/dl

[h]<1 mg/dl

[i]<0.5 mg/dl

gathered during the dry season. This difference might reflect different activity levels in the chimpanzees between seasons, although Bossou chimpanzees are typically less active during the rainy season than in dry season months (December–February) (see Chap. 33).

Ketones are metabolic parameters that are excreted from subjects with diabetes and starvation. In wild orangutans, urinary ketones increase in association with negative energy balance (Knott 1998). In the present study, 20–30% of samples were positive for ketones, and seasonal differences were unclear. A negative energy balance might explain this high presence of ketones.

Although glucose is normally excreted into urine in small amounts, the amount increases when blood sugar levels increase or the threshold of sugar excretion in the kidneys is reduced. For a juvenile male (Peley, 6 years old), relatively high levels (100 mg/dl) of urinary glucose were noted twice between June and August 2004. The occurrence of measurable glucose is not normal, and diabetes mellitus could be suspected. However, the occurrence of urinary glucose is not diagnostic of the condition because this factor may be observed in many other cases. For example, urinary glucose may be observed after large amounts of foods containing sugar are eaten, in cases of acute emotional strain, and after exercise. Further tests are therefore needed for a more detailed diagnosis.

Hematuria appears mainly in subjects with bladder problems such as inflammation, calculi, and malignancy. In this study, all subjects that presented hematuria were cycling females, presumably menstruating. However, hemoglobin may also be excreted in the urine by subjects with hemolytic anemia and infectious diseases such as filariasis, babesiosis, or malaria. A juvenile female (Fanle, 4 years old) presented slight hemoglobinuria (0.06 mg/dl) in the dry season.

The presence of nitrites and leukocytes in urine can indicate urinary tract infection. In the present study, although chimpanzees were tested only in the wet season, seven samples from six subjects (an adult male, three adult females, a young female, and a juvenile male) were positive for both nitrites and leukocytes. These subjects were suspected to suffer from a urinary tract infection.

Urobilinogen and bilirubin show positive values in subjects with hepatic damage. A specimen from a juvenile male (Jeje, 4 years old) had above-normal levels of urobilinogen (1 mg/dl). Bilirubin, although only tested in subjects during the wet season, was positive in nine samples from seven subjects.

All subjects in the present study had no obvious medical condition. The subjects in which some parameters in the urinary test strip were positive could present subclinical illnesses. In combination with direct observation of each individual's behavior, such as inactiveness, lameness, hair coat abnormalities, and fecal condition, urinalysis may provide a simple and useful in situ tool for health monitoring of wild chimpanzees (Kaur and Huffman 2004; Krief et al. 2005; Lonsdorf et al. 2006).

36.4   Conclusion

The present study has yielded preliminary data on health monitoring of the wild chimpanzees of Bossou, although none of the subjects showed any clinical signs of illness. Although further research and assessments are needed, the simple and noninvasive techniques used in this study might be useful in the monitoring of health risks in habituated wild chimpanzee populations.

Acknowledgments I sincerely thank the Direction Nationale de la Recherche Scientifique et Technologique (DNRST), and the Institut de Recherche Environnementale de Bossou (IREB) of the Republic of Guinea, for permission to study the chimpanzees at Bossou; and the Tanzania Commission for Science and Technology (COSTECH), the Tanzania Wildlife Research Institute (TAWIRI), the Tanzania National Parks (TANAPA), the Mahale Wildlife Research Centre (MWRC), and the Mahale Mountains National Park (MMNP) for permission to study the chimpanzees at Mahale. This study was financed by a grant from the Research Fellowships of the Japan Society for the Promotion of Science for Young Scientists (No. 4502, 2628), and the Ministry of Education, Culture, Sports, Science and Technology, Japan (No. 12375003).

Chapter 37
Green Corridor Project: Planting Trees in the Savanna Between Bossou and Nimba

Tetsuro Matsuzawa, Gaku Ohashi, Tatyana Humle, Nicolas Granier, Makan Kourouma, and Aly Gaspard Soumah

37.1 The Aim of the Green Corridor Project

A unique feature of the Bossou community is the coexistence between humans and chimpanzees. Chimpanzees often cross roads with a special sociospatial organization in which adult males take the division of roles to protect females and their offspring (Hockings et al. 2006; see Chap. 23). Bossou chimpanzees also learned to get papaya fruits from the village and use them as a gift to females (Hockings et al. 2007; Ohashi 2007; see Chap. 23). This kind of human tolerance to chimpanzees results from the religious belief of the local people: Manon people at Bossou believe that the chimpanzees are the reincarnation of their ancestors.

Chimpanzee society is characterized by male philopatry. Males typically stay in their natal community while females emigrate to adjacent communities. Data from Bossou indicate that all females born in Bossou disappeared before giving birth for the first time or soon after (see Chap. 3).

During the past three decades, there has been no record of females immigrating into the Bossou community. The isolation of the Bossou community, resulting from its close proximity to the village and the presence of savanna stretching in all directions beyond the chimpanzees' core area, may explain this pattern. The chimpanzees born in the community can emigrate, but the nonhabituated chimpanzees born in other neighboring communities cannot easily immigrate into the Bossou community. More young adult males >14 years old than females of that age have remained in Bossou.

T. Matsuzawa (✉) and G. Ohashi
Primate Research Institute, Kyoto University, 41-2 Kanrin, Inuyama, Aichi 484-8506, Japan
e-mail: matsuzaw@pri.kyoto-u.ac.jp

T. Humle
School of Anthropology and Conservation, The Marlowe Building, University of Kent, Canterbury CT2 7NR, UK

N. Granier
Behavioral Biology Unit, Department of Environmental Sciences, University of Liège, Quai Van Beneden, 22, 4020 Liège, Belgium

M. Kourouma and A.G. Soumah
Institut de Recherche Environnementale de Bossou, Bossou, Republic of Guinea

T. Matsuzawa et al. (eds.), *The Chimpanzees of Bossou and Nimba*,
DOI 10.1007/978-4-431-53921-6_38, © Springer 2011

This pattern is in line with the typical pattern of male philopatry characteristic of chimpanzee society. All females and some males emigrated (or disappeared to be more precise) at the average age of 10.2 years ($n=12$; range, 7–16 years old) (see Chap. 3).

For decades, the village of Bossou numbered about 1,500 inhabitants. Then, during the years following the civil war in Liberia, which started in 1990, the population almost doubled. Finally, after the end of the civil war in Liberia, the human population gradually decreased to about 2,500 (Kabasawa, personal communication).

According to the IUCN (2007), there are about 187,000 chimpanzees living in tropical rainforest and surrounding savanna. The chimpanzee is an endangered species, and its numbers are rapidly decreasing (see Chaps. 39 and 40 for more details of the main threats). There seem to be three major threats to their survival. The first threat is deforestation, mostly consequent to human population growth, which accentuates the need for arable land and leads to the shrinking of chimpanzee habitat. The second threat is poaching and the bush-meat trade. Although the chimpanzees of Bossou are never hunted or poached, the story does not apply to chimpanzees in other neighboring localities where the villagers are of different ethnic origins and hold religious or cultural beliefs different from those of the Manon people of Bossou. The third major threat is contagious diseases (see Chaps. 32 and 36 for more details on the threat of diseases to chimpanzees).

The size of the Bossou chimpanzee community has ranged, since 1976, between 12 and 23 individuals. It numbers at present 13 members (May 2010). The data clearly show that (1) the number of chimpanzees was stable over many years, at around 20, but has drastically decreased in recent years, and (2) the community currently consists of many old chimpanzees and very few young (see Chap. 3). To preserve the Bossou community, we need to expand their habitat and also connect it to that of adjacent communities to facilitate gene exchange.

The "Green Corridor Project" started in January 1997. This project aims to expand the forests of Bossou to the east to connect them to the forest of the Nimba Mountains. This corridor may increase the opportunities for the chimpanzees to travel back and forth between the two areas. The government of Guinea founded the Institute for Environmental Research at Bossou (IREB) in October 2001. The Green Corridor Project is the product of a collaborative effort among researchers, governmental staff, and local villagers.

37.2   The History of the Green Corridor Project

The Nimba Mountains are located about 10 km east of Bossou. The massif borders three countries: Guinea, Côte d'Ivoire, and Liberia. The Guinean and Ivorian portions of the Nimba Mountains are recognized by UNESCO as a World Natural Heritage Site (WNH) in danger (see Chap. 39 for more details).

The first author, T.M., reached the highest summit of the Nimba Mountains on March 3, 1986. The view from the summit clearly showed the importance of this area from the strategic point of view of chimpanzee conservation in West Africa.

Matsuzawa and Yamakoshi (1996) conducted the first preliminary surveys of the Nimba chimpanzees by the Kyoto University Primate Research Institute (KUPRI) team on the Ivorian side in 1993. Surveys of the Nimba chimpanzees were then undertaken by Makoto Shimada, Tatyana Humle, Kathelijne Koops, and Nicolas Granier (Shimada et al. 2004; Humle and Matsuzawa 2001, 2004; Koops and Matsuzawa 2006; Koops et al. 2007; Granier et al. 2007a) (see Chaps. 27–29, 34, and 39 for more details). Thanks to these continuous efforts, we know that there is at least one community of chimpanzees in the Seringbara area on the Guinean side of the Nimba Mountains. This area thus represents the nearest adjacent community to the Bossou community.

There is savanna stretching for about 4 km between Bossou and Seringbara. There are also small gallery forests along small streams. The Green Corridor Project aims to connect Bossou and Nimba by planting trees in the savanna (Fig. 37.1).

According to the old people of Bossou, historically this expanse of savanna was forested. We thus hope to retransform the savanna into forest by planting trees 5 m apart in a 120-ha area (4 km long and 300 m wide). For this, we need to plant a total of 48,000 trees. Based on this plan, we have been growing 7,000–10,000 young trees annually in a tree nursery.

The unique aspect of this plan is that we utilize chimpanzee feces, thus favoring chimpanzee plant foods. If the purpose had been to simply create a forest, we could have chosen tree species more suitable for transplantation such as *Eucalyptus*. An alternative approach would have been to plant fruiting trees such as mangoes and oranges, which are also beneficial to humans. However, we thought it best to reforest with tree species utilized by chimpanzees rather than by humans, thus minimizing future incursions by humans into the area.

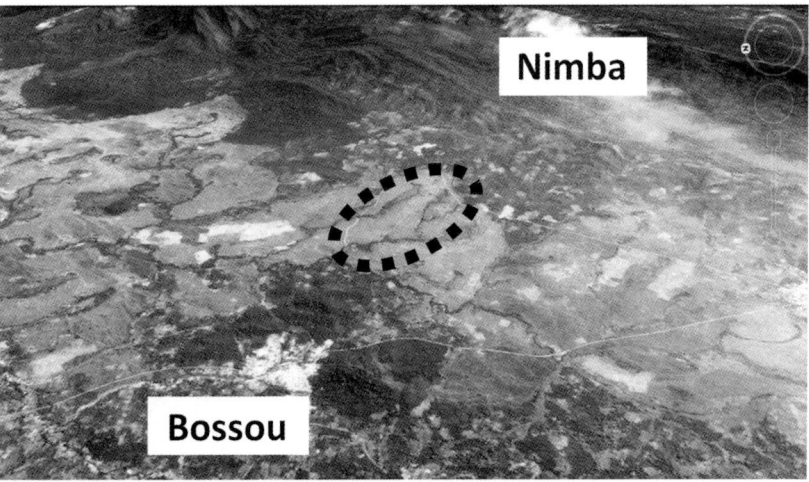

Fig. 37.1 Map of Bossou and Nimba: the Green Corridor Project aims to connect the two habitats by planting trees in the savanna (adapted from google map)

37.3   Tree Nursery: Utilization of Chimpanzee Feces

We have established our unique way of planting trees. It consists of three stages: nursing the young trees, planting them in the savanna, and protecting the growing trees.

The first stage is the tree nursery. We primarily use seeds gathered from chimpanzee feces. These seeds have a better germination rate than those simply shed by the tree onto the ground (Takemoto, unpublished data). The young trees are cared for in the tree nursery, where each sapling is kept in a plastic bag (Fig. 37.2).

We sometimes also collect small saplings that we find in the forest and bring them back to the tree nursery. While in the forest, researchers often encounter clusters of saplings that originated from chimpanzee feces. Chimpanzees are a key species for seed dispersal and for the maintenance of the diversity of tropical rainforests. In a sense, the Green Corridor Project is an attempt to artificially enhance the seed dispersal skills of chimpanzees.

Second, we transplant the young trees to the savanna. After the terrain has been cut clear of poaceous species, the young nursery saplings are transplanted in 5-m intervals. We have also tried an agro-forestry approach, by encouraging the villagers to cultivate the savanna by planting cassava (*Manihot esculenta*) or rice (*Oryza* sp.) and to transplant saplings between the crops. Then, we ask them to abandon the field after the harvest. This combination of cassava/rice cultivation and transplantation of young trees motivates local farmers to cultivate in the savanna rather than in the forest. The concept of "forestation" is a little hard for the local people to understand because traditionally they are used to practicing slash-and-burn agriculture, whereby

Fig. 37.2 Young trees growing in the tree nursery, where each sapling is kept in a plastic bag filled with humus (photograph by Tatyana Humle)

trees are cut down rather than planted. Therefore, a planting tree project combined with cultivating crops has proved to be a rather successful approach.

Third, we made an effort to protect the trees that were growing. The main reasons why young trees do not survive in the savanna are threefold: (1) the nonnutritious soil and dry conditions, (2) grazing by domesticated animals and damage by insects, and (3) bushfires. There are several places in the savanna where no trees grow because iron-packed soil covers the ground surface. Domestic animals such as sheep, goats, and cows also often intrude into the savanna to forage on grass and young leaves. In addition, cattle breeders often start bushfires in the savanna to stimulate the growth of young and fresh grasses to provide natural fodder for their cattle. Some people also use fire to hunt cane rats (*Thryonomys* sp.) that live in underground tunnels. These fires often go out of control and cause bushfires. To protect the planted young trees, we set up a 20-m-wide firebreak on both sides of the corridor. We employ two local people to maintain and patrol the firebreak area, to extinguish fires, and to prevent invasion by domesticated animals. The recent attempt to use "hexatubes" to protect the young saplings is explained below.

37.4   Assessment of Planting Activity

The initial effort of the Green Corridor Project was to create a small botanical garden ("Petit Jardin Botanique"). This pilot study aimed to evaluate the plant species that would best thrive in a savanna environment. The garden was constructed on 0.36 ha (about 60×60 m) in the periphery of the savanna bordering the Bossou chimpanzees' habitat. Several local assistants cleared the bush and then planted nursery trees from 28 species, all present in the core area of the Bossou chimpanzees (Sugiyama and Koman 1987, 1992). The total number of trees planted was 250. After 1.5 years (July 1998), the trees in the garden were inspected by Hirata and Morimura (Hirata et al. 1998a): 125 planted trees were still alive (50.0%). Then, 8 years later, in January 2005, a second inspection was performed (Matsuzawa 2010).

During these 8 years, we did not transplant more tree saplings to the garden. During the second inspection, we cut back poaceous species and identified all the trees that had grown. Among the initial transplanted trees, only nine species totaling 62 trees (24.8% survival) had survived. We noted the high survivorship in the following four species: *Uapaca heudelotii*, *Parkia bicolor*, *Craterispermun laurinum*, and *Albizia zygia*. The tallest tree was a *P. bicolor* tree, 9.3 m in height. The second tallest tree was a *U. heudelotii* tree, 9.2 m in height (Fig. 37.3).

In addition to the planted trees, we found 386 initially nontransplanted young trees that had grown naturally. The wind or animals such as birds must have carried the seeds. Thus, the second inspection indicated a total of 448 trees in the initial plot, 79.2% more than the number of trees that had initially been transplanted. However, 86.2% of these trees were the product of a natural regeneration process that occurred over a period of 8 years, from 1997 until 2005. Among the naturally grown trees, we identified 30 species. This small pilot plot therefore presented a

Fig. 37.3 The savanna transformed into a secondary forest. You can still recognize the planted *Uapaca* trees at 5-m intervals (photograph by Tetsuro Matsuzawa)

wide variety of species, dominated by three species: *Harungana madagascariensis* ($n=55$), *Nauclea latifolia* ($n=55$), and *Dychrostachys glomerata* ($n=40$).

Based on our efforts during the past years, we can conclude the following. First, some transplanted tree species, such as *U. heudelotii* and *P. bicolor*, can survive well in the savanna. We have since therefore selected tree species that can be utilized by chimpanzees and that can thrive in a savanna environment. Second, planting efforts are not the main contribution to reforestation, as natural regeneration of tree species can readily be promoted through guarding efforts and prevention of bushfires. Third, savanna can be transformed into forest through our reforestation program. Based on the initial attempt in the Petit Jardin Botanique, we can estimate that 8 years is sufficient to grow trees to a height of about 10 m by transplanting nursery-grown trees to the savanna.

37.5   Hexatubes: A New Way to Protect the Young Trees

In this section, we introduce our recent attempt of using "hexatubes" to protect the young trees in the Green Corridor. A hexatube is a hexagon-shaped tube, 1.4 m tall, made from polypropylene. These tubes sustain a microclimate favoring plant growth, maintaining adequate temperature and humidity conditions. The tubes also protect the trees from strong winds and also from grazed animals (Figs. 37.4 and 37.5).

Fig. 37.4 A hexatube is a hexagon-shaped tube, 1.4 m tall, made from polypropylene. The tubes were set to protect the young trees transplanted to the savanna (photograph by Gaku Ohashi)

Fig. 37.5 (**a**) The hexatube maintains a microclimate, favoring adequate temperature and humidity, and protecting the young trees from strong winds and also from grazed animals (photograph by Tatyana Humle). (**b**) This young tree grew above the tube's height within a year (photograph by Ryo Hasegawa)

We first brought from Japan 1,200 tubes in September 2005. All of them were set by December 2005. Then we shipped 5,000 tubes in 2006, and 3,000 more in 2007. Among the 3,000 tubes set up so far, about 70% of the young trees in the tubes survived after a year. The tubes were also set up in the courtyard of a school to encourage pupils observe the growth of young trees.

37.6 New Attempt of Planting Trees: Local Hangars and Tree Cuttings

For a decade, we have nursed saplings and transplanted these into savanna. During transplantation, the environment around the saplings is drastically changed. The saplings are forced to survive under conditions of strong sunshine, and for this reason, many die soon after transplantation. In 2007, we constructed three "hangars" in the savanna to protect saplings of *Uapaca* trees, a deciduous species, against strong sunshine.

These hangars are made of natural materials, such as local people use for their temporary encampments: bamboo are used as columns, leaf stalks of *Raphia* as beams, and oil-palm fronds as roofing (Fig. 37.6). The roof allows the passage of water but blocks off most of the sunlight. Under each hangar, we transplanted 25 *Uapaca* saplings.

Fig. 37.6 The local hanger: a new way of growing the young trees on site in the savanna. The tree nursery was built in the central part of the savanna; this helps the young trees because they are not transplanted from the tree nursery to the severe climate of the savanna (photograph by Gaku Ohashi)

One year later, we checked the condition of the *Uapaca* trees under each of the three hangars. Thirteen *Uapaca* trees suffered damage from termites; the other 62 trees survived and matured. The ground beneath was already covered by large *Uapaca* leaves. By September 2008, we had constructed a total of 23 hangars in the savanna. The dimensions of the hangars are not very large, but this approach allows us to create many forest patches in the center of savanna where it was previously difficult to plant young saplings.

In 2007, we experimented with another planting method using tree cuttings. When we observed traditional fences in villages and around agricultural fields, we noticed that some sticks had sprouted. We identified a total of 8,998 sticks, and found that 176 sticks were sprouting. Fifty-one of the 176 sprouting sticks were *Spondias cytherea* cuttings (Fig. 37.7). We thus proceeded to collect 1,523 cuttings of *S. cytherea* from the forest, and planted these around gallery forest and small forest patches in the savanna. Three weeks after planting, 891 of the 1,523 cuttings already presented new shoots.

This novel approach drew the interest of the local people, because their local knowledge could help advance the Green Corridor Project. Some villagers repaired the hangars voluntarily, and others started to spontaneously plant *Spondias* cuttings in the savanna. This strategy has clearly stimulated environmental awareness among the local people who are now realizing that they themselves can help change the savanna into forest by employing their own traditional methods and techniques.

Fig. 37.7 Planting of a cutting of Guei-buna (*Spondias cythera*): a new method of directly planting trees into the corridor (photograph by Gaku Ohashi)

37.7 Conclusion

Thanks to the Green Corridor Project, the forests of Bossou have gradually been expanding eastward. Bossou chimpanzees have started to use this area more regularly, heading toward the Nimba Mountains. The number of observations of Bossou chimpanzees in the Seringbara area has significantly increased in recent years. Researchers and students, as well as villagers, have reported the presence of Bossou chimpanzees in the Seringbara area for several days or even a week at a time.

A reforestation program such as the Green Corridor Project has two components. One is the active process of planting trees to transform savanna into forest. The other is environmental education. Reforestation in itself is not the primary goal of the project. Through this activity, we aim to change the local people's attitude toward the fauna and flora surrounding them in their everyday lives. We thus aim to pursue our wildlife conservation efforts by encouraging and helping activities initiated by the local people for the local people. Our activities are always to be based on respect for the local Manon people, who will continue to protect the chimpanzees and their forest during future generations.

Acknowledgments The Green Corridor Project has been promoted by the collaboration of researchers (KUPRI-International team), the governmental institute (IREB), and the local people of Bossou and Seringbara. The project of planting trees is partly supported by a grant from the Ministry of the Environment to Dr. Toshisada Nishida. Special thanks are due to Mr. Ryo Hasegawa, Mr. Paquilé Chérif, Mr. Gouano Goumi, Mr. Tino Zogbila, Mr. Boniface Zogbila, and others for their help with the different phases of the Green Corridor Project. The details of this long-term project are available on the following website: http://www.greenpassage.org/indexE.rhtml.

Chapter 38
Environmental Education and Community Development in and Around Bossou

Tatyana Humle

38.1 Environmental Education in Villages in and Around Bossou

Members of the Kyoto University Primate Research Institute (KUPRI) research team have helped organize and implement local environmental education initiatives ever since the early 1990s. These initiatives began with public screenings of videos about Bossou chimpanzees in the village of Bossou. The videos used over the years include "The Tool-makers of Bossou" (© BBC, UK, narrated in English), "A Hard Nut to Crack" (© NHK, Japan, narrated in French), "The Green Corridor" (© NHK, Japan, narrated in Japanese), and "Jokro: The Death of a Chimpanzee" (© KUPRI, Japan, narrated in French). Since 2001, we have also regularly been producing pamphlets in English, French, and Japanese, as well as other educational materials, including badges and T-shirts (pamphlets and other printed educational materials mentioned in this chapter can be downloaded from http://www.greenpassage.org/green-corridor/education/indexE.html, and copies of all videos are available upon request), and organizing regular events in villages and schools in and around Bossou (Fig. 38.1). Pamphlets in English are aimed at Liberian refugees and settlers to the region and non-French-speaking tourists or officials, whereas those in French are widely distributed to villagers and village officials in and around Bossou, as well as French-speaking tourists and nongovernmental organizations (NGOs) and officials nationally. Some pamphlets were developed specifically to provide general information on the activities of KUPRI and Institut de Recherche Environnementale de Bossou (IREB), as well as to explain rules and guidelines to be respected by tourists wishing to see the chimpanzees of Bossou. One of the pamphlets, entitled "Jokro: The Death of an Infant Chimpanzee," as well as the complementary video of Jokro's story, also aims to educate people about the risks of disease transmission and about the strong bond that exists between chimpanzee mothers and their offspring.

T. Humle (✉)
School of Anthropology and Conservation, The Marlowe Building,
University of Kent, Canterbury CT2 7NR, UK
e-mail: t.humle@kent.ac.uk

T. Matsuzawa et al. (eds.), *The Chimpanzees of Bossou and Nimba*,
DOI 10.1007/978-4-431-53921-6_39, © Springer 2011

Fig. 38.1 Example of educational materials for environmental education. See http://www.green-passage.org/green-corridor/education/indexE.html for further details on materials and activities (© KUPRI)

Between June 2003 and March 2004, we ran a series of three environmental education campaigns across nine villages in the locality including Bossou, Thuo, Nion, Seringbara, Gbénémou, Gbah, Solméta I, Solméta II, and Thiassou. Each campaign involved public screenings of one or two videos and the distribution of pamphlets and badges (Fig. 38.2). These campaigns were ran by volunteer-trained local youth group members, including on occasion university students, members of a local NGO (UVODIZ: Union des Volontaires pour le Développement Integré de Zantompiézo), and staff from IREB.

The pamphlets developed in 2003 for these three campaigns provided basic information on chimpanzee socioecology, social and material intelligence, and life history. These pamphlets also explained the current status of chimpanzees in Africa and in Guinea and the national and international laws concerning wildlife and chimpanzees. Further, they importantly described behaviors to adopt or not to adopt when encountering wild chimpanzees. The pamphlets provided useful written supporting material to members of local youth groups, and UVODIZ and IREB staff, who led the three environmental education campaigns in the nine villages.

After a public video screening, the first campaign involved the presentation of a questionnaire to three age groups (young, adults, and elders). This questionnaire aimed to evaluate people's perceptions and understanding of chimpanzees and traditional and national laws pertaining to hunting, including what to do and not to do when you encounter a chimpanzee. Young students from youth groups were trained in presenting these questionnaires to the villagers and in recording answers and comments made. Three months later, the second campaign aimed to informally educate villagers about national laws concerning wildlife, chimpanzees, and hunting, as well as behaviors to adopt or avoid when faced with a chimpanzee. Presentation of the questionnaire was repeated 6 months later across all nine villages during the third and last environmental education campaign. This third campaign aimed to

Fig. 38.2 Public video screening for environmental education (photograph by Susana Carvalho)

evaluate the level of awareness and understanding retained by the people since the first two campaigns. The results revealed that villagers, especially the young, retained much of the information that was conveyed to them (Humle et al. 2004). Interestingly, during the first questionnaire round, the group of adults and elders performed significantly better than youth on questions pertaining to traditional hunting laws that disallow, for example, killing young mammals, mothers with young, and animals drinking from natural water sources. In conjunction with the strongly held taboo of killing or eating a selected set of animal taxa (up to five taxa in some families, a list dependent on family name), these laws perpetuate intrinsic conservation values that have long helped regulating of hunting and conserving wildlife populations. Although none of our education materials contained information on traditional hunting laws and none of the educators were asked to discuss these laws during any of the sessions, youth performed at 100% on these questions 9 months later, indicating that adults and elders in villages were prompted by the questions to convey this knowledge to the young people.

In addition, since 2004, we have been helping a local youth association, mainly comprised of university students native to the locality, coordinate a youth festival. This festival lasts one week in August and has now become an annual event. It includes a football tournament attended by youth from several surrounding villages. During that week, we help and encourage the youth association to develop and perform short theater sketches addressing themes such as hunting, poaching, and forest and chimpanzee conservation, as well other topics including acquired immunodeficiency syndrome (AIDS), hygiene, and literacy. These sketches are usually performed outdoors and are attended by villagers from Bossou and surrounding villages or towns. We also provide the organizers with educational materials for distribution.

38.2   Environmental Education in Schools

In 1999, we distributed a first small booklet in French entitled "Kikeimi" aimed at local schools. This booklet, illustrated with pictures, relates the fictional story of a young female chimpanzee born into the Bossou community until her emigration at adolescence into a neighboring community in the Nimba Mountains.

In 2004, our outreach to primary and secondary schools in the locality involved the distribution of a new bilingual (French–English) book. This book, entitled "Juru the Chimpanzee: Young and Curious," presents a fictional but factual story (illustrated with photographs and drawings) about a young female chimpanzee. This story touches on many different aspects of chimpanzee behavior, reflecting their social and material intelligence, their similarity to humans, and their strong social affiliations. It also provides examples of the intricate links between the animal and plant world. In addition, it emphasizes how habitat destruction has a wide-ranging impact on wildlife, as well as on humans.

Before the dissemination of this second book across 16 primary and secondary schools in the locality, we organized two workshops, attended by schoolteachers, headmasters, local NGO staff involved in environmental education (UVODIZ), and scientists and students on site involved in studies of chimpanzee behavior and wildlife ecology. During the workshops, ideas were exchanged on how best to make use of the book for educational purposes. Because no formal guidelines were provided, leeway was given to the imagination and the initiative of school-teachers, most of whom showed great enthusiasm for using this new tool to promote environmental education. Some copies of the book were also allotted to local and national organizations for use in their own conservation awareness campaigns or programs.

We have also conducted informal classroom interventions aimed at educating pupils and students about the environment and chimpanzees. Since 2006, such interventions, often animated by our local field assistants, have involved drawing competitions among high school students (Fig. 38.3). The theme of the drawings is chosen by the pupil and should illustrate what they have learned about threats to chimpanzees. Some of the drawings and future ones have been and will be used in developing our educational materials (Fig. 38.4). Members of KUPRI-International are also now carrying out interventions in schools and distributing educational materials to other regions, including Yealé in Côte d'Ivoire and Diecké in Guinea (Fig. 38.5).

Fig. 38.3 Local field assistant from Bossou animates an environmental education session in the primary school of Thuo, 4 km from the village of Bossou, using a PowerPoint projection (photograph by Susana Carvalho)

Fig. 38.4 Examples of drawing produced by secondary school pupils at Bossou (photograph by Celestin Niamy)

Fig. 38.5 Schoolteacher from a village near Diécké reading booklet to young children in the village (photograph by Susana Carvalho)

38.3   Signposts

On the front walls of IREB's buildings in Bossou, two mural paintings draw the attention of visitors and local people to the importance of protecting chimpanzees. The famous sacred hill of Gban, which represents the core of the chimpanzees' home range, is depicted in the background, illustrating the unique coexistence between humans and chimpanzees at Bossou (Fig. 38.6).

In 2005, a signpost painted by a Guinean artist and targeted at passing trucks, cars, motorcycles, and pedestrians was also strategically posted at the side of the largest and only Bossou road leading toward the Liberian border, which is frequently crossed by the chimpanzees (see Chap. 23). This signpost, which depicts a mother chimpanzee carrying an infant crossing the road, says "Laissez nous passer" ("Let us cross") (Fig. 38.7). In addition, 22 signposts were manufactured in November 2003. Several of the larger signposts were posted at the periphery of towns in the prefecture of Lola and N'Zo. Smaller ones were positioned alongside smaller roads and paths along the forest edge to help demarcate the boundaries of strictly protected core areas of the Nimba Mountains Biosphere Reserve, which includes the Nimba Mountains and the forests of Déré (see Chaps. 29 and 39) and Bossou. These signposts stress wildlife protection and the unlawfulness of hunting and poaching. Depending on their emplacement, these signposts are aimed at either passing vehicles or local pedestrians frequenting these areas.

Fig. 38.6 Mural painting on the facade of IREB (photograph by Tatyana Humle)

Fig. 38.7 Signpost placed strategically on Bossou road leading to Liberia. This road is frequently crossed by Bossou chimpanzees (see Chap. 23) (photograph by Tatyana Humle)

Fig. 38.8 School of Seringbara built in 2001 by KUPRI and villagers (photograph by Tatyana Humle)

38.4 Contributions to Local Development: Schools, Health, and Ecotourism

Our school outreach program has also involved the construction of a school in Yealé in Côte d'Ivoire (see Chap. 27) and Seringbara in Guinea (see Chap. 28) and yearly material donations to schools in the locality, including benches, tables, books, pens, and chalk (Fig. 38.8). In addition, KUPRI has helped fund the construction of a well

Fig. 38.9 Four latrines were constructed in the village of Koronhan in the region of Diécké with the support of the Brisitish Embassy in Guinea and in collaboration with KUPRI-International. In the photo on the left, Susana Carvalho is standing in front of one of the newly built latrines with the first elder of the village (on her right) and the chief of the village (on her left). (photograph by Susana Carvalho)

in the village of Seringbara and a health center in Bossou, as well as several latrines in schools and districts of the village located at the forest edge (see Chap. 32). In 2009, with the support of the British Embassy in Guinea, four latrines were also constructed in the village of Koronhan in the region of Diécké (Fig. 38.9). This village has been the focus of etho-archeological studies by Susana Carvalho (see Chaps. 15 and 30).

Finally, in 2004, we helped set up a committee of co-management of ecotourism revenues between the village and IREB. This committee's role is to set fees and decide on the allocation of the incoming funds, thus ensuring that ecotourism revenues also benefit the village.

38.5   Conclusions and Lessons Learnt

Our environmental education initiatives have taught us several useful lessons. Adolescents and young adults are often more receptive than adults to awareness-raising campaigns. However, we learned that, if prompted, older adults can play a significant role in imparting traditional conservation habits to youth. Before initiating a village-based environmental campaign, it is therefore vital to identify a priori and understand local traditional laws or habits that are directly or indirectly relevant to the preservation of wildlife and flora. Finally, we are convinced that community involvement and a multifaceted approach can effectively sustain environmental education programs or conservation activities over the long term.

Since KUPRI initiated its environmental education efforts in combination with local aid to villages and schools, good progress has been made in the Nimba region to increase awareness of villagers, refugees, and settlers to the region about the plight of chimpanzee and forest conservation. Through these campaigns, alongside the Green Corridor Project, we have been able to make contact with people of all ages across several local villages and towns. This connection has been achieved thanks to a close collaboration with the local people, especially youth groups and students, and staff from local NGOs and IREB. Through the schoolbooks and school interventions, we have also been able to specifically target children, adolescents, and young adults and teachers. The Green Corridor Project (see Chap. 37) has also inspired young people in the locality to take conservation issues seriously and to organize their own initiatives. Progress achieved especially since 2003 is encouraging, but our efforts need to be sustained if we are to effectively and positively change people's perceptions, attitudes, and behavior towards conservation matters and the plight of chimpanzees in the area and nationwide.

Acknowledgments I wish to acknowledge the contributions of all KUPRI international members, especially Tetsuro Matsuzawa and Yukimaru Sugyama for setting the example, and Gen Yamakoshi, Dora Biro, Susana Carvalho, Nicolas Granier, Misato Hayashi, Kim Hockings, Kathelijne Koops, Laura Martinez, Gaku Ohashi, and Claudia Sousa for their engagement and help. We thank the staff of IREB, our local assistants, and the staff of UVODIZ, especially Soh Pletah Bonimy, and all the volunteer students and youths for their help and assistance. We would also like to acknowledge the support of all traditional and governmental authorities and women and youth associations that have helped us make environmental education and conservation initiatives a reality in and around Bossou. Finally, we are in particular grateful to GRASP-Japan (Great Apes Survival Project), the U.S. Fish and Wildlife Services (Great Apes Conservation Fund), Conservation International (Primate Action Fund), the British Embassy in Guinea, and Houston Zoo for their financial support.

Chapter 39
Conservation Issues in the Nimba Mountains

Nicolas Granier and Laura Martinez

39.1 A Tri-National Biogeographic and Anthropological Entity

The Nimba Mountains exhibit a particularly rich and unique biodiversity, which results from a highly specific conjunction of multiple biological factors and distinctive geographic, geological, climatological, and ecological patterns. Their tri-national location at the crossroads of several ethnic influences and migratory fluxes also contributes to their uniqueness. Yet, the Nimba Mountains can be defined as much by their intrinsic diversity as by their global homogeneity.

39.1.1 Biogeomorphology

The Nimba Mountains (7°25′–7°42′ N and 8°20′–8°40′ W) peak at 1,752 m, constituting the second highest relief in West Africa. The massif forms a 40-km-long barrier oriented northeast–southwest, which marks the border between Guinea, Liberia, and Côte d'Ivoire (Fig. 39.1). Rising abruptly more than 1,000 m above the surrounding plains, it presents a thin crest with steep and rocky slopes, which exceed 75° inclination in some places. In its particular location at the crossroads of three climatic influences (Equatorial-Guinean, Libero-Guinean, and Sub-Sudanian) and of two major tropical winds (the monsoon, a humid wind blowing from the south, and the dry trade wind, or Harmattan, blowing from the north), this relief constitutes an important climatic barrier. Pluviometry varies from 1,500 to 4,000 mm³ across areas and years, with generally more rain at the highest altitudes

N. Granier (✉)
Behavioral Biology Unit, Department of Environmental Sciences, University of Liège,
Quai van Beneden, 22, 4020 Liège, Belgium
e-mail: nicogranier@yahoo.fr

L. Martinez
Research Institute of EcoScience, Ewha Womans University, B 365, Science Building,
11-1 Daehyun-Dong, Seodaemun-Gu, Seoul 120-750, Republic of Korea
e-mail: lauramatlan@gmail.com

T. Matsuzawa et al. (eds.), *The Chimpanzees of Bossou and Nimba*,
DOI 10.1007/978-4-431-53921-6_40, © Springer 2011

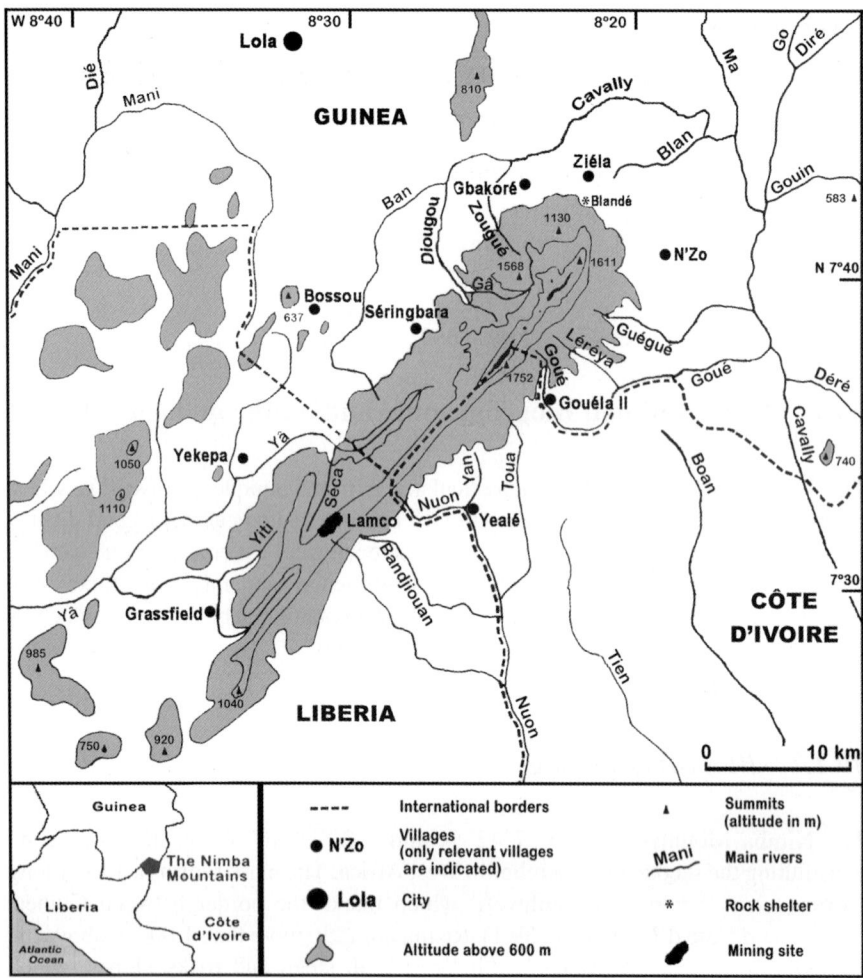

Fig. 39.1 Map of the Nimba Mountains along the tri-national border between Guinea, Côte d'Ivoire, and Liberia, West Africa. This map shows the main geomorphological patterns and hydrographic network of the Nimba region, as well as important human settlements surrounding the massif

and in the southeast (Lamotte 1998a; Soropogui et al. 2008). These original climatological features, added to the steepness of the relief, the complexity of the geological structure, and peculiar edaphic characteristics, have favored the emergence of an important diversity of microclimates and ecological niches populated by a highly diverse and endemic wildlife. Almost all vegetation types of the West African region are represented, which led Schnell (1998) to describe the Nimba Mountains as a "West African crossroads of floras."

In the northeastern (Guinean) end, the massif's top is covered by altitude grasslands from 800 m in elevation. This ecosystem, which harbors a highly endemic orophyte vegetation composed of Poaceae, terrestrial orchids, heathers, and Euphorbiaceae, is an essential component of the massif's uniqueness (Adam 1971–1983). The slopes and foothills of the relief are covered by altitude and lowland Guineo-Equatorial rainforest, and the surrounding plain presents lowland forest with sprinkled stretches of savannas. A fifth habitat type described by Schnell (1998) is the secondary vegetation, which consists of a low tree density habitat with ground heliotropic vegetal species. Mainly found at forest edges, it has a growing importance because of human activities and uncontrolled bushfires. Toward the southwestern part of the mountain range (Liberian side), the ridge progressively descends from 1,752 m to 1,000 m, and the forest rises over the crest to cover the entire massif. The floral and habitat type richness is accompanied by a particularly diverse and important fauna, which has been among the most studied in West Africa (Lamotte and Roy 2003). An important and regularly developed hydrographic network drains the Nimba Mountains with deep and steep ravines shaped by watercourses (Lamotte and Rougerie 1998). The numerous streams originating in the massif feed three main rivers: the Cavally, flowing southward into Côte d'Ivoire; the Ya, flowing southeastward into Liberia; and the Nuon, flowing south into Liberia (see Fig. 39.1). Finally, the substratum of the relief is composed of old granitic and gneiss formations, superimposed with layers of green schist inlaid with highly concentrated and pure iron ore (Pascual 1988).

39.1.2 Cultural Influences

The oldest traces of human settlements ever discovered around the Nimba Mountains were found in the northern-end foothill of the massif, in a rock shelter named the Blandé Cave (see Fig. 39.1). The site was explored and studied from 1949 by French anthropologists (Holas 1952b; Mauny and Holas 1953), who collected more than 2,000 pottery fragments and 100 lithic pieces such as rough flints (hatchets, knives, and points). According to them, occupation of the Blandé Cave lasted approximately from the sixth century BC to 1000 AD, and resulted from migratory influxes starting in the Sahara and the Sudan. Subsequently, knowledge of occupancy of the Nimba region from these dates relies more on oral tradition than on archeological data, and the first traces recorded after this period are reported from about 1750. Nowadays, the three main ethnic groups settled around the Nimba Mountains are the Manon, Kono, and Yakuba. An important feature that structures the social system in these animist populations is their clanistic organization. Each clan or family is composed of a group of individuals who possess a common ancestor and follow the same prohibitions. The most common are the food prohibitions, which concern proscribed animals or plants called totems. There is an intricate mixing between clans, some of them being absorbed or assimilated by others, with alliances being formed, all of which result in a highly intertwined social network.

   According to Germain (1984), the early creation legends of the Manon group say that the oldest unit was formed by the alliance of two clans: the Nia and the Ma. After the formation of this first nucleus in the Diécké area, the community was dispersed. One part of the Ma clan (whose totems are the chimpanzee, the goat, and the snail) emigrated to the region of Man in Côte d'Ivoire, and then moved to settle around N'Zo. The rest of the community walked around the Nimba Mountains to the plain located north (Bossou region), passing either by the east (Vépo region) or by the west (today constituting Liberian territories). Later, the vast Mandingo islamization movements of the seventeenth century forced the Manon people to limit their occupation to the Diécké and the Nimba Mountains regions. Mixing between the already established forest populations and the newly arrived Mandingos led to the creation of the Kono ethnic group. An alliance was later concluded between the Manon and Kono groups, reinforcing the cultural and genetic interconnections. In sum, although intragroup traditional characteristics have been preserved throughout generations, complex migratory fluxes have led to a continuous intergroup mixing.

39.2 Conservation Keystones

39.2.1 Conservation Status of the Nimba Mountains

The unique biological characteristics of the Nimba Mountains led to the early protection of their Guinean and Ivorian parts in June 1944. In contrast to the four national parks of Guinea (which were managed by local forestry administration), the Mount Nimba Full Nature Reserve (Réserve Naturelle Intégrale du Mont Nimba) was placed under scientific management of the Museum National d'Histoire Naturelle (MNHN, Paris) and the Institut Français d'Afrique Noire (IFAN, Dakar). The latter was in charge of scientific and anthropological studies in the former West African French Territories (Lamotte et al. 2003; Brugière and Kormos 2009). After their independence, the administrations of both Guinea and Côte d'Ivoire maintained the massif under protective status in their legislation. Table 39.1 shows chronological landmarks of the Nimba Mountains in the three countries over the past 70 years.

In 1980, the Guinean side of Nimba was classified as a Biosphere Reserve by the Man and Biosphere (MAB) Program of UNESCO. In 1981–1982, the Guinean part (8,520 ha) and the Ivorian part (6,482 ha) were established as Strict Nature Reserves by the International Union for Conservation of Nature (IUCN) and as a Natural World Heritage Site (NWHS) by the WHS Program of UNESCO (Hartley et al. 2008; WHS-UNESCO 2008). The Nimba Mountains Strict Nature Reserve is assigned to IUCN category Ia, which corresponds to protected areas "managed mainly for science, possessing some outstanding or representative ecosystems, geological or physiological features and/or species, available primarily for scientific research and/or environmental monitoring" (IUCN 2009). In 1992, the Guinean part of the Nimba Mountains was labeled an "Endangered World Heritage Site" because of potential mining activity and the increasing human pressure caused by successive waves of refugees from Liberia and Sierra Leone. In 1993, the Biosphere

Table 39.1 Chronological landmarks of the Nimba Mountains in Guinea, Côte d'Ivoire, and Liberia

Year	Guinea	Côte d'Ivoire	Liberia
1939	First visit of a scientific team to NM		
1942	Scientific studies of NM begin		
1944	Strict Nature Reserve		
1963			Mining activities launched
1976	Study of Bossou chimpanzees begins		
1980	NMBR		
1981	NWHS		
1982		NWHS	
1992	Endangered NWHS		Armed conflict begins
1993	NMBR enlarged Mining concession	Study of Nimba chimpanzees begins	
2001	Armed conflict ends Tri-national Program for the Protection of NM		
2002		Armed conflict begins	
2003			Nature Reserve
2004	PCBMN begins		
2006	Mining activities launched (drilling)		

NM Nimba Mountains, *NMBR* Nimba Mountains Biosphere Reserve, *NWHS* Natural World Heritage Site, *PCBMN* Program for Biodiversity Conservation of the Nimba Mountains

Reserve was enlarged from 12,700 to 22,000 ha by inclusion of two additional core areas: the Bossou hills and Déré Forest, with an extended buffer zone (Fig. 39.2). However, no Guinean legal text ratified the new protected status of these two core areas, leading to difficulties in protection and management. The same year, an enclave of 1,550 ha was withdrawn from the core area of the Biosphere Reserve in the Gbakoré region (northeast of the massif), to become an iron-ore mining concession (Debonnet and Collin 2007). The civil war that began in 2002 in Côte d'Ivoire has induced political instability, which put a complete stop to environmental research and management activities in the country. This situation facilitated many kinds of illicit activities such as poaching, deforestation, and settlement of rebel forces. Today the armed conflict is over, and the government shows a renewed interest in environmental issues.

The Liberian part of Nimba has suffered from extensive logging activities in the East Nimba National Forest, Grassfield region (Verschuren 1983), and from iron-ore mining in the Yekepa area. Finally, in 2003, the Liberian authorities showed their willingness to officially protect the Nimba Mountains by publishing a legal act establishing the Nimba Nature Reserve (13,500 ha) and stipulating their wish to include it in the NWHS complex. This reserve incorporates the former Nimba East National Forest extending up to the border with Guinea and Côte d'Ivoire (Act for the Establishment of the East Nimba Nature Reserve 2003; Beamont and Suter 2004).

Taken together, the three countries cover more than 31,000 ha of protected areas that would greatly benefit from being considered as a complete and single ecological unit.

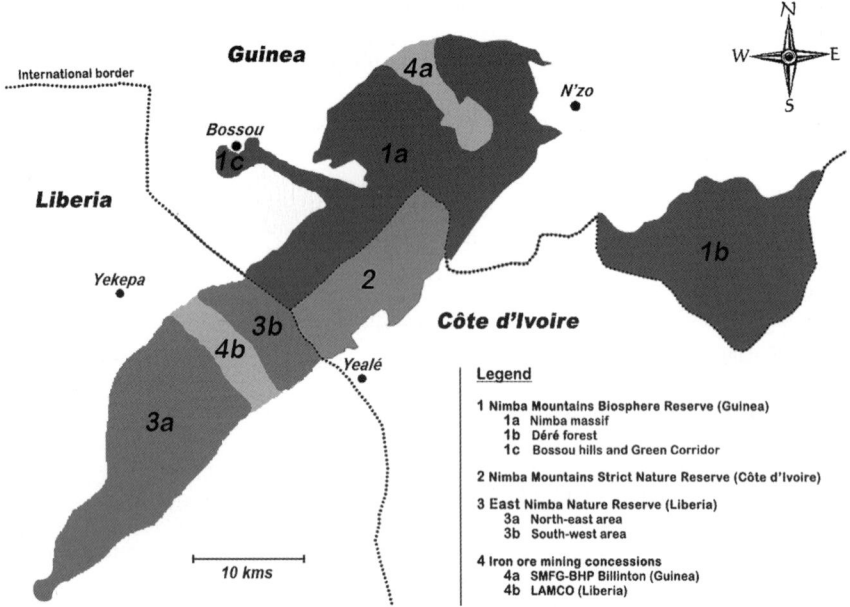

Fig. 39.2 Protected areas in the Nimba Mountains region. This map shows the mosaic of areas with different status of protection across the countries of Guinea, Côte d'Ivoire, and Liberia. Despite an ecological and geographic continuum, the conservation status is "discontinuous"

39.2.2 Tri-National Perspective of Conservation

Despite their formal conservancy status, the Nimba Mountains have suffered from a severe lack of law enforcement, rational utilization of forest resources, and global biodiversity management. As a consequence, the ecological entity of the Nimba Mountains is split into a mosaic of areas exhibiting different levels of preservation and conservation status (see Fig. 39.2). Although large areas of well-preserved forest are still present in the three countries (e.g., areas 1a, 2, 3b in Fig. 39.2), other areas have been severely damaged by logging exploitation and subsequent habitat destruction (areas 1b, 3a; see Fig. 39.2). In the same way, although long-term conservation activities such as the Kyoto University Primate Research Institute's (KUPRI) efforts in the Bossou hills and the Green Corridor Project (see Chap. 37) have permitted the protection, reforestation, and connection of conservation key areas (area 1c), iron-ore mining sites and their surroundings have already been heavily damaged in Liberia, and we fear similar future developments around the Guinean mining enclave (areas 4a, 4b).

Aware of this problematic situation, concerned authorities from the three countries initiated in 2001 a "Tri-national Program for the Protection of the Nimba Mountains." This program has two objectives. The first is to elaborate a legal context that would ensure consistency in the administrative rules and protective status of the

Guinean, Liberian, and Ivorian parts of the Nimba Mountains. The second objective is to increase and update the scientific knowledge on Nimba wildlife and environment, by monitoring climate, hydrometry, fauna, and flora. Two meetings were organized in 2001 and 2002 to launch administrative and field activities in the three countries (Touré and Suter 2002), but the armed conflict that started in Côte d'Ivoire has frozen the whole initiative. It is to be hoped that the newly peaceful situation in the subregion would favor the restoration of this tri-national initiative.

39.3   Contribution of Scientific Studies

39.3.1   History of Scientific Studies in the Nimba Mountains

The Nimba Mountains has been one of the most studied sites in West Africa in terms of the number of scientific investigations (Lamotte and Rougerie 1998). In 1939, a first scientific mission explored the Nimba Mountains area on the border between Guinea and Côte d'Ivoire and reported an unusual level of unknown species as well as a spectacular landscape (Brugière and Kormos 2009). To facilitate further investigations, a scientific station was established in the early 1940s in Ziéla (northern end of the massif; see Fig. 39.1). Until 1957, a long series of scientific studies were carried out from there by Maxime Lamotte, Roger Roy, and many IFAN scientists on geology, geomorphology, fauna and flora, and climate (Lamotte et al. 2003). Later, another research station was built by miners in Grassfield, Liberia (see Fig. 39.1), catering for a new study area, from which, among others, an extended floral description of Nimba was completed by Adam (1971–1983).

Since the 1980s, a number of scientific missions were supported by the UNESCO, mainly to assess anthropic pressures affecting Nimba biodiversity and to propose suitable conservation recommendations (Debonnet and Collin 2007). Different NGOs, organizations, and universities have also investigated this biodiversity. For example, the Royal Botanic Garden of Kew conducted botanical studies (Hawthorne and Jongkind 2006), Würzburg University studies on amphibians (Hillers et al. 2008), MNHN studies on carnivores (Gaubert et al. 2002), and BirdLife International ornithological surveys (Borrow and Demey 2001). In 1942, Maxime Lamotte highlighted the research potential of Bossou as a field site for the study of wild chimpanzees (Kortlandt 1986). Adrian Kortlandt visited Bossou briefly several times during the early 1960s and was the first primatologist to conduct research on this chimpanzee community (Kortlandt 1986, 1989; Kortlandt and Holzhaus 1987). It was not until 1976 that Yukimaru Sugiyama from KUPRI initiated the long-term research on the wild chimpanzees of Bossou, and Tetsuro Matsuzawa extended the research on chimpanzees to the Nimba Mountains in 1993 (Table 39.1; see Chap. 1).

Today, the recent increase in scientific studies designed and conducted by Guinean, Ivorian, and Liberian researchers and students, sometimes in collaboration with internationally recognized researchers, heralds a new era in Nimba biodiversity studies (e.g., Kourouma et al. 2008).

39.3.2   From Research to Conservation

Results of scientific studies can lead to the emergence of conservation measures. By providing ecological data and designing tools to evaluate biodiversity and endemism, systematic scientific accounts can constitute the best ally for setting up concrete and adaptable programs of biodiversity preservation.

The establishment of biodiversity hotspots is a good example of the great worldwide impact that can be reached through this kind of scientific work (Myers et al. 2000; Fa and Funk 2007). The Nimba Mountains are actually located in one of these hotspots, called the Guinean Forests of West Africa (Conservation International 2008). In addition, identification of more than 2,000 plant species including 16 endemics has contributed to the classification of the Nimba Mountains as a center of plant diversity (Hartley et al. 2008). In the same way, following ornithologists' discoveries the site was classified as an Important Bird Area by BirdLife International (BirdLife 2008). A further example is provided by the ongoing long-term studies on chimpanzees, which have led to the classification of the Nimba Mountains, together with Bossou and Déré Forests in Guinea and Tiapleu Forest in Côte d'Ivoire, as one of the six exceptionally important priority areas for the conservation of West African chimpanzees in the IUCN/SSC/PSG-CI Action Plan (Kormos et al. 2003a). Finally, the IUCN Red List of Endangered Species is another example of conservation statements based on the long-term efforts undertaken by scientists. In a bibliographical study, Rondeau and Lebbie (2007) listed 3,384 animal species in the Guinean ecosystems of the Nimba Mountains. More than 500 animal species new to science were discovered across Nimba, including 200 endemics, among which 13 species of mammals, birds, and amphibians are classified as Critically Endangered (CR), Endangered (EN), or Vulnerable (VU) in the IUCN Red List (Hartley et al. 2008; IUCN 2009). Endangered flagship species such as the viviparous Nimba toad (*Nectophrynoides occidentalis*–CR), the only viviparous bufonid known in the world, the Mount Nimba otter shrew (*Micropotamogale lamottei*–EN), or the West African chimpanzee (*Pan troglodytes verus*–EN), by attracting focus of public opinion, can play important roles in fundraising and become weighty arguments in favor of Nimba wildlife preservation.

39.4   Threats to Nimba Chimpanzees

Chimpanzee conservation cannot be dissociated from natural habitats and biodiversity protection. In Bossou, despite the long-term traditional and scientific protection, chimpanzees have become increasingly isolated from neighboring communities and are nowadays threatened by an intricate web of pressures (see Chaps. 37 and 40). In Nimba, the two main types of threats exerted on biodiversity are anthropic pressures, characterized by a domestic and relatively limited environmental exploitation, and industrial pressures, characterized by large-scale use of the environment. Both these threats lead to habitat modifications, which jeopardize chimpanzee survival.

39.4.1 Anthropic Pressures

Forest habitats supply an important part of the local people's domestic needs. Since the hunter-gatherer era, forester ethnic groups have been using forest by-products such as fruits, leaves, seeds, roots, and bark as food, medicine, or construction materials. In the same way, the fauna of the forest has always constituted the main source of protein. More recently, small-scale agriculture has enlarged food resources by providing a staple diet based on cereals and tubers, especially rice and cassava. This way of exploiting natural resources has apparently been sustainable over centuries, but nowadays, the balance has been upset: the natural habitat cannot regenerate rapidly enough to continue fulfilling the needs of a growing resident population. The demographic overgrowth and increase in human densities in the Nimba region have been aggravated by the successive armed conflicts that occurred during the course of the past 20 years in Sierra Leone, Liberia, and Côte d'Ivoire. These conflicts have fostered important instabilities in the countries of the Mano River Union (MRU; i.e., alliance and economical collaboration between Sierra Leone, Liberia, and Guinea, recently joined by Côte d'Ivoire), which pushed their residents to settle in Guinea.

As a result of the demographic increase, agricultural yield has been intensified to keep up with local needs. The slash-and-burn agricultural technique, used in combination with a practice of fallowing land, always requires more arable land, pushing communities to deforest new parcels every year. As a consequence, the Nimba forests are becoming more and more isolated, surrounded by a fragmented habitat composed of deteriorated patches of forest, perennial cultures (palm tree, coffee, cocoa, pineapple, and banana), seasonal cultures (rain-fed rice, cassava, tarot, yam, peanut, and corn), and fallow lands covered by grasses and shrub species.

Bushfires lit by villagers are among the most threatening pressures for Nimba ecosystems. People are accustomed to burning the herbaceous vegetation of savannas, altitude grasslands, and secondary forests for purposes related to cultivation, poaching, clearing of trails, and the promotion of new growth for grazing cattle. Fires are central to the ecology of African tropical savannas and are commonly used as a tool for managing the tree–grass balance in protected areas (Clerici 2006). However, the excessive and uncontrolled use of fire occurring in Nimba has a strong negative impact on habitats. Every year during the dry season, fires lit in herbaceous ecosystems penetrate into the forest's driest edges, leaving partially burned trees that finally fall down, creating large holes in the canopy, favoring in this way the development of ground heliotropic vegetation. Natural regeneration of the forest is then particularly long and difficult.

Poaching and excessive hunting with guns and snares are reported across the whole Nimba Mountains Biosphere Reserve, despite international and national regulations. Hunting appears to be rather unselective, and chimpanzee poaching is sometimes reported, although it seems to have greatly declined thanks to the long-term conservation efforts undertaken by KUPRI in the area (Granier, in preparation). Several reasons can lead to the killing of chimpanzees. First, their meat is eaten, although traditional taboos in some ethnic groups, such as among the Manon people, appear to limit its consumption. Second, in certain ethnic groups, chimpanzee meat is believed to hold medicinal virtues, and prized parts of the body can be sold at high prices. Third, seasonality in food availability, fragmentation, and reduction of habitat can cause chimpanzees to feed on human cultivars (Hockings and Humle 2009). Crop-raiding may drive cultivators to kill chimpanzees, as has happened in other regions of Guinea (Granier and Martinez 2004a, b). Fourth, nonselective snaring and trapping could also represent a fatal danger for chimpanzees owing to injuries that can cause infection and gangrene. One case of death from a snare was recorded in the Taï Forest (Côte d'Ivoire) during the past 29 years, and four cases are known from both snare and trap in the Budongo Forest (Uganda) during the past 18 years (Boesch and Boesch-Achermann 2000; Reynolds 2005b). Finally, the pet trade also represents a menace, because capturing a baby chimpanzee necessitates killing members of the community who will protect their group. This menace can potentially cause much more severe damage to the population than poaching.

39.4.2 Mining and Logging

Industrial views on iron exploitation have been the sword of Damocles hanging over Nimba biodiversity since the 1960s. Today we know that mining can indeed have significant local and regional negative impacts on ecosystems (habitat loss and quality decrease, waste discharge, pollution of rivers and soil, modifications of the hydrographic regime and network). In fact, an iron-mining project operated between 1963 and 1992 in the site of Yéképa, in the Liberian Nimba, just 10 km southeast of Bossou (Verschuren 1983) (see Fig. 39.1). Exploitation was under control of a consortium called Liberian American Swedish Minerals Company (LAMCO). Early in 2007, the exploitation lease was bought out by another multinational mining company, ArcelorMittal. In addition, the Guinean government signed in 2003 an agreement for iron exploitation in the Guinean side of the Nimba Mountains with EuroNimba, a consortium held by three companies: BHP Billiton, Newmont, and Areva. This consortium, in association with the Guinean government, has created the SMFG company (Société des Mines de Fer de Guinée), which operates the mining site. Aware of the irreversible negative impacts of LAMCO mining and constraint by international lobbies, the SMFG program has undertaken an environmental commitment, which includes long-term impact studies of mining on wildlife, hydrographic regime, and climate.

Logging also exists in the Nimba region. The example of Déré forest, which was logged in 1999–2000, shows how commercial exploitation of timber can be destructive even over a short timescale (see Chap. 29). In addition to deforestation and direct collateral damage caused by falling trees, the construction of infrastructure and roads in the forest has favored human settlement for cultivation and subsequently increased poaching. Similarly, the Liberian part of the Nimba Mountains (especially the southern and western ridges of the East Nimba Nature Reserve) has suffered from extensive timber exploitation by a company named Nimbaco, from the 1970s until at least 1983 (Verschuren 1983). Although exploitation has now stopped, this part of the forest is severely deteriorated and may contain lower faunal abundance than the northern area of the reserve (see Fig. 39.2, areas 3a, 3b).

The highly recurrent demands of communities for local development show how much the financial spinoffs of industrial activities cannot be ignored or simply rejected. Industrial companies judiciously invest in infrastructures (roads, schools, hospitals), create local employment, and favor local/national dynamism of the economy. Even if such activities are probably not sustainable and are certainly harmful to the environment, the majority of locals see the arrival of an industrial activity as a great opportunity. Thus, logging and mining companies easily benefit from support of local populations.

39.5   Toward Integrated Conservation

Many poor countries today are trapped in a cycle of poverty and environmental degradation (Ehmke and Shogren 2008). To turn the corner, they often concentrate efforts on the poverty problem to the detriment of environmental issues. The biggest challenge for the conservation of the Nimba Mountains undoubtedly consists of the trade-off between biodiversity preservation and local development. A strong argument that could be put forward to meet this challenge is the potential long-term economic value of such an exceptional and unique natural heritage, if rigorously protected (Verschuren 1983; Debonnet and Collin 2007). However, threats leading to its degradation, such as the local anthropic pressures and industrial activities, perpetually keep this perspective at a distance. Consequently, these threats should be considered as inescapable and difficult-to-change components of the complex matter of natural resource management. In Nimba, the problem is becoming even more complicated because of the absolute necessity of a tri-national harmonization of approaches. In this context, the inability to efficiently carry out conservation actions without a collaborative and integrated multipartite approach is salient. Thus, we propose some suggestions for a successful and sustainable conservation program of the Nimba Mountains.

Local populations are undoubtedly the key actors in the sustainable protection of their heritage (Bajracharya et al. 2005; Danielsen et al. 2005). Conservation programs may not be successful in the long run without integrating communities in a central position. Additional anthropological investigations are necessary to obtain a better understanding of local concepts of natural heritage and human–wildlife coexistence, as well as the ongoing changes in their beliefs consequent to environmental evolution. Progression of customary and legal environmental practices and laws also need to be more carefully examined and must be better taken into account in conservation programs.

Permanent collaboration between conservationists and researchers appears essential to enhance suitable conservation actions. All conservation stakeholders should consider the most updated outcomes from scientific studies. To facilitate this process, researchers should be keen to present their work in easily comprehensible formats and to ensure that the potential conservation actions are in agreement with their scientific findings. From the same perspective, researchers should keep in mind conservation issues when designing and conducting their studies. For example, the numerous threats jeopardizing the Nimba Mountains chimpanzees and the almost nonexistent management of the Reserve may lead researchers to consider very carefully the benefits of habituating apes to the presence of human observers, because it is very difficult to guarantee proper protection to wildlife over the long term (Jenkins 2008; Köndgen et al. 2008).

Bio-monitoring programs focusing on Nimba wildlife should be developed on a long-term basis. A perennial follow-up of flagship and good bio-indicator species appears to be the most appropriate strategy as it would provide information on trends in global population and ecosystems evolution, which are still insufficiently known (Hortal and Lobo 2005).

The Nimba Mountains form an essential water catchment area, which has great regional importance because it contains the headwaters of three major river systems flowing into the Atlantic Ocean: the Cavally, Ya, and Nuon Rivers. These rivers irrigate vast zones of Guinea, Liberia, and Côte d'Ivoire, where they constitute important water resources for domestic, agricultural, and industrial needs. Thus, degradation of the Nimba ecosystems, by affecting the headwaters of these rivers, would have a long-term negative impact on their entire hydrographic network, with sanitary/economic consequences in the three countries. Indeed, the sustained protection of the Nimba Mountains has important regional correlates.

As a conclusion, the biggest challenge for the sustainable conservation of the Nimba Mountains is to integrate all these issues in a tri-national coherent initiative, which should be thoughtfully designed and strictly enforced. Such an attempt is embodied by a newly launched biodiversity conservation program, which unfortunately focuses only on the Guinean part of the Biosphere Reserve (Programme de Conservation de la Biodiversité des Monts Nimba). Funded by the United Nations Environment Program and the Global Environment Facility, this program was initiated in 2004 for a 9-year period with the objective of conciliating research, conservation, mining activities, and local development. Our biggest hope is to see this initiative backed, durably reinforced, and enlarged to the entire massif so as to efficiently tackle the most critical conservation issues and to ensure a sustainable future for the habitats, wildlife, chimpanzees, and humans of the Nimba Mountains.

Acknowledgments We wish to express our sincere gratitude to Tetsuro Matsuzawa for his support and advice. We are grateful to Dora Biro and Andrew MacIntosh for their help with English corrections to this chapter. The research was financially supported by grants from the Japanese Ministry of Education, Culture, Sports, Science and Technology, the Japanese Ministry of Environment, and the Japanese Society for the Promotion of Science to T. Matsuzawa, and by a grant from IUCN/SSC Primate Specialist Group, Great Ape Emergency Conservation Action Fund to N. Granier.

Chapter 40
Chimpanzees in Guinea and in West Africa

Tatyana Humle and Rebecca Kormos

40.1 Introduction

40.1.1 General Conservation Status of Chimpanzees

Many species of large mammals throughout Africa are threatened with extinction because of the destruction of their habitat and unsustainable levels of hunting and capture. The common chimpanzee (*Pan troglodytes*) is no exception (IUCN 2009). Taxonomists generally agree that *P. troglodytes* can be divided into four subspecies that exhibit mutually exclusive geographic ranges: (1) the eastern chimpanzee (*P. t. schweinfurthii*), living in Rwanda, Tanzania, Uganda, and the Democratic Republic of Congo, with relict populations in Sudan and Burundi; (2) the central chimpanzee (*P. t. troglodytes*), living in Angola, Central African Republic, Equatorial Guinea, Cameroon, the Democratic Republic of Congo, and Gabon; (3) the Nigerian chimpanzee (*P. t. ellioti*; previously referred to as *P. t. vellerosus*) in eastern Nigeria and western Cameroon north of the Sanaga River; and (4) the western chimpanzee (*P. t. verus*) (IUCN 2009). The Nigerian subspecies was only recently recognized as a separate subspecies (Gonder et al. 1997) and, to date, we still know little of its behavior and ecology (Fowler and Sommer 2007; Schöning et al. 2007). The subspecies *P. t. verus* and *P. t. ellioti* have the fewest individuals (IUCN 2009).

Wild chimpanzees (*P. troglodytes*) inhabit primarily evergreen forest, but some populations also persist in deciduous woodland and grassland biotopes interspersed with gallery forest. Most wild chimpanzees live between 13° N and 7° S of the equator. Their populations have declined by more than 66% in the past 30 years, from around 600,000 to fewer than 200,000 individuals (Kormos et al. 2003a). This tendency is all the more concerning as chimpanzees are extremely vulnerable to

T. Humle (✉)
School of Anthropology and Conservation, The Marlowe Building, University of Kent, Canterbury CT2 7NR, UK
e-mail: t.humle@kent.ac.uk

R. Kormos
1170 Grizzly Peak Blvd, Berkeley, CA 94708, USA

T. Matsuzawa et al. (eds.), *The Chimpanzees of Bossou and Nimba*,
DOI 10.1007/978-4-431-53921-6_41, © Springer 2011

demographic decline and are unable to recover as rapidly as other species. Female chimpanzees indeed typically only give birth to a single offspring every 5–6 years (Goodall 1986; Sugiyama 2004). This severe population decline is additionally all the more alarming as chimpanzees represent our evolutionary cousins, with whom we shared a common ancestor about 5–6 million years ago (Goodman et al. 1998) and with whom we share today many aspects of behavior, physiology, and genetics.

40.1.2 Status of Pan troglodytes verus

The Upper Guinea forests, which extend from southern Guinea into Sierra Leone and eastward into through Liberia, Côte d'Ivoire, and Ghana into Western Togo, are home to the majority of *P. t. verus* (Sayer et al. 1992; Kormos et al. 2003a). Although these forests contain a high diversity of mammals and are among the most biologically diverse in the world, they are unfortunately also among the most threatened (Myers et al. 2000). The forests of Upper Guinea have already been reduced to 15% of their original size, and much of the remaining forest is severely fragmented (Mittermeier et al. 1999). The largest remaining blocks of tropical rainforest that harbor chimpanzees in West Africa include the Taï National Park in Côte d'Ivoire, the Grebo Forest and the Sapo National Park in Liberia, and the Ziama and Diécké Forests in Guinea and the Nimba Mountains at the crossroad of the three countries (see Chaps. 27–31 for further details on some of these forests).

P. t. verus is the second most endangered subspecies of the four recognized to date. This subspecies is patchily distributed and numbers between 21,300 and 55,600 individuals (Kormos et al. 2003a). It is very rare or close to extinction in four West African countries: Burkina Faso, Ghana, Guinea-Bissau, and Senegal (IUCN 2009). It has already disappeared from the wild in Togo (Campbell and Radley 2006) and the Gambia. The subspecies is possibly also now extinct in Benin (IUCN 2009). *P. t. verus*, therefore, survives mainly in Côte d'Ivoire, Guinea, Liberia, Mali, and Sierra Leone (IUCN 2009).

40.2 Current and Future Threats to Chimpanzees

40.2.1 Habitat Destruction and Degradation

Deforestation across West Africa has severely reduced suitable habitats for chimpanzees. It is estimated that more than 80% of the region's original forest cover has been lost (Kormos et al. 2003a). A range of activities may yield varying degrees of habitat destruction and degradation (Fig. 40.1). Agricultural practices, such as slash-and-burn agriculture, and unsustainable land use management are among the most widespread causes of habitat destruction and degradation in some regions; and rapid human population growth across Africa is expected to lead to

Fig. 40.1 The main threats to chimpanzees include deforestation from an increase in agricultural pressure (*foreground:* slash-and-burn agriculture), large-scale extractive industrial activities (*background:* remains of an open air iron-ore mine in the Nimba Mountains, Liberia) (photograph by Tatyana Humle)

continued widespread conversion of forest and woodland to agricultural land. Chimpanzees are therefore increasingly forced into a human interface, which often results in increased resource competition between the two species, and ultimately chimpanzee population decline (Hockings and Humle 2009).

Extractive activities such as logging and mining are also responsible for extensive and sometimes irreversible habitat destruction and degradation. Such activities, when practiced on an industrial scale, additionally often involve road building for access to remote areas (Fig. 40.2). Road building or the establishment of an access network in and out of the forest or mining concession often result in habitat degradation and fragmentation. The threat to chimpanzees is also accrued by encouraging the bush-meat trade in providing access to hunters and poachers to areas previously naive to such anthropic pressures. Human incursions into less accessible or remote areas of forest may also exacerbate the risk of disease transmission.

Fig. 40.2 Large-scale logging (Liberia) is responsible for extensive and sometimes irreversible habitat destruction and degradation. The creation of large roads tend to exacerbate the bush-meat trade as hunters and poachers are more readily able to sell and readily deliver bush meat to city markets (photograph by Jeremy Holden)

Logging generally, but not always, has a negative impact on chimpanzee density as a consequence of habitat alteration (removal of important food trees) and disturbance (Plumptre and Johns 2001; White and Tutin 2001). Mining for gold and bauxite are widespread in West Africa. Iron-ore mining is increasingly being developed, especially in Guinea, which harbors some of the highest grade iron in the world, as well as some of the largest reserves of bauxite. The impact of these mining activities on chimpanzee behavior, ecology, and density has not yet been evaluated precisely; however, considering the high levels of disturbance and habitat alteration (including erosion and water and noise pollution) resulting from such activities and the ensuing significant human influx into these region, we can confidently predict that chimpanzee populations in such areas will negatively be affected. Indeed, some chimpanzee communities may be forced to shift their ranges, which could severely threaten their long-term survival, either through increased intra- and intercommunity competition for resources, which could result in warfare and death of chimpanzees, or in a greater susceptibility to diseases caused by stress on the immune system.

40.2.2 Hunting and Poaching

Hunting and poaching have different causes, as well as different consequences on chimpanzee populations. In some regions, because of the low population density of chimpanzees and their slow reproductive rates, hunting and poaching may lead to rapid local disappearance of chimpanzee populations. One of the motivations for hunting and poaching is meat consumption. For example, chimpanzees currently constitute 1–3% of the bush meat sold in urban markets in Côte d'Ivoire (Caspary et al. 2001). Commercial hunting, often facilitated by logging, is also responsible for declines in chimpanzee populations in some areas (Tutin et al. 2005; Wilkie and Carpenter 1999).

Chimpanzees are also sometimes captured for commercial purposes. In some regions, such practices may even be carried out by people with religious and cultural taboos on their consumption. Although the pet trade of chimpanzees is illegal in all range countries that are signatories to Convention on International Trade of Endangered Species (CITES), it persists illegally across Africa, including West Africa (see Chap. 5). The capture of an infant chimpanzee usually implies the death of its mother, as well as often other members of the community, thus representing a huge threat to individual communities.

In some localities, chimpanzees may also be killed and/or consumed solely for traditional medicinal purposes or traditional remedies. Some range countries, such as Guinea in West Africa, also have legal loopholes that officially permit the capture of chimpanzees for scientific research. People may also hurt or kill chimpanzees intentionally to protect their crops. In addition, chimpanzees may be maimed or killed when caught in snares set for other animals, such as baboons or cane rats, considered problem species in some areas because of their crop-raiding habits.

40.2.3 Disease

The main cause of death in chimpanzees at Gombe, Mahale, Taï, and Bossou is infectious disease (Goodall 1986; Nishida et al. 2003; see Chap. 32). Because chimpanzees and humans are so similar, chimpanzees may contract many pathogens that are either simply carried by humans or which afflict humans (Köndgen et al. 2008). The frequency of encounters between chimpanzees and humans and/or human waste is increasing as human populations expand and habitat fragmentation and degradation continues, leading to higher risks of disease transmission between humans and chimpanzees (Fig. 40.3). If not properly managed, research and ecotourism also present an elevated risk of disease transmission for both species (see Chap. 32). As presented in Chap. 32, chimpanzees are extremely vulnerable to respiratory diseases, which have become a major cause of death, especially among habituated populations. In the past 15 years, repeated epidemics of Ebola hemorrhagic fever have caused dramatic declines of chimpanzees in remote protected areas in Gabon and the Republic of Congo (Huijbregts et al. 2003; Walsh et al. 2005; P. Walsh, unpublished data). Although recent surveys have not always distinguished between the nests of chimpanzees and sympatric gorillas, the pooled density of apes in several large areas has declined by 50–90% following Ebola epidemics (Bermejo et al. 2006; Lahm et al. 2007; Tutin et al. 2005; P. Walsh, unpublished

Fig. 40.3 Yolo (15-year-old adult male from Bossou) walks by houses in the village to raid papaya trees growing at the forest edge; this close proximity with human habitations could favor disease transmission between humans and chimpanzees (photograph by Tatyana Humle)

data). In contrast to the situation in Central Africa, so far only one Ebola epidemic has ever been reported in West Africa (see Chap. 32).

40.3   Status of Wild Chimpanzees in Guinea

The Republic of Guinea is probably the country in West Africa harboring the greatest number of chimpanzees, with approximately 17,582 (8,113–29,011) chimpanzees nationwide (Ham 1998). The majority (estimated at more than 90%) of chimpanzees in Guinea live outside protected areas (Kormos et al. 2003a). Guinea harbors the smallest protected areas network in West Africa, whether in terms of the number of protected areas or the percentage of the country that is protected (only 2.9% of the country's surface area) (Brugière and Kormos 2009). Guinea has only two national parks, the Parc National de Badiar (PNB) near the border with Senegal, and the Parc National du Haut Niger (PNHN) in the center of Guinea, near the town of Faranah. These two parks are managed as Biosphere Reserves, which include a strictly protected core area, a buffer zone, and a transition zone. They respectively cover 284,000 ha (core area, 113,800 ha) and 647,000 ha (core area, 55,400 ha). The Nimba Mountains (total area, 145,200 ha; core area, 21,780 ha) and the classified forest of Ziama (total area, 116,170 ha; core area, 42,547 ha) are the only other two biosphere reserves in Guinea (see Chaps. 27–29).

Several areas either overlapping or entirely within Guinea were classed as priority sites for the conservation of chimpanzees in 2002 during a workshop aimed at evaluating our current understanding of the status of chimpanzees in West Africa (Kormos et al. 2003a). One of these areas is the Madingue Plateau, a transnational area overlapping Guinea, Mali, and Senegal. This area, the largest expanse of savanna-woodland in West Africa, includes the Mali's Wongo and Korofin National Parks, components of the Bafing Biosphere Reserve in Mali, Senegal's Niokolo-Koba National Park, and Guinea's Badiar National Park. Although chimpanzees in this area generally are traditionally not hunted for meat, they are increasingly suffering from habitat loss and fragmentation (Granier and Martinez 2004). The Madingue Plateau is estimated to harbor approximately 1,500 chimpanzees and represents one of the largest populations of savanna-dwelling chimpanzees in West Africa (Kormos et al. 2003a). The Fouta Djallon highlands extending into Guinea-Bissau is another priority site, home to probably more than 3,000 chimpanzees, a significant proportion of the chimpanzee population of Guinea (Ham 1998; Kormos et al. 2003a). The forest of this region is very dry and highly fragmented. Chimpanzees are common in this region as they are not hunted by the local people for traditional, cultural, and religious reasons.

However, habitat degradation and resource competition are of increasing concern for chimpanzees in this region of Guinea. The habitat fragmentation prevalent in this region does imply that an increasing number of communities are at risk of inbreeding if connectivity between forest blocks or areas is not preserved or restored. A third site of exceptional priority for the conservation of chimpanzees is the Parc National du Haut Niger (PNHN). Fleury-Brugière and Brugière (2002) estimated the chimpanzee population within the PNHN to exceed 500 individuals. Between 1995 and 2005, thanks to European Union funding, through the AGIR project (Programme Régional d'Appui à la Gestion Intégrée des Ressources Naturelles des Bassins du Niger et de la Gambie), the park benefited from some protection and law enforcement. The presence of chimpanzee bush meat in villages surrounding the Park was nevertheless recorded during this period, albeit infrequently (Brugière and Magassouba 2009). Since the end of the AGIR project in 2005, hunting, poaching, and logging activities have unfortunately markedly increased within the park's strictly protected zones (Humle, personal observation). Humle began a survey in the Mafou area in 2007 that has yielded data for comparison with the AGIR data and has indicated that the wild chimpanzee population has remained relatively stable since previous surveys conducted by Fleury-Brugière and Brugière (2010; Humle 2007). Finally, the cross-border region of the Nimba Mountains, including the whole massif, and the Bossou, Déré, and Tiapleu ecosystems, is the final and fourth site of exceptional priority for chimpanzee conservation in Guinea and is estimated to harbor several hundred chimpanzees (see Chaps. 27, 28, 29, and 39). This population faces heavy direct and indirect threats from open air iron-ore mining activities, already in their planning and prospection stages, as the Nimba massif contains some of the highest grade iron ore in the world. The mining concession activities will directly affect the World Heritage Site established by UNESCO in 1981. Although chimpanzees are traditionally protected in the region (see Chaps. 1 and 4), increased habitat encroachment and degradation in this region are likely to seriously aggravate resource competition between humans and chimpanzees (Hockings and Humle 2009; see Chaps. 22 and 23) and the potentiality for disease transmission (see Chap. 32). Finally, the border regions with Guinea-Bissau and Sierra Leone are also known to harbor chimpanzees and are potentially important for the nationwide conservation of the species.

40.4   Implications and Conclusion

Chimpanzees are listed under Appendix A of CITES and as Class A under the African Convention on the Conservation of Nature and Natural Resources. Chimpanzees are additionally officially protected by separate national laws in most countries. Some chimpanzee populations also occur in protected areas such as National Parks. However, in spite of this legal protection, chimpanzee populations continue to decline throughout their range, especially because most chimpanzees occur outside protected areas, and law enforcement within protected areas is all too often either insufficient or deficient. It is clear that stricter enforcement of endangered species laws and more effective management of protected areas are urgently needed in many countries in West Africa. In addition, conservation education and the promotion of sustainable and ecologically friendly economic alternatives to hunting and slash-and-burn agricultural practices should be developed and encouraged to ensure better coexistence between humans and chimpanzees. Improving the hygiene and health standards of humans in and around areas inhabited by chimpanzees should also contribute to chimpanzee conservation by limiting the risks of disease transmission (see Chap. 32). These indirect benefits of chimpanzee conservation for the local human communities should help outweigh the costs, which the communities all too often perceive as imposed on them by external governmental or nongovernmental bodies or organizations.

The conservation of chimpanzees in Guinea and Africa as a whole depends mostly on our ability to convey and apply our knowledge and understanding of their behavior, ecology, and demographics with the support and will of the people living in proximity to chimpanzees. As large extractive industrial conglomerates continue to implant themselves in regions of high value for the conservation of chimpanzees and biodiversity as a whole, it is also becoming increasingly urgent to assess the true impact of their activities on chimpanzees, including other fauna and flora, and the long-term reality and effectiveness of their proposed mitigation strategies and of their contributions to the development of local human communities and of their host country as a whole. We urgently recommend the elaboration of a national strategic plan for mining activities in Guinea, aimed at examining the cumulative impact of large-scale extractive activities, especially mining, on both chimpanzees and biodiversity. "No-go" zones should be created and trust funds set up by the mining companies to support the protection of these areas and local sustainable development.

Finally, a combination of factors has led to a poor understanding of the current population status of *P. t. verus*. Much of the range has not been surveyed, survey methods have been inconsistent, and many of the surveys are outdated. Older survey data are particularly unreliable because disease, commercial hunting, and extractive industries are known to have since caused dramatic declines in some areas (Tutin et al. 2005). New surveys using consistent methods are greatly needed throughout most of the range of *P. t. verus* (Kühl et al. 2008). In addition, however, as the urgency of chimpanzee conservation is making itself felt, first and foremost, in those regions where threats are clearly identified, efforts should be made to conserve chimpanzees with a long-term perspective of population viability in mind within a framework favorable to chimpanzee–human coexistence.

Acknowledgments We are particularly grateful to Tetsuro Matsuzawa, Yukimaru Sugiyama, Hiroyuki Takemoto, Makoto Shimada, David Brugière, Marie-Claude Fleury-Brugière, Janis Carter, Christelle Colin, Saliou Diallo, Christine Sagno, and Elhadj Ousmane Tounkara for all their valuable contributions to our current understanding of the status of chimpanzees in Guinea. We also wish to thank Christophe Boesch, Mohamed Bakarr, Tom Butynski, John Oates, and all the participants of the workshop organized in Abidjan in September 2002 for helping develop the 2003 Action Plan for the West African Subspecies of Chimpanzee, upon which much of the information provided in this chapter is based. Finally, we thank all the people throughout West Africa who endeavor to conserve chimpanzees and to ensure their viability in the region.

Appendices

T. Matsuzawa et al. (eds.), *The Chimpanzees of Bossou and Nimba*,
DOI 10.1007/978-4-431-53921-6, © Springer 2011

Appendix A   Lineage of the Bossou Community as of January 2010

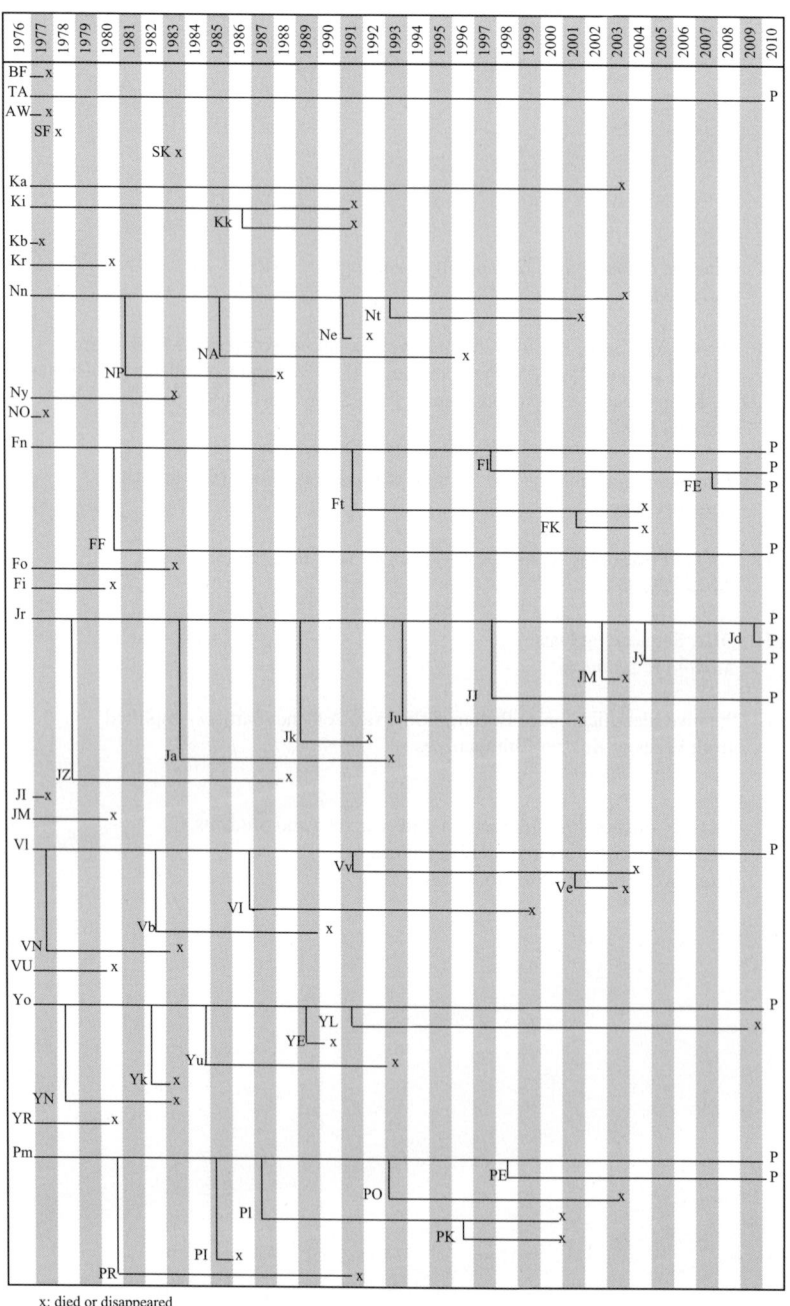

x: died or disappeared
P: present

Name	Abbr.	Sex	Birth day/month/year	Dissapear day/month/year
BAFU	BF	♂	1950 (estimated)	Dissapeared (Post May-77)
TUA	TA	♂	1957 (estimated)	P
AIWA	AW	♂	1955 (estimated)	Dissapeared (Post May-77)
SAFI	SF	♂	1945 (estimated), Immigrated (21/01/1977)	Dissapeared (9/2/1977)
SAKAI	SK	♂	1963 (estimated), Immigrated (Mar-83)	Dissapeared (Post Mar-83)
Kai	Ka	♀	1950 (estimated)	Died (3-5/12/2003)
Kie	Ki	♀	1976 (estimated)	Dissapeared (5/3/1991)
Kakuru	Kk	♀	1986 late	Dissapeared (5/3/1991)
Kubo	Kb	♀	1972 (estimated)	Dissapeared (22/2/1977)
Kure	Kr	♀	1969 (estimated)	Dissapeared (Post Mar-80)
Nina	Nn	♀	1954 (estimated)	Dissapeared (19/11/2003)
Nto	Nt	♀	1993 early	Dissapeared (Dec-01)
Nyele	Ne	♀	27/03/1991	Dissapeared (5-8/1991)
NA	NA	♂	1985	Dissapeared (3-5/1996)
NPEI	NP	♂	1981 early	Died (1/1/1988)
Nyu	Ny	♀	1976 (estimated)	Dissapeared (Post Mar-83)
NON	NO	♂	1969 (estimated)	Dissapeared (Post May-77)
Fana	Fn	♀	1956 (estimated)	P
Fanle	Fl	♀	08-18/10/1997	P
FLANLE	FE	♂	14/09/2007	P
Fotaiu	Ft	♀	1991 middle	Dissapeared (Nov-04)
FOKAIYE	FK	♂	Jul-01 middle	Dissapeared (Nov-04)
FOAF	FF	♂	1980 late	P
Fon	Fo	♀	1976 (estimated)	Dissapeared (Post Mar-83)
Fino	Fi	♀	1971 (estimated)	Dissapeared (Post Mar-80)
Jire	Jr	♀	1958 (estimated)	P
Jodoamon	Jd	♀	18/11/2009	P
Joya	Jy	♀	14:00 2/9/2004	P
JIMATO	JT	♂	4-12/10/2002	Died (27/11/2003-03/12/2003)
JEJE	JJ	♂	Dec-97	P
Juru	Ju	♀	19/11/1993	Dissapeared (Dec-01)
Jokuro	Jk	♀	1989 early	Dissapeared (25/1/1992)
Ja	Ja	♀	1983 late	Dissapeared (Feb-93)
JIEZA	JZ	♂	1978	Dissapeared (Post April-88)
JI	JI	♂	1975 (estimated)	Dissapeared (Post May-77)
JIMA	JM	♂	1972 (estimated)	Dissapeared (Post Mar-80)
Velu	Vl	♀	1959 (estimated)	P
Vuavua	Vv	♀	1991 middle	Dissapeared (Mar-04)
Veve	Ve	♀	May-01 late	Died (29-30/12/2003)
VUI	VI	♂	1986 late	Dissapeared (Post Jul-99)
Vube	Vb	♀	1982	Dissapeared (Post Mar-90)
VUNA	VN	♂	1977	Dissapeared (Post Mar-83)
VU	VU	♂	1972 (estimated)	Dissapeared (Post Mar-80)
Yo	Yo	♀	1961 (estimated)	P
YOLO	YL	♂	1991 middle	Disappeared (Nov-09)
YERA	YE	♂	1989 middle	Dissapeared (Mar/Nov-90)
Yunro	Yu	♀	1984 late	Dissapeared (Feb-93)
Yaka	Yk	♀	1982 early	Dissapeared (5/3/1983)
YANA	YN	♂	1978	Dissapeared (Post Apr-83)
YIRI	YR	♂	1974 (estimated)	Dissapeared (Post Mar-80)
Pama	Pm	♀	1967 (estimated)	P
PELEY	PE	♂	Apr-98	P
PONI	PO	♂	4/2/1993	Died (19/11/2003-3/12/2003)
Pili	Pl	♀	1987 early	Dissapeared (Sept-00/May-01)
POKURU	PK	♂	Aug-96	Dissapeared (Sept-00/May-01)
PIRE	PI	♂	1985 early	Dissapeared (10/1/1986)
PRU	PR	♂	1980 late	Dissapeared (Nov-91)

Appendix B   Plant Food List of Bossou Chimpanzees Including Vernacular Manon Name, as Well as Plant Type and Part Consumed (Updated and Compiled by T. Humle, K. Koops, and P. Chérif and adapted from Sugiyama and Koman 1987, 1992)

	Scientific name	Family	Vernacular name (Manon)	Plant type	Part consumed
1	*Acacia pennata*	Mimosaceae	Dan	Vine	Bark
2	*Acioa tenuiflora*	Chrysobalanaceae		Tree	Seed
3	*Adenopus breviflorus*	Passifloraceae		Vine	Fruit
4	*Aframomum citratum*	Zingiberaceae	Sin	Herb	Fruit, Pith
5	*Aframomum excapum*	Zingiberaceae	Sin	Herb	Fruit, Pith
6	*Aframomum latifolium*	Zingiberaceae	Douandi	Herb	Fruit, Pith
7	*Aframomum longiscapum*	Zingiberaceae	Yièlasin	Herb	Fruit, Pith
8	*Aframomum melegueta*	Zingiberaceae	Taasuo	Herb	Fruit, Pith
9	*Aframomum sceptrum*	Zingiberaceae		Herb	Fruit, Pith
10	*Aframomum strobilaceum*	Zingiberaceae		Herb	Fruit, Pith
11	*Aframomum subsericeum*	Zingiberaceae		Herb	Fruit, Pith
12	*Aframomum sulcatum*	Zingiberaceae		Herb	Fruit, Pith
13	*Albizia adianthifolia*	Mimosaceae	Kpanforo	Tree	Gum
14	*Albizia ferruginea*	Mimosaceae	Safouklé	Tree	Gum
15	*Albizia zygia*	Mimosaceae	Kpanti	Tree	Gum
16	*Alchornea cordifolia*	Euphorbiaceae	Fanalè	Shrub	Fruit, Pith
17	*Allophyllus africanus*	Sapindaceae	Weilazana	Tree	Fruit
18	*Ampelocissus macrocirrha*	Ampelidaceae	Leilè	Vine	Fruit
19	*Ananas comosus*	Bromeliaceae	Yérabibi	Herb	Fruit, Pith
20	*Ancilobotrys scandens*	Apocynaceae		Vine	Fruit
21	*Aningeria altissima*	Sapotaceae	Yara	Tree	Fruit
22	*Anthonotha macrophylla*	Caesalpiniaceae	Gbèké	Tree	YoLeaf
23	*Antiaris africana*	Moraceae	Kpo	Tree	Fruit, YoLeaf
24	*Artabotrys jollyanus*	Annonaceae	Tiédong	Vine	Fruit
25	*Baphia* sp.	Papilionaceae		Tree	YoLeaf
26	*Belschiemeidia mannii*	Lauraceae	Kwè	Tree	Fruit
27	*Bersama abyssinica*	Sapindaceae	Wuison-yiri	Tree	Fruit
28	*Blighia sapida*	Sapindaceae	Gleinpourou	Tree	Fruit
29	*Blighia unijugata*	Sapindaceae	Glein	Tree	Fruit
30	*Blighia welwitschii*	Sapindaceae	Glein	Tree	Fruit
31	*Bombax buonopozense*	Bombacaceae	Guèdéré	Tree	YoFruit
32	*Bosquea angolensis*	Moraceae	Pâa	Tree	Bark, Leaf
33	*Bridelia ferruginea*	Euphorbiaceae	Gwan	Tree	Fruit
34	*Bridelia micrantha*	Euphorbiaceae	Lougwan	Tree	Fruit
35	*Bussea occidentalis*	Caesalpiniaceae	Kpakélé	Tree	Seed
36	*Calpocalyx aubrevillei*	Mimosaceae		Tree	Fruit

(continued)

Appendix B (continued)

	Scientific name	Family	Vernacular name (Manon)	Plant type	Part consumed
37	*Canarium schweinfurthii*	Burseraceae	Biin	Tree	Fruit, Petiole of Leaf
38	*Canthium horizontale*	Rubiaceae	Ni-inwéléyiri	Tree	Fruit
39	*Carapa procera*	Meliaceae	Gbon	Tree	YoLeaf
40	*Carica papaya*	Caricaceae	Gblanghin	Tree	Fruit, Leaf, Pith
41	*Ceiba pentandra*	Bombacaceae	Guè	Tree	YoLeaf, Flower
42	*Celtis adolfi-frederici*	Ulmaceae	Kosingwan	Tree	Fruit
43	*Celtis brownii*	Ulmaceae	Nionoziigwan	Tree	Leaf
44	*Celtis mildbraedii*	Ulmaceae	Sowéléyiri	Tree	Fruit
45	*Chlorophora excelsa*	Moraceae	Gué-i	Tree	Fruit, Flower, Petiole
46	*Chlorophora regia*	Moraceae	Gué-i	Tree	Fruit, Flower
47	*Chytranthus longiracemosus*	Sapindaceae	Taangban	Shrub	Fruit
48	*Cissus aralioides*	Vitaceae	Kanso	Vine	Fruit
49	*Citrus aurantium*	Rutaceae	Boiguein	Tree	Fruit
50	*Citrus grandis*	Rutaceae	Gléfou	Tree	Fruit
51	*Citrus reticula*	Rutaceae	Kélébosoka	Tree	Fruit
52	*Citrus sinensis*	Rutaceae	Gein	Tree	Fruit
53	*Cola caricaefolia*	Sterculiaceae	Lougho	Tree	Fruit
54	*Cola cordifolia*	Sterculiaceae	Boba	Tree	Fruit, Bark
55	*Cola reticulata*	Sterculiaceae	Guérapourouyiri	Tree	Fruit
56	*Costus afer*	Zingiberaceae	Zin	Herb	Pith
57	*Costus deistelii*	Zingiberaceae	Zin	Herb	Pith
58	*Costus dubius*	Zingiberaceae	Zin	Herb	Pith
59	*Costus lucanusianus*	Zingiberaceae	Zin	Herb	Pith
60	*Craterispermum laurinum*	Rubiaceae	Gbékè	Shrub	Fruit
61	*Cyrtosperma senegalense*	Aracae		Herb	Rhizome
62	*Dacryodes klaineana*	Burseraceae	Wéwé	Tree	Fruit
63	*Dalbergia* sp.	Papilionaceae	Tounoula	Vine	YoLeaf
64	*Deinbollia pennata*	Sapindaceae	Loubonbon	Tree	Fruit
65	*Dialium dinklagei*	Caesalpiniaceae	Kpèi	Tree	Fruit
66	*Dialium guineensis*	Caesalpiniaceae	Kpèi	Tree	Fruit
67	*Dicranolepis laciniata*	Thymeleaceae	Torô	Shrub	Fruit
68	*Dioscorea minutiflora*	Dioscoreaceae	Bhélé	Vine	Fruit
69	*Dioscorophyllum cumminsii*	Menispermaceae	Bonlégwan	Vine	Fruit
70	*Dissotis jacquesii*	Melastomataceae	Silamounugbo	Herb	YoLeaf
71	*Dracaena arboreus*	Agavaceae	Ziri	Tree	Petiole
72	*Elaeis guineensis*	Palmae	Ton-yiri	Tree	Petiole, Seed, Fruit, Flower (woody tissue)
73	*Erythrina mildbraedii*	Papilionaceae	Kédoh	Tree	YoLeaf
74	*Euclinia longiflora*	Rubiaceae		Shrub	Fruit
75	*Fagara leprieurii*	Rutaceae	Ménéyiri	Tree	YoLeaf
76	*Fagara macrophylla*	Rutaceae	Gueinyiri	Tree	YoLeaf

(continued)

Appendix B (continued)

	Scientific name	Family	Vernacular name (Manon)	Plant type	Part consumed
77	*Ficus annomani*	Moraceae	Gorolékènin	Tree	Fruit, Whole Leaf
78	*Ficus asperifolia*	Moraceae	Goro	Tree	Fruit
79	*Ficus barteri*	Moraceae	Goro	Tree	Fruit
80	*Ficus bignonifolia*	Moraceae	Blôléboiboi	Tree	Fruit
81	*Ficus capensis/sur*	Moraceae	Blô	Tree	Fruit
82	*Ficus dectekena*	Moraceae	Goro	Tree	Fruit
83	*Ficus eriobotrioides*	Moraceae	Goro	Tree	Fruit
84	*Ficus eriota*	Moraceae	Goro	Tree	Fruit
85	*Ficus exasperata*	Moraceae	Nyanalé	Tree	Fruit, YoLeaf, MatLeaf
86	*Ficus leprieurii*	Moraceae	Goro	Tree	Fruit
87	*Ficus macrosperma*	Moraceae	Goro	Tree	Fruit
88	*Ficus mucuso*	Moraceae	Soroblo	Tree	Fruit, YoLeaf
89	*Ficus ovata*	Moraceae	Goro	Tree	Fruit
90	*Ficus polita*	Moraceae	Goro	Tree	Fruit
91	*Ficus sagittifolia*	Moraceae	Goro	Tree	Fruit
92	*Ficus thonningii*	Moraceae	Goro	Tree	Fruit
93	*Ficus umbellata*	Moraceae	Gorolèkérékéré/ Goroboa	Tree	Fruit, YoLeaf
94	*Ficus vallis-choudae*	Moraceae	Goro	Tree	Fruit
95	*Ficus variifolia*	Moraceae	Soroblogwan	Tree	Fruit, YoLeaf
96	*Funtumia elastica*	Apocynaceae	Sékélé	Tree	YoPetiole
97	*Gambeya gigantia*	Sapotaceae	Bomo	Tree	Fruit
98	*Gambeya perpulchrum*	Sapotaceae	Weingbeinléyiri	Tree	Fruit
99	*Gambeya taiense*	Sapotaceae	Bomogwan	Tree	Fruit
100	*Garcinia kola*	Guttiferae	Soniyiri	Tree	Fruit
101	*Glyphae brevis*	Tiliaceae	Kwè-i	Shrub	Leaf
102	*Gongronema latifolium*	Asclepiadaceae	Noyblé	Vine	Fruit, Bark, Pith (Stem), Leaf
103	*Grewia barombiensis*	Tiliaceae	Diéti	Vine	Fruit
104	*Grewia pubescens*	Tiliaceae	Yirizan	Vine	Fruit
105	*Halopegia azurea*	Marantaceae		Herb	Pith
106	*Hannoa klaineana*	Siimaroubaceae	Fâa	Tree	Fruit, Leaf
107	*Harungana madagascariensis*	Hypericaceae	Lorô	Tree	Fruit, YoLeaf
108	*Hibiscus esculentus*	Malvaceae	Boobhé	Herb	Fruit, Flower, Leaf
109	*Hibiscus rostellatus*	Malvaceae	Gban-nana	Herb	Leaf
110	*Hibiscus sabdariffa*	Malvaceae	Bhomi	Herb	Flower, Leaf
111	*Hippocratea paniculata*	Celastraceae	Kpané	Vine	Fruit, Leaf
112	*Hypselodelphis poggeana*	Marantaceae	Gomo	Herb	YoLeaf, Fruit, Pith
113	*Hypselodelphis violaceae*	Marantaceae	Gomo	Herb	Fruit, Pith
114	*Ituridendron bequaertii*	Sapotaceae	Lougin	Tree	Fruit
115	*Justicia tenella*	Acanthaceae	Tongdilé	Herb	Leaf
116	*Khaya ivorensis*	Meliaceae	Kpitirizoro	Tree	Fruit

(continued)

Appendix B (continued)

	Scientific name	Family	Vernacular name (Manon)	Plant type	Part consumed
117	*Landolphia dulcis*	Apocynaceae	Bovouakara	Vine	Fruit
118	*Landolphia hirsuta*	Apocynaceae	Siensien	Vine	Flower, Fruit
119	*Landolphia incerta*	Apocynaceae	Dékpolo	Vine	Fruit
120	*Landolphia owariensis*	Apocynaceae	Sénédé	Vine	Fruit
121	*Lecaniodiscus cupanioides*	Sapindaceae	Gleinkaba	Tree	Fruit
122	*Leea guineensis*	Ampelidaceae		Herb	Pith
123	*Leptoderris brachyptera*	Papilionaceae	Toublégwan	Vine	Bark (Look for larvae)
124	*Leptoderris fasciculata*	Papilionaceae	Toublé	Vine	Bark
125	*Lippia* sp.	Verbenaceae		Herb	Flower
126	*Macaranga barteri*	Euphorbiaceae	Béghou	Tree	Fruit
127	*Macaranga heterophylla*	Euphorbiaceae	Gbinlago	Tree	Fruit
128	*Maesobotrya barteri*	Euphorbiaceae	Kin	Tree	Fruit
129	*Mangifera indica*	Anacardiaceae	Mangolo	Tree	Fruit
130	*Manihot esculenta*	Euphorbiaceae	Bé-i	Herb	Tuber
131	*Maranthochloa macrophylla*	Marantaceae	Yorô	Herb	Pith, Fruit
132	*Marantochloa congensis*	Marantaceae	Yorôlékénin	Herb	Pith
133	*Marantochloa cuspidata*	Marantaceae		Herb	Pith
134	*Marantochloa filipes*	Marantaceae	Beralé	Herb	Pith
135	*Marantochloa flexuosa*	Marantaceae		Herb	Pith
136	*Marantochloa leucantha*	Marantaceae	Beralézoro	Herb	YoPith
137	*Marantochloa purpurea*	Marantaceae		Herb	YoPith
138	*Megaphrynium macrostachyum*	Marantaceae	Ghaa	Herb	Pith, YoLeaf, Fruit
139	*Milletia zechiana*	Papilionaceae	Kpétuan	Tree	Leaf
140	*Momordica cabraei*	Curcurbitaceae	Louguo	Tree	Fruit
141	*Monodora tenuifolia*	Annonacea	Kpanayiri	Tree	Fruit, YoLeaf
142	*Morinda longiflora*	Rubiaceae	Sorogbo	Vine	Fruit
143	*Morinda morindoides*	Rubiaceae	Sorogbo	Vine	Fruit
144	*Morus mesozygia*	Moraceae	Ghangbe	Tree	Fruit, Flower, YoLeaf
145	*Musa paradisiaca*	Musaceae		Herb	Fruit, Pith
146	*Musa sinensis*	Musaceae	Blo	Herb	Fruit, Pith
147	*Musaenda erithrophylla*	Rubiaceae	Tokpèrélé	Vine	Fruit, Leaf
148	*Musanga cecropioides*	Moraceae	Wolo	Tree	Fruit, YoLeaf, Flower
149	*Myrianthus arboreus*	Moraceae	Gbale	Tree	Fruit, Yoleaf
150	*Myrianthus libericus*	Moraceae	Gbalo	Tree	Fruit, Yoleaf
151	*Myrianthus serratus*	Moraceae	Gbalogwan	Tree	Fruit
152	*Napoleona leonensis*	Lecythidaceae	Nimo	Tree	Fruit
153	*Napoleona vogelii*	Lecythidaceae	Dole	Tree	Fruit
154	*Nauclea latifolia*	Rubiaceae	Yéilaweinyiri	Shrub	Fruit
155	*Nephrolepis bisserata*	Nephrolepidaceae	Klaklalé	Herb	Leaf, Pith, Rhizome
156	*Newbouldia laevis*	Bignoniaceae	Dian	Tree	YoLeaf
157	*Oryza* sp.	Graminae	Bou	Herb	Pith

(continued)

Appendix B (continued)

	Scientific name	Family	Vernacular name (Manon)	Plant type	Part consumed
158	*Oxyanthus formosus*	Rubiaceae	Wôlô-yiri	Tree	Fruit
159	*Pachystella pobeguiniana*	Sapotaceae		Tree	Fruit
160	*Palusota hirsuta*	Commelinaceae	Kiekopalé	Herb	Fruit, Pith, Flower, Whole Leaf
161	*Parinari excelsa*	Rosaceae	Koin	Tree	Fruit
162	*Parkia bicolor*	Mimosaceae	Komi	Tree	Fruit
163	*Pennisetum purpureum*	Gramineae	Ka	Tree	Pith
164	*Persea gratissima/americana*	Lauraceae	Avoca	Tree	Leaf Petiole
165	*Persia americana*	Lauraceae	Avoca	Tree	Petiole, YoLeaf
166	*Phyllanthus discoideus*	Euphorbiaceae	Tié	Tree	Fruit
167	*Phyllanthus margariana*	Euphorbiaceae	Loutié	Tree	Fruit
168	*Piper guineense*	Piperaceae	Zemlé	Herb	Fruit
169	*Piper umbellatum*	Piperaceae	Win-ilé	Herb	YoFruit
170	*Platycerium angolense*	Polypodiaceae		Epiphyte	YoLeaf
171	*Polycephalium capitatum*	Icacinaceae	Bonlé	Vine	Fruit, Whole Leaf
172	*Popowia klainii*	Annonaceae		Vine	Fruit
173	*Premna hispida*	Verbenaceae	Bosorolé	Shrub	Fruit
174	*Pseudospondias microcarpa*	Anacardiaceae	Poni	Tree	Fruit, YoLeaf, Bark
175	*Pteridium aquilinum*	Polypodiaceae	Kpokoulou	Herb	YoLeaf, Pith
176	*Pterocarpus santalinoides*	Papilionaceae	Gbano	Tree	YoLeaf
177	*Pycnanthus angolensis*	Myristicaceae	Dini	Tree	Fruit
178	*Pyrenacantha reticulata*	Icacinaceae	Deisoblégwan	Vine	Leaf, Pith
179	*Raphia gracilis*	Palmae	Duo	Tree	Pulp, Sap
180	*Renealmia maculata*	Zingiberaceae		Herb	Pith
181	*Rhaphiostylis beninensis*	Icacinaceae	Ploplo	Vine	Fruit
182	*Rhigiocarya racemifera*	Menispermaceae	Wuiburulé	Vine	Fruit
183	*Ricinodendron heudelotii*	Euphorbiaceae	Kôh	Tree	Fruit
184	*Rothmania longiflora*	Rubiaceae	Loupilalé	Shrub	Fruit
185	*Rutidea parviflora*	Rubiaceae		Shrub	Fruit
186	*Saccharum officinarum*	Poaceae	Kwiyoo	Herb	Pith
187	*Salacia owabiensis*	Celastraceae	Sokologbé	Vine	Fruit
188	*Santiria trimera*	Burseraceae	Goo	Tree	Fruit
189	*Sarcophrynium brachystachyum*	Marantaceae	Sanbela	Herb	Pith, Fruit, YoLeaf
190	*Sarcophrynium prionogonium*	Marantaceae	Bera	Herb	Pith, Fruit, YoLeaf
191	*Scleria barterii*	Cyperaceae	Pépé	Herb	Leaf Swallowing
192	*Sherbournia bignoniflora*	Rubiaceae	Kpanéti	Vine	Fruit
193	*Sherbournia calycina*	Rubiaceae	Monbè	Vine	Fruit
194	*Smilax kraussiania*	Smilacaceae	Kpeinkho	Vine	YoLeaf, MatLeaf (with Meat/ Egg/Larvae)
195	*Spondias cythera*	Anacardiaceae	Gueibuna	Tree	Bark

(continued)

Appendix B   (continued)

	Scientific name	Family	Vernacular name (Manon)	Plant type	Part consumed
196	*Spondias mombin*	Anacardiaceae	Buna	Tree	Fruit
197	*Sterculia tragacantha*	Sterculiaceae	Tou	Tree	YoLeaf, Yoseed, Bark
198	*Strophantus sarmantosus*	Apocynaceae	Noyblégwan	Vine	Seed
199	*Strychnos* sp.	Loganiaceae	Kouinkana	Vine	Fruit
200	*Taraktogenus* sp.	Flacourtiaceae		Shrub	Bark
201	*Terminalia glautescens*	Combretaceae	Walan	Tree	YoLeaf
202	*Terminalia ivorensis*	Combretaceae	Béi	Tree	YoLeaf
203	*Tetrapleura tetraptera*	Mimosaceae	Zian	Tree	YoLeaf
204	*Tetrorchidium didymostemon*	Euphorbiaceae	Longlô	Tree	Fruit
205	*Thaumatococus daniellii*	Marantaceae	Saa	Herb	Fruit, Pith, YoLeaf
206	*Theobroma cacao*	Sterculiaceae	Cacao	Tree	Fruit
207	*Tiliacora dinklagei*	Icacinaceae	Deisoblé	Vine	Fruit
208	*Trema guineensis*	Ulmaceae	Wama	Tree	Fruit
209	*Trichilia heudelotii*	Meliaceae	Waa	Tree	Fruit, YoLeaf, Petiole
210	*Triplochiton scleroxylon*	Sterculiaceae	Jokoro	Tree	YoLeaf
211	*Uapaca guineensis*	Euphorbiaceae	Sona	Tree	Fruit
212	*Uapaca heudelotii*	Euphorbiaceae	Sona	Tree	Leaf
213	*Urera cameroonensis*	Urticaceae	Yobli	Vine	Pith, Bark
214	*Uvaria afzelii*	Annonaceae	Goun-gbé	Vine	Fruit
215	*Uvaria chamae*	Annonaceae	Goun-gbégwan	Vine	Fruit
216	*Uvariopsis guineensis*	Annonaceae	Gbleingpourou	Tree	Fruit
217	*Venguealla campyllacantha*	Rubiaceae	Tofonghénè	Shrub	Fruit
218	*Vitex cienkowskii*	Verbenaceae		Tree	Fruit
219	*Vitex doniana*	Verbenaceae	Bon	Tree	Fruit
220	*Vitex ferruginea*	Verbenaceae		Tree	Fruit
221	*Vitex grandifolia*	Verbenaceae		Tree	Fruit
222	*Vitex madiensis*	Verbenaceae		Tree	Fruit
223	*Vitex micrantha*	Verbenaceae	Bouyiri	Tree	Fruit
224	*Vitex oxycuspis*	Verbenaceae		Tree	Fruit
225	*Xylopia staudtii*	Annonaceae	Gbanzoro	Tree	Fruit
226	*Zea mays*	Poaceae	Kpei	Herb	Fruit

Appendix C   Temperature, Humidity, and Rainfall at Bossou Between 1995 and 1998

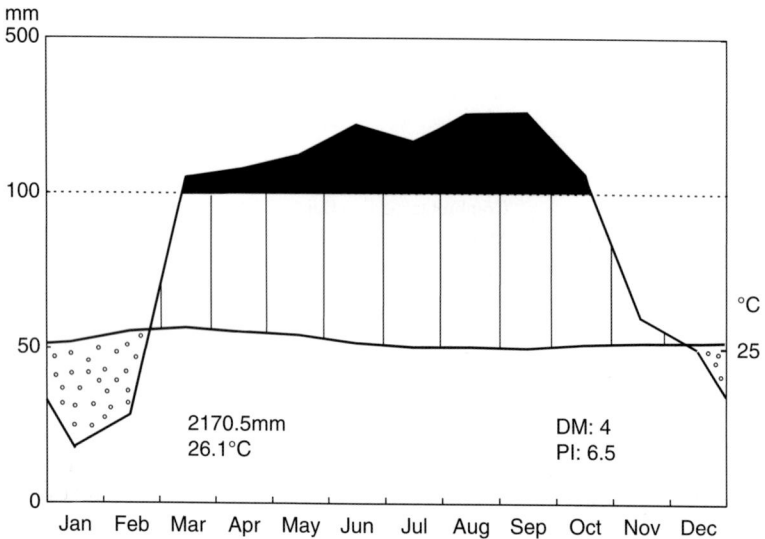

Fig. 1 Climate diagram at Bossou (averages from 1995 to 1998). "In this diagram, regimes of monthly rainfall and monthly temperature are displayed on a single graph, in which the temperature scale (from 0 to 50°C) is twice as large as the rainfall scale (from 0 to 100 mm); to save space the rainfall scale above 100 mm is reduced by a factor of 10. Three types of month are distinguished: 'wet' months, in which rainfall exceeds 100 mm (taken as a rough guide to potential evapotranspiration in the tropics); 'drought' months, in which rain falls below the plotted temperature graph; and "dry" or 'intermediate' months, in which there is a less serious deficit of rainfall below potential evapotranspiration" (Richards 1996, p. 165). The data are based on Table 2. The mean annual rainfall (mm) and mean annual temperature (°C) are indicated on the left-hand side of the graph. PI, per humidity index (the degree of continuity of wetness of a tropical climate); DM, number of dry months (<100 mm)

Table 1 Mean monthly temperature (°C) and humidity (%) in the forest (Mont Gban, 650 m) at 1.5 m above ground between 1996 and 1998, measured with a digital thermo-hygrometer with a determination range (± precision) of −10 to 60°C (±1°C) and 10–98% RH (±6%)

	1996				1997				1998			
	Max. temperature	Min. temperature	Max. humidity	Min. humidity	Max. temperature	Min. temperature	Max. humidity	Min. humidity	Max. temperature	Min. temperature	Max. humidity	Min. humidity
Jan					26.8	20.8	95.0	61.8	27.6	20.8	72.6	41.3
Feb					28.0	21.1	85.5	32.8	29.5	22.1	93.3	44.2
Mar					28.4	21.8	94.6	49.6	29.1	22.3	94.2	51.3
Apr					26.0	20.8	98.0	84.5	27.0	21.6	98.0	87.0
May					25.3	20.5	98.0	92.1	26.1	21.7	98.0	94.3
Jun					23.9	20.1	98.0	98.0	24.9	20.9	98.0	96.8
Jul	22.8	20.4	98.0	98.0	23.3	19.8	98.0	97.5				
Aug	23.3	20.2	98.0	97.2	23.8	20.0	98.0	96.9				
Sep	23.6	20.3	98.0	97.9	24.1	20.5	98.0	98.0				
Oct	24.2	19.9	98.0	91.6	25.2	20.8	98.0	94.4				
Nov	26.3	20.6	97.3	73.3	25.8	20.6	98.0	88.3				
Dec	25.7	20.7	97.4	74.2	26.0	20.5	86.9	66.4				

Table 2 Mean monthly temperature (°C) and humidity (%) at 1.5 m above ground and mean monthly rainfall (mm), measured at the field station (550 m) between 1995 and 1998, using a maximum and minimum mercury thermometer and a wet-and-dry-bulb hygrometer

	1995				1996				1997				1998			
	Max. temperature	Min. temperature	Humidity (17:00)	Rain fall	Max. temperature	Min. temperature	Humidity (17:00)	Rain fall	Max. temperature	Min. temperature	Humidity (17:00)	Rain fall	Max. temperature	Min. temperature	Humidity (17:00)	Rain fall
Jan					34.4	17.8	79.3	18.3	33.9	18.4	70.8	58.0	36.5	14.0	59.8	0.0
Feb					34.9	18.9	71.3	125.9	38.0	15.9	66.2	0.0	39.4	18.4	57.3	9.9
Mar					34.2	19.2	78.2	166.5	37.5	19.1	84.4	88.0	38.4	19.9	74.1	103.5
Apr					33.9	20.0	83.5	124.6	34.2	19.6	93.4	168.8	35.1	21.2	83.9	212.2
May					32.4	20.9	–	172.7	32.6	20.3	86.2	241.6	34.5	20.7	94.8	191.5
Jun					30.8	19.9	80.1	316.8	30.1	20.6	85.5	432.2	32.7	19.9	98.4	266.9
Jul					29.9	19.9	81.6	300.9	29.2	20.1	84.7	315.5				
Aug					30.3	20.2	80.4	286.1	30.1	19.9	86.1	428.5				
Sep					30.2	20.1	83.2	460.4	30.6	19.7	94.0	298.9				
Oct					31.1	19.5	90.3	188.3	31.4	19.3	89.6	143.3				
Nov					33.0	17.9	79.7	19.4	31.7	18.3	84.6	96.5				
Dec	32.7	17.1	85.8	0.0	32.5	17.4	76.1	60.0	32.6	15.7	74.0	33.6				

Appendix D   Global Positioning System (GPS) Data of Specific Locations

Place	Latitude	Longitude
Bossou		
IREB Institute	7°38′ 40.0″ N	8°30′ 09.1″ W
Gban Hill	7°38′ 32.4″ N	8°29′ 27.5″ W
Guein Hill	7°38′ 41.5″ N	8°29′ 50.0″ W
Breton Hill	7°38′ 04.9″ N	8°30′ 39.3″ W
Ueyaton Hill	7°39′ 16.5″ N	8°30′ 08.8″ W
Colin de Concasser (Nut-cracking Hill)	7°38′ 39.2″ N	8°29′ 58.5″ W
Zono Hill	7°37′ 52.4″ N	8°31′ 21.5″ W
Gbah (Gba) village	7°37′ 05.4″ N	8°32′ 24.0″ W
Maru (Riverine Forest)	7°38′ 43.6″ N	8°29′ 36.6″ W
Mobli (Riverine Forest)	7°38′ 22.9″ N	8°30′ 40.6″ W
Go-yigba (Riverine Forest)	7°39′ 40.2″ N	8°30′ 01.7″ W
Godingba (Riverine Forest)	7°39′ 21.0″ N	8°29′ 32.6″ W
Bureau (Outdoor Laboratory)	7°38′ 41.5″ N	8°29′ 50.0″ W
Maison de KUPRI (KUPRI Facilities)	7°38′ 43.2″ N	8°30′ 22.4″ W
Maison du Sous-Prefet (Sous-Prefet House)	7°39′ 01.8″ N	8°30′ 10.9″ W
Petit Jardin Botanique (Botanical Garden Project)	7°38′ 18.2″ N	8°29′ 09.5″ W
Nimba (Guinea)		
Seringbara Village	7°37′ 50.1″ N	8°27′ 44.7″ W
Nion Village	7°37′ 22.8″ N	8°28′ 52.8″ W
Gahtoy Camp	7°39′ 36.2″ N	8°25′ 10.2″ W
Kiépa Camp	7°37′ 47.3″ N	8°26′ 04.8″ W
Madei Camp	7°38′ 58.6″ N	8°25′ 23.6″ W
Yiéton Hill	7°39′ 21.3″ N	8°25′ 00.8″ W
Tongbongbon Hill	7°37′ 58.3″ N	8°25′ 41.1″ W
Tonzoro Hill	7°38′ 21.3″ N	8°24′ 49.2″ W
Glouton-1 Hill	7°39′ 03.6″ N	8°24′ 37.2″ W
Niénéton Hill	7°38′ 54.6″ N	8°24′ 33.6″ W
Gahtoyton Hill	7°39′ 15.0″ N	8°23′ 34.8″ W
Gouoton Hill	7°39′ 34.8″ N	8°24′ 01.2″ W
Nimba (Cote d'Ivoire)		
Yealé Village	7° 31′21.8″ N	8° 25′29.1″W
Yanleu Camp	7° 32′50.09″N	28′03.01″W
Diécké		
Diécké Town	7°20′ 51.6″ N	8°57′ 12.0″ W
Yossono Village	7°33′ 11.4″ N	8°48′ 52.2″ W
Nonah Village	7°31′ 50.9″ N	9°04′ 27.7″ W
Korohouan Village	7°26′ 09.6″ N	8°59′ 22.2″ W
Liberia		
New-Yekepa Village	7°35′ 38.6″ N	8°33′ 34.0″ W
Bonla Village	7°34′ 24.6″ N	8°39′ 34.4″ W
Kpayee-Lepula Village	7°13′ 24.8″ N	8°41′ 29.5″ W

Appendix E   DNA Sequencing Data

Table S1 IDs of mitochondrial DNA (mtDNA) sequences of chimpanzees from Bossou and Nimba registered in DDBJ/EMBL/GenBank

Sequence ID	Sample ID	Accession ID	Length (bp)
Bs-TJFK	Bs_T/Bs_F/Bs_J/Bs_K	AB189231	604
Bs-P	Bs_P	AB189232	604
Bs-V	Bs_V	AB189233	605
Bs-Y	Bs_Y	AB189234	605
Bs-N	Bs_N	AB189235	604
S-1	S1/S7/S10/S11/S12/S14/S19/S22	AB189236	603
S-2	S2/S8/S17/S18/S21/Y5	AB189237	604
S-3	S3/S13	AB189238	605
S-4	S4/G1	AB189239	604
S-5	S5/S15/S23	AB189240	603
S-6	S6	AB189241	604
S-7	S9/S16/S20	AB189242	605
Y-1	Y1	AB189243	604
Y-2	Y3	AB189244	605
Y-3	Y6/Y11	AB189245	604
Y-4	Y7	AB189246	605
Y-5	Y9	AB189247	603
Y-6	Y12	AB189248	604
Y-7	Y13	AB189249	604
Y-8	Y14	AB189250	603
Y-9	Y20	AB189251	605

Bs from Bossou, *S* Seringbara, *Y* Yealé, *G* Gouéla

Table S2 Variable site of mitochondrial DNA (mtDNA) sequences found in this study

Nucleotide position corresponding to Anderson Sequence (J01415)

SeqID	Sequence
	atacctgcacctgcc-ccttatccccttaactctgcatccc-aagtctccctcactcctcacgttggcgtc
S-1	..t..a...c..t-.t....cc.g..t.ta...tc..a.....g..a..aat..
S-2	.ctt.cat...ttct..cg....ccggt..c...g..t.cg....ct..c.c.c....gt...aa..a.
S-3	..t....t..a..-..........t.......c................c...........
S-4	...................t...........-.........t.....
S-5	...................................t.....
S-6	..t....t..a.-............tc........c.g.....c.......
S-7	.ctt.cat...ttct..cg.....tccggt..c...g..t.cg....ct..c.c.c....gt...aa..a.
Y-1	..t....a.-..t....t......-...............c.......t.....
Y-2	.ctt.ca....ttct..cg.t...ccggt..c...g.tt.cg..tct..c.c.ctc....t.t..aa..at
Y-3	..t....cat.-........c.g.ctc.....c...c.g......tt....t.c..cg...ac.a.t..
Y-4	.ctt.ca....ttct..cg.....tccggt..c...g..t.cg....ct..c.c.c.c....gt...aa..a.
Y-5	................................c................t......
Y-6	..t....t..a.-..........c...............c.....
Y-7	..t....a.-...............t.....................c.....
Y-8	tcttt.a.c...attct.cc....t..cg....g...g......-g.....t..ctt.ctcc-....gt..caa.aa.
Y-9	.ctt.ca....ttct..cg.....tccggt..c...g..t.cg....ct..c.c.c.c....gt...aa..at
Bs-TJFK	..t.c..t..a.-..........t...........c...........c.....
Bs-P	..t.a...c..t-.t.....cc.g..t.ta...tc..a.......g..a..aat...
Bs-V	.ctt.a...t.a.tc.t.c.c.tt.ccgg....c.atg....cg....cg....c.c.c..t...gt...aa..a.
Bs-Y	.ctt.cat...ttct..cg.....tccggt..c...g..t.cgg..ct..c.c.c.c....gt...aa..a.
Bs-N	.cttt.a....attct.cc....t..cg.......-g.....t..ctt.ctcc....gt..caa.aa.

Table S3 Summary of observed haplotypes in each sample collected from Bossou

RM lineage[a]	Individual name[b]	Sample type	No. sample[c]	No. majority[d]	No. inconsistency[e]	RM haplotype[f]	No. expected/no. total[g]
F	Fana +	Hair from night bed	3(2)	3(2)	0	Bs-TJFK	6/6
		Feces	1(0)	1	0		
	FOAF	Hair from ground	2(0)	2	0	Bs-TJFK	
J	Jire +	Hair from night bed	2(0)	2	0	Bs-TJFK	5/5
		Hair from ground	1(0)	1	0		
	Juru	Hair from ground	1(0)	1	0	Bs-TJFK	
		Feces	1(0)	1	0		
K	Kai	Hair from night bed	5(4)	2(2)	3(2)	Bs-TJFK	16/26
		Hair from ground	11(6)	5(2)	6(4)		
		Urine	11(0)	10	1		
N	Nina	Hair from night bed	4(2)	4(2)	0	Bs-N	8/9
		Hair from ground	1(0)	1	0		
	Nto	Hair from night bed	2(2)	2(2)	0	Bs-N	
		Hair from ground	1(1)	0	1		
		Feces	1(0)	1	0		
P	Pama	Hair from ground	3(0)	3	0	Bs-P	5/5
		Feces	1(0)	1	0		
	Pili +	Hair from night bed	1(0)	1	0	Bs-P	
T	TUA	Hair from ground	1(1)	1	0	Bs-TJFK	4/4
		Urine	3(0)	3	0		

V	Velu	Hair from ground	3(2)	0	3(2)	Bs-V	16/19
		Urine	5(0)	5	0		
	Vuavua	Urine	11(0)	11	0	Bs-V	
Y	Yo	Hair from night bed	3(0)	3	0	Bs-Y	8/8
		Feces	1(0)	1	0		
	YOLO	Urine	4(0)	4	0	Bs-Y	

[a] Recorded maternal (RM) lineages based on field observation

[b] The first capital character of an individual name indicates its RM lineage. Names of males are represented in capital letters. A plus sign represents females with dependent young

[c] Number of samples whose mtDNA sequences were obtained. Number in parentheses shows the number of results obtained by two-round polymerase chain reaction (PCR)

[d] Number of samples yielding a majority of detected haplotypes in each RM lineage (RM haplotype), which is the authentic haplotype determined by selecting majority when more than two haplotypes are detected for a RM. Number in parentheses shows the number of results obtained by two-round PCR

[e] Number of samples not yielding a RM haplotype. Number in parentheses shows number of results obtained by two-round PCR

[f] Sequence ID of the RM haplotype determined by selecting majority of detected haplotypes in each RM lineage

[g] Number of samples yielding a RM haplotype / number of samples yielding a mtDNA sequence

Appendix F Putative Identification of Bacterial SSU rRNA Gene (Partial) Amplified from Feces of Bossou Chimpanzees

Number of detected individuals ($N_{total}=12$)	OTU[a] Name	Putative identification by BLAST[b] at Family level	Nearest known species suggested by BLAST[b]
12	#3	Bacteroidetes; Bacteroidetes (class); Bacteroidales; Prevotellaceae	*Prevotella* sp.
12	#142	Firmicutes; Mollicutes; Anaeroplasmatales; Erysipelotrichaceae	*Holdemania filiformis*
10	#2	Bacteroidetes; Bacteroidetes (class); Bacteroidales; Prevotellaceae	*Prevotella* sp.
10	#372	Firmicutes; Mollicutes; Anaeroplasmatales	*Anaeroplasma varium*
9	#42	Bacteroidetes; Bacteroidetes (class); Bacteroidales; Prevotellaceae	*Prevotella nigrescens*
7	#13	Bacteroidetes; Bacteroidetes (class); Bacteroidales; Prevotellaceae	*Prevotella* sp.
7	#61	Bacteroidetes; Bacteroidetes (class); Bacteroidales; Prevotellaceae	*Prevotella* sp.
7	#40	Proteobacteria; Gammaproteobacteria; Pseudomonadales; Moraxellaceae	*Acinetobacter* sp.
7	#311	Firmicutes; Clostridia; Clostridiales; Lachnospiraceae	*Catonella morbi*
7	#55	Firmicutes; Clostridia; Clostridiales; Clostridiaceae	*Anaerotruncus colihominis*
6	#23	Firmicutes; Mollicutes; Anaeroplasmatales; Anaeroplasmataceae	*Anaeroplasma varium*
5	#295	Proteobacteria; Gammaproteobacteria; Aeromonadales; Succinivibrionaceae	*Succinivibrio dextrinosolvens*
5	#64	Bacteroidetes; Bacteroidetes (class); Bacteroidales; Prevotellaceae	*Prevotella* sp.
5	#218	Firmicutes; Clostridia; Clostridiales; Lachnospiraceae	*Ruminococcus obeum*
5	#562	Firmicutes; Clostridia; Clostridiales; Clostridiaceae	*Clostridium leptum*
4	#187	Bacteroidetes; Bacteroidetes (class); Bacteroidales; Prevotellaceae	*Prevotella albensis*
4	#513	Bacteroidetes; Bacteroidetes (class); Bacteroidales; Porphyromonadaceae	*Dysgonomonas gadei*
4	#197	Firmicutes; Clostridia; Clostridiales; Eubacteriaceae	*Eubacterium infirmum*
4	#26	Firmicutes; Clostridia; Clostridiales; Eubacteriaceae	*Eubacterium dolichum*

(continued)

Appendix F (continued)

Number of detected individuals ($N_{total}=12$)	OTU[a] Name	Putative identification by BLAST[b] at Family level	Nearest known species suggested by BLAST[b]
4	#78	Firmicutes; Clostridia; Clostridiales; Clostridiaceae	*Clostridiaceae str.*
4	#432	Actinobacteria; Actinobacteridae; Bifidobacteriales	*Bifidobacterium* sp.
4	#219	Firmicutes; Mollicutes; Anaeroplasmatales; Anaeroplasmataceae	*Anaeroplasma abactoclasticum*
3	#10	Proteobacteria; Deltaproteobacteria; Syntrophobacterales; Syntrophaceae	*Syntrophus buswellii*
3	#200	Firmicutes; Lactobacillales; Streptococcaceae	*Streptococcus gallolyticus*
3	#33	Bacteroidetes; Bacteroidetes (class); Bacteroidales; Prevotellaceae	*Prevotellaceae bacterium*
3	#502	Bacteroidetes; Bacteroidetes (class); Bacteroidales; Prevotellaceae	*Prevotella veroralis*
3	#124	Bacteroidetes; Bacteroidetes (class); Bacteroidales; Prevotellaceae	*Prevotella* sp.
3	#4	Bacteroidetes; Bacteroidetes (class); Bacteroidales; Prevotellaceae	*Prevotella salivae*
3	#45	Bacteroidetes; Bacteroidetes (class); Bacteroidales; Prevotellaceae	*Prevotella oulora*
3	#216	Bacteroidetes; Bacteroidetes (class); Bacteroidales; Prevotellaceae	*Prevotella nigrescens*
3	#387	Bacteroidetes; Bacteroidetes (class); Bacteroidales; Prevotellaceae	*Prevotella nigrescens*
3	#356	Bacteroidetes; Bacteroidetes (class); Bacteroidales; Prevotellaceae	*Prevotella buccae*
3	#323	Bacteroidetes; Bacteroidetes (class); Bacteroidales; Prevotellaceae	*Prevotella albensis*
3	#86	Actinobacteria; Actinobacteridae; Act inomycetales;Propionibacterineae; Nocardioidaceae	*Nocardioides* sp.
3	#143	Firmicutes; Lactobacillales; Lactobacillaceae	*Lactobacillus vitulinus*
3	#313	Firmicutes; Clostridia; Clostridiales; Lachnospiraceae	*Ruminococcus obeum*
3	#361	Firmicutes; Clostridia; Clostridiales; Lachnospiraceae	*Ruminococcus callidus*
3	#104	Firmicutes; Clostridia; Clostridiales; Lachnospiraceae	*Coprococcus eutactus*
3	#140	Firmicutes; Mollicutes; Anaeroplasmatales; Erysipelotrichaceae	*Holdemania filiformis*
3	#282	Firmicutes; Mollicutes; Anaeroplasmatales; Erysipelotrichaceae	*Holdemania filiformis*

(continued)

Appendix F (continued)

Number of detected individuals (N_{total}=12)	OTU[a] Name	Putative identification by BLAST[b] at Family level	Nearest known species suggested by BLAST[b]
3	#416	Actinobacteria; Coriobacteridae; Coriobacteriales; Coriobacterineae; Coriobacteriaceae	*Slackia heliotrinreducens*
3	#319	Firmicutes; Clostridia; Clostridiales; Clostridiaceae	*Clostridium sporosphaeroides*
3	#50	Firmicutes; Clostridia; Clostridiales; Clostridiaceae	*Clostridium spiroforme*
3	#80	Bacteroidetes; Bacteroidetes (class); Bacteroidales; Bacteroidaceae	*Porphyromonas-like* sp.
3	#59	Firmicutes; Clostridia; Clostridiales; Acidaminococcaceae	*Dialister pneumosintes*
3	#220	Firmicutes; Mollicutes; Acholeplasmatales; Acholeplasmataceae	*Acholeplasma morum*
2	#298	Proteobacteria; Gammaproteobacteria; Aeromonadales; Succinivibrionaceae	*Succinimonas amylolytica*
2	#43	Firmicutes; Lactobacillales; Streptococcaceae	*Streptococcus pleomorphus*
2	#30	Spirochaetes; Spirochaetales; Spirochaetaceae	*Treponema* sp.
2	#391	Bacteroidetes; Bacteroidetes (class); Bacteroidales; Prevotellaceae	*Prevotella veroralis*
2	#403	Bacteroidetes; Bacteroidetes (class); Bacteroidales; Prevotellaceae	*Prevotella veroralis*
2	#1	Bacteroidetes; Bacteroidetes (class); Bacteroidales; Prevotellaceae	*Prevotella* sp.
2	#62	Bacteroidetes; Bacteroidetes (class); Bacteroidales; Prevotellaceae	*Prevotella* sp.
2	#352	Bacteroidetes; Bacteroidetes (class); Bacteroidales; Prevotellaceae	*Prevotella* sp.
2	#208	Bacteroidetes; Bacteroidetes (class); Bacteroidales; Prevotellaceae	*Prevotella* sp.
2	#119	Bacteroidetes; Bacteroidetes (class); Bacteroidales; Prevotellaceae	*Prevotella salivae*
2	#7	Bacteroidetes; Bacteroidetes (class); Bacteroidales; Prevotellaceae	*Prevotella ruminicola*
2	#192	Bacteroidetes; Bacteroidetes (class); Bacteroidales; Prevotellaceae	*Prevotella ruminicola*
2	#211	Bacteroidetes; Bacteroidetes (class); Bacteroidales; Prevotellaceae	*Prevotella nigrescens*
2	#17	Bacteroidetes; Bacteroidetes (class); Bacteroidales; Prevotellaceae	*Prevotella multiformis*
2	#446	Bacteroidetes; Bacteroidetes (class); Bacteroidales; Prevotellaceae	*Prevotella multiformis*
2	#8	Bacteroidetes; Bacteroidetes (class); Bacteroidales; Prevotellaceae	*Prevotella denticola*

(continued)

Appendix F (continued)

Number of detected individuals (N_{total}=12)	OTU[a] Name	Putative identification by BLAST[b] at Family level	Nearest known species suggested by BLAST[b]
2	#390	Bacteroidetes; Bacteroidetes (class); Bacteroidales; Prevotellaceae	*Prevotella buccalis*
2	#194	Bacteroidetes; Bacteroidetes (class); Bacteroidales; Prevotellaceae	*Prevotella buccae*
2	#74	Bacteroidetes; Bacteroidetes (class); Bacteroidales; Prevotellaceae	*Prevotella albensis*
2	#121	Bacteroidetes; Bacteroidetes (class); Bacteroidales; Prevotellaceae	*Prevotella albensis*
2	#189	Bacteroidetes; Bacteroidetes (class); Bacteroidales; Prevotellaceae	*Prevotella albensis*
2	#191	Bacteroidetes; Bacteroidetes (class); Bacteroidales; Prevotellaceae	*Prevotella albensis*
2	#330	Bacteroidetes; Bacteroidetes (class); Bacteroidales; Prevotellaceae	*Prevotella albensis*
2	#484	Bacteroidetes; Bacteroidetes (class); Bacteroidales; Porphyromonadaceae	*Bacteroides distasonis*
2	#100	Actinobacteria; Actinobacteridae; Act inomycetales;Propionibacterineae; Nocardioidaceae	*Nocardioides* sp.
2	#144	Firmicutes; Lactobacillales; Lactobacillaceae	*Lactobacillus vitulinus*
2	#105	Firmicutes; Clostridia; Clostridiales; Lachnospiraceae	*Ruminococcus* sp.
2	#103	Firmicutes; Clostridia; Clostridiales; Lachnospiraceae	*Ruminococcus obeum*
2	#196	Firmicutes; Clostridia; Clostridiales; Lachnospiraceae	*Ruminococcus obeum*
2	#81	Firmicutes; Clostridia; Clostridiales; Lachnospiraceae	*Ruminococcus callidus*
2	#85	Firmicutes; Clostridia; Clostridiales; Lachnospiraceae	*Ruminococcus callidus*
2	#172	Firmicutes; Clostridia; Clostridiales; Lachnospiraceae	*Roseburia faecalis*
2	#79	Firmicutes; Clostridia; Clostridiales; Lachnospiraceae	*Lachnospiraceae bacterium*
2	#307	Firmicutes; Clostridia; Clostridiales; Lachnospiraceae	*Catonella morbi*
2	#22	Firmicutes; Clostridia; Clostridiales; Lachnospiraceae	*Lachnospiraceae oral clone MCE9*
2	#314	Firmicutes; Clostridia; Clostridiales; Lachnospiraceae	*Lachnobacterium* sp.
2	#19	Proteobacteria; Deltaproteobacteria; Desulfuromonadales; Geobacteraceae	*Geobacter* sp.
2	#333	Bacteroidetes; Sphingobacteria; Sphingobacteriales; Flexibacteraceae	*Cytophaga* sp.

(continued)

Appendix F (continued)

Number of detected individuals (N_{total} = 12)	OTU[a] Name	Putative identification by BLAST[b] at Family level	Nearest known species suggested by BLAST[b]
2	#336	Bacteroidetes; Sphingobacteria; Sphingobacteriales; Flexibacteraceae	*Cytophaga* sp.
2	#25	Firmicutes; Clostridia; Clostridiales; Eubacteriaceae	*Eubacterium* sp.
2	#112	Firmicutes; Mollicutes; Anaeroplasmatales; Erysipelotrichaceae	*Holdemania filiformis*
2	#118	Firmicutes; Mollicutes; Anaeroplasmatales; Erysipelotrichaceae	*Holdemania filiformis*
2	#286	Firmicutes; Mollicutes; Anaeroplasmatales; Erysipelotrichaceae	*Holdemania filiformis*
2	#72	Firmicutes; Mollicutes; Anaeroplasmatales; Erysipelotrichaceae	*Erysipelothrix rhusiopathiae*
2	#238	Firmicutes; Mollicutes; Anaeroplasmatales; Erysipelotrichaceae	*Erysipelothrix rhusiopathiae*
2	#109	Firmicutes; Mollicutes; Anaeroplasmatales; Erysipelotrichaceae	*Bulleidia moorei*
2	#111	Firmicutes; Mollicutes; Anaeroplasmatales; Erysipelotrichaceae	*Bulleidia moorei*
2	#51	Actinobacteria; Coriobacteridae; Coriobacteriales; Coriobacterineae; Coriobacteriaceae	*Slackia heliotrinreducens*
2	#39	Firmicutes; Clostridia; Clostridiales	*Epulopiscium* sp.
2	#304	Firmicutes; Clostridia; Clostridiales; Clostridiaceae	*Eubacterium formicigenerans*
2	#239	Firmicutes; Clostridia; Clostridiales; Clostridiaceae	*Clostridium* sp.
2	#345	Firmicutes; Clostridia; Clostridiales; Clostridiaceae	*Clostridium scindens*
2	#254	Firmicutes; Clostridia; Clostridiales; Clostridiaceae	*Clostridium paraputrificum*
2	#267	Firmicutes; Clostridia; Clostridiales; Clostridiaceae	*Clostridium paraputrificum*
2	#269	Firmicutes; Clostridia; Clostridiales; Clostridiaceae	*Clostridium paraputrificum*
2	#270	Firmicutes; Clostridia; Clostridiales; Clostridiaceae	*Clostridium paraputrificum*
2	#281	Firmicutes; Clostridia; Clostridiales; Clostridiaceae	*Clostridium paraputrificum*

(continued)

Appendix F (continued)

Number of detected individuals ($N_{total}=12$)	OTU[a] Name	Putative identification by BLAST[b] at Family level	Nearest known species suggested by BLAST[b]
2	#150	Firmicutes; Clostridia; Clostridiales; Clostridiaceae	*Clostridium disporicum*
2	#383	Firmicutes; Clostridia; Clostridiales; Clostridiaceae	*Clostridium botulinum*
2	#349	Firmicutes; Clostridia; Clostridiales; Clostridiaceae	*Acetivibrio cellulolyticus*
2	#350	Firmicutes; Clostridia; Clostridiales; Clostridiaceae	*Acetivibrio cellulolyticus*
2	#351	Firmicutes; Clostridia; Clostridiales; Clostridiaceae	*Acetivibrio cellulolyticus*
2	#206	Firmicutes; Bacillales; Bacillaceae	*Bacillus* sp.
2	#76	Firmicutes; Mollicutes; Anaeroplasmatales; Anaeroplasmataceae	*Asteroleplasma anaerobium*
2	#167	Firmicutes; Mollicutes; Anaeroplasmatales; Anaeroplasmataceae	*Anaeroplasma bactoclasticum*
2	#363	Proteobacteria; Betaproteobacteria; Burkholderiales; Alcaligenaceae	*Sutterella wadsworthensis*
2	#173	Firmicutes; Clostridia; Clostridiales; Acidaminococcaceae	*Selenomonas sputigena*
2	#107	Firmicutes; Mollicutes; Acholeplasmatales; Acholeplasmataceae	*Acholeplasma granularum*
2	#24	Firmicutes; Mollicutes; Acholeplasmatales; Acholeplasmataceae	*Acholeplasma axanthum*

In total 1,074 *E. coli* clones were subjected to Restriction Fragment Length Polymorphism (RFLP) Analysis. Of about 600 OTU identified, 117 OTU that were shared by at least two individuals were subjected to further sequence analyses

For details for PCR amplification of eubacterial SSU rRNA genes, please see Uenishi et al. (2007)

[a]OTU Operational Taxonomic Unit

[b]Basic Local Alignment Search Tool

Appendix G List of Collaborators: International Researchers and Students, Head Administrators and Staff in Guinea

Table 1 Researchers and students

Name	Most recent affiliation[a]	Topic of research or status	Research site[b]	Years of visit[c]
BAMAMOU Cécé	ISAV & IREB, G	Student	*B, S, R*	2005–2008
BAMY Fampé	Guinée Ecologie, G	Researcher	*B*	2004
BARRY Yaya	ISAV, G	Researcher	*B*	2007
BILIVOGUI Papa	ISAV, G	Student	*B*	2004, 2008
Dr. BIRO Dora	Oxford Univ, UK	Cognition, Social learning and tool use	*B*	1998–2005
CAMARA Muctar	ICCRDG, G	Researcher	*B*	1985
CAMARA Nawa	ISAV, G	Researcher	*E*	2007
CARVALHO Susana	Cambridge Univ, UK	Stone-tool use, Archeology	*B, D*	2006–2009
Dr. CELLI Maura	KUPRI, J	Tool use	*B*	1998
CONDÉ Néma	UC, G	Student	*B*	2003
COULIBALY Siliman	ICCRDG, G	Researcher	*B*	1991
COUMBASSA Abdoulaye	INRDG, G	Researcher	*B*	1976–1977
DIALLO Abdou	DNRST, G	Researcher	*B*	1986
DIALLO Sarabailo	ISAV, G	Researcher	*B, GC*	2004, 2006
DIALLO Ousmane	UC, G	Student	*B*	2006
DORÉ Togna	ISAV, G	Student	*GC*	2007
DOUMOBOUYA Mariam	ISAV, G	Student	*GC*	2005
DRAMÉ Gausu	DNRST, G	Researcher	*B*	1986–1987
FOFANA Mory	CUZ, G	Student	*S, R*	2006
Dr. FUJISAWA Michiko	WRC-KU, J	Geriatrics, Enrichment in captivity	*B*	2008–2010
Dr. FUJITA Shiho	Yamaguchi Univ, J	Reproduction, and hormones	*B*	1999–2004
Dr. FUSHIMI Takao	Kitasato Univ, J	Field experiment, Stone-tool use	*B*	1989
GAMYS Cé	CUZ, G	Student	*S, R*	2007
GAMYS Joël	CI, Liberia	Student	*B, S*	2003–2005
GBONIMI Pé	ISAV, G	Researcher	*E*	2007
GOTO Ryutaro	KU, J	Plant-insect interaction	*B*	2005
GRANIER Nicolas	Liège Univ, Belgium	Habitat-use, Conservation	*B, G, GC, R, Y*	2002–2010

(continued)

Table 1 (continued)

Name	Most recent affiliation[a]	Topic of research or status	Research site[b]	Years of visit[c]
GUÉMI Essaie	CUZ, G	Student	*S, R*	2006
HABA Yoki	ISAV, G	Student	*GC*	2005
HASEGAWA Ryo	Phytoculture Control Co. Ltd., J	Hexatubes, Maintenance	*GC*	2005–2007
Dr. HAYAKAWA Sachiko	KUPRI, J	Genetic analysis	*B*	1999
Dr. HAYASHI Misato	KUPRI, J	Field experiment, Stone-tool use	*B*	2002–2009
Dr. HIRATA Satoshi	Great Ape Research Institute, J	Tool use, social interactions	*B, D*	1996–2000
HIROSAWA Mari	KUPRI, J	Enrichment in captivity	*B*	2008
Dr. HOCKINGS Kimberley	New Univ of Lisbon, Portugal	Human–chimpanzee interactions, Behavioral adaptations	*B*	2004–2008
Dr. HUFFMAN Michael	KUPRI, J	Medicinal plants use	*B*	1997
Dr. HUMLE Tatyana	Kent Univ, UK	Tool use, Ecology and culture, Conservation	*B, S, Y*	1995–2010
Dr. HUYNEN Marie-Claude	Liège Univ, Belgium	Ecology, Conservation	*B*	2001
Dr. INOUE-NAKAMURA Noriko	Showa Women's Univ, J	Field experiment, Stone-tool use	*B*	1992–1996
ITO Miho	AAAS-KU, J	Raphia-palm utilization by local people	*S*	2004–2007
KABA Mory	ISAV, G	Researcher	*B*	2004
Dr. KABASAWA Asami	AAAS-KU, J	Conservation, Chimpanzee sanctuaries	*B*	2003–2004
KADOTA Chiemi	Wanpark Kochi Animal Land, J	Enrichment in captivity	*B*	2001
KASAHARA Yokiro	School of Medicine-KU, J	Geriatrics	*B*	2008
Dr. KATO Akino	KUPRI, J	Veterinary medicine, Behavior	*B*	2004
KEITA Alpha	SSMNZ, G	Researcher	*B*	1976–1977
KIYONO Mieko	KU, J	Behavior	*B*	2006
Dr. KOBAYASHI Shigeo	AAAS-KU, J	Fuel wood consumption and water utilization by humans	*B*	2004–2007

(continued)

Table 1 (continued)

Name	Most recent affiliation[a]	Topic of research or status	Research site[b]	Years of visit[c]
KOMAN Jérémy	SSMNZ, G	Researcher	*B, S*	1976–1990
KOOPS Kathelijne	Cambridge Univ, UK	Elementary technology, Feeding ecology	*B, S*	2003–2008
KOUROUMA Abdoulaye	ISAV, G	Researcher	*B*	2004
Dr. KOUROUMA Makan	IREB, G	Researcher	*GC*	2003–2008
KOUROUMA Sékou	ISAV, G	Student	*B*	2004–2005
KPOGHOMOU Elie	CEGENS, G	Researcher	*B, R, G*	2003–2008
KUMAZAKI Kiyonori	KUPRI, J	Enrichment in captivity	*B*	1991
LENO Arnaud	UC, G	Student	*B*	2006
MAOMY Nyan	CUZ, G	Student	*B, GC*	2007
Dr. MARCHANT Linda	Miami Univ, USA	African apes behavior, Hominid evolution,	*B*	2004
MARTIN Christopher	KUPRI, J	Behavior, Field experiment	*B*	2007
Dr. MARTINEZ Laura	Ewha Womans Univ, Korea	Vocal communication, Conservation	*B, Y*	2002–2009
Dr. MATSUZAWA Tetsuro	KUPRI, J (Field Site Director)	Tool use, Field experiment	*B, S, Y, D*	1986–2010
Dr. McGREW William	Cambridge Univ, UK	Socio-ecology, Evolution of material culture	*B, S*	2004–2007
MIWA Nobukatsu	KUPRI, J	Enrichment in captivity	*B*	1991
Dr. MIZUNO Kazuharu	AAAS-KU, J	Landscape ecology, Geology	*B*	2009
Dr. MORIMURA Naruki	WRC-KU, J	Enrichment in captivity, Behavior	*B*	1998, 2009–2010
Dr. MYOWA-YAMAKOSHI Masako	KU, J	Playing behavior, Mother–infant interaction	*B*	1996–1999
NAKAMURA Miho	ANC Productions Inc. & WRC-KU, J	Video recording	*B*	1990, 1996, 2000
Dr. NAKAMURA Michio	WRC-KU, J	Behavioral comparison with Mahale	*B*	2002
NAKATSUKA Masahiro	School of Medicine-KU, J	Geriatrics	*B*	2008

(continued)

Table 1 (continued)

Name	Most recent affiliation[a]	Topic of research or status	Research site[b]	Years of visit[c]
NOGAMI Etsuko	WRC-KU, J	Behavior, Enrichment in captivity	B	2005
NYAMY Issac	ISAV, G	Student	B	2007
OCHIAI-OHIRA Tomomi	KUPRI, J	Behavior, Time budget	B	1998
OHASHI Gaku	KUPRI, J	Fission–fussion and mating, Culture	GC, B, L	1999–2010
OHNO Hisato	Railway Technical Research Institute, J	Behavior	B	1989
Dr. OKAMOTO Sanae	Maastricht Univ, The Netherlands	Cognitive development, Mother–infant interaction	B	2010
PHILLIPS Caroline	Cambridge Univ, UK	Elementary technology, Feeding ecology	S	2007
SACKO Kalil	ISAV, G	Student	GC	2005
SACKO Mory	ISAV, G	Student	B	2004
SAGNO Mory	ISAV, G	Student	GC	2005
SAID Nawab	ISAV, G	Researcher	B	2007
Dr. SAKURA Osamu	Tokyo Univ, J	Party composition, Tool use	B	1987–1990
SANGBÉ Nyan	CUZ, G	Student	GC	2005
Dr. SHIMADA Makoto	Institute of Genetics and Biological Research, J	Genetic analysis	B, S, Y, G	1999
SONOMY Labilé	ISAV, G	Researcher	GC	2007
Dr. SOUMAH Aly-Gaspard	UC, G	Site management, Conservation	B	1982–1983, 2008–2010
SOUMAHORO Ibrahima	ISAV, G	Student	B	2004
Dr. SOUSA Claudia	New Univ of Lisbon, Portugal	Tool use and social learning, Vocal communication	B, D	2000–2006
SOW Mamadou Bhoye	DNEF, G	Researcher	B	1987
Dr. SUGIYAMA Yukimaru	KUPRI, J (Field Site Founder)	Population dynamics, Socio-ecology and culture	B, S, G	1975–2008
Dr. TAKEMOTO Hiroyuki	KUPRI, J	Forest and feeding ecology interactions	B	1995–2008

(continued)

Table 1 (continued)

Name	Most recent affiliation[a]	Topic of research or status	Research site[b]	Years of visit[c]
THÉA Emmanuel	ISAV, G	Student	R	2007
Dr. TONOOKA Rikako	Tokai Women's Univ, J	Leaf-folding tool use	B	1992–1995
TRAORÉ Gopou	CUZ, G	Student	GC	2006
UCHIDA Akiko	Waseda Univ, J	Urinary hormones	B	1996
UENISHI Gentaro	KPU, J	Intestinal microbiology	B	2005
Dr. USHIDA Kazunari	KPU, J	Intestinal microbiology	B	2004–2007
Dr. WALKER Polly	Oxford Univ, UK	Behavior	B	2000
Dr. YAMAKOSHI Gen	AAAS-KU, J	Landscape ecology, tool use and feeding	B	1992–2009
YAMANASHI Yumi	KUPRI, J	Time budget, Enrichment in captivity	B	2009
Dr. YAMAMOTO Shinya	KUPRI, J	Tool use, social interaction	B	2004
YOKOTA Naoto	Oita Junior College, J	Topography, vegetation	B	1992
ZEOMY Kolapé	ISMVD, G	Student	B	2006

[a]*Abbreviations for affiliations*: **AAAS-KU**, Asian and African Areas Studies – Kyoto University; **CEGENS**, Centre d'Etude et de Gestion de l'Environnement du Nimba et de Simandou (Center for the Study and the Management of Nimba and Simandou Environments); **CI**, Conservation International; **CUZ**, Centre Universitaire de N'Zérékoré (N'Zérékoré University); **ISMVD**, Institut Supérieur de Médecine Vétérinaire de Dalaba (Higher Institute of Veterinarian Medecine of Dalaba); **DNEF**, Direction Nationale des Eaux et Forêts (National Department of Forestry); **DNRST**, Direction Nationale de la Recherche Scientifique et Technologique (National Department of Scientific and Technological Research); **G**, Republic of Guinea; **ICCRDG**, Institut Central de Coopération pour la Recherche et Documentation de Guinée (Guinean Central Institute of Cooperation for Research and Documentation); **INRDG**, Institut National pour la Recherche et Documentation de Guinée (Guinean Institute for Research and Documentation); **IREB**, Institut de Recherche Environnementale de Bossou (Environmental Research Institute of Bossou); **ISAV**, Institut des Sciences Agronomique et Vétérinaire (Institute of Agronomic and Veterinarian Sciences); **KPU**, Kyoto Prefectural University; **J**, Japan; **KU**, Kyoto University; **KUPRI**, Kyoto University Primate Research Institute; **SSMNZ**, Station Scientifique du Mont Nimba de Ziéla (Scientific Station of the Nimba Mountains – Ziéla); **UC**, Université de Conakry (Conakry University); **Univ**, University; **WRC-KU**, Wildlife Research Center – Kyoto University
[b]*Abbreviations*: **B**, Bossou; **D**, Diecké; **E**, Leyba plain; **G**, Gouéla; **GC**, Green Corridor; **L**, Liberia (Nimba County); **R**, Déré; **S**, Séringbara; **Y**, Yealé
[c]Comma: years of visit up to 2010; hyphen: period of regular visits up to 2010

Table 2 Head administrators at Conakry and staff at Bossou–Nimba sites

Name	Affiliation[a]	Function[b]. Site[c]	Years in function[d]
Dr. BALAMOU Jean-Pierre	ICCRDG	Ad National Director. C	1982–1983
BONIMY Soh Pletah	KUPRI	Environmental Educator. B, L	2005–2010†
CAMARA Fodé	IREB	Maintenance. B	2003–p
Dr. CAMARA Vassydan	DNRST	Ad National Director. C	2008–p
CHÉRIF Paquilé	IREB, KUPRI	Field Botanist. B, S	1995–p
	ENATEF		2006–2007
CONDÉ Iba	IREB	Ad Director. B	2001–p
Dr. COULIBALY Bakary	DNRST	Ad National Director. C	1991–2000
DELAMOU Ouo-Ouo	IREB	Genetic Resources Dept. B	2001–p
Dr. DIAKITÉ Mamadou	IREB	Director. B	2001–2003
DORÉ Fromo	KUPRI	Field Assistant. S	1999–p
DORÉ Gilles	KUPRI	Field Assistant. B	1998–p
DORÉ Kassié	IREB, KUPRI	Field Assistant. S	1999–p
DORÉ Marcel	KUPRI	Field Assistant. B	2000–2007
DORÉ Paquilé	IREB	Accountant. B	2001–p
DORÉ Seraphin	KUPRI	Camp Cook. S	2007–p
DOUNAHARA Delphine	IREB	Secretary. B	2004–p
DROH David	KUPRI	Field Assistant. Y	1993–p
GBÉREGBÉ C. Henry	KUPRI	Field Assistant. B, S	2003–p
	ENATEF		2009–p
GBIAN Pierre	KUPRI	Driver. B	2003–p
GOGO Anatole	KUPRI	Field Assistant. Y	1993–p
GONDO Pascal	KUPRI	Field Assistant. Y	2003–p
GOPOU Anthony	KUPRI	Field Assistant. Y	1999–p
GOUMY Cé	IREB, KUPRI	Poaching Surveillance. B, GC	2003–p
GOUMY Guano	KUPRI	First Field Assistant, Consultant. B	1976–2000 2001–2006†
GOUMY Pascal	IREB, KUPRI	Field Assistant. B, L	1993–p
HABA Michel	IREB	Technician. B	2001–p
KOLIÉ Cécé	IREB	Primatology Dept. B	2001–p
Prof. KANTÉ Kabiné	INRDG DNRST	National Director. C	1991–2008
Dr. KEITA Mamby	DNRST	National Director. C	2008–p
Dr. KEITA Sidiki	ICCRDG DNRST	National Director. C	1982–1983 1985–1986
Dr. KOUROUMA Makan	IREB	Director. B	2003–2008
PAHON Philibert	KUPRI	Field Assistant. Y	1993–p
SAMY Dagouka	IREB, KUPRI	Guardian, Housekeeping & Temperature records. B	1993–p
SACKO Djémory	IREB	Scientific Secretary. B	2001–p
Dr. SOUMAH Aly-Gaspard	IREB	Director. B	2009–p

(continued)

Table 2 (continued)

Name	Affiliation[a]	Function[b]. *Site*[c]	Years in function[d]
Dr. SOUMAH Fodé	DNRST	National Director. *C*	1987–1991
Dr. TAGBINO Tamba	DNRST	Ad National. *C*	2000–2008
		Director	2008–2010
TOUNKARA Jean	IREB	Computer Specialist. *B*	2006–p
TRAORÉ Nyonko	KUPRI	Cook. *B*	1989–2006†
TRAORÉ Rémy	KUPRI	Surveillance. GC	2003–p
TONGA Ferdinand	KUPRI	Cook & Logistics. *Y*	2007–p
		Assistant Guide	1993–1999
WONSEU Alexis	KUPRI	Field Assistant. *Y*	2001–p
ZOGBÉLÉMOU Michel	KUPRI	Field Assistant. *G*	2006–p
ZOGBILA Boniface	KUPRI	Field Assistant. *B, GC*	2000–p
ZOGBILA Buna	IREB, KUPRI	Surveillance & Maintenance. GC	2003–p
ZOGBILA Cé	IREB, KUPRI	Maintenance. *B, GC*	2002–2009†
ZOGBILA Fokayé	KUPRI	Field Assistant. *S*	2003–p
ZOGBILA Souah	IREB, KUPRI	Driver. *B*	2003–p
ZOGBILA Tino	KUPRI	Second Field Assistant. *B*	1982–1997
		Consultant	1998–2005†
ZOUMANIGUI Kognon	IREB	Documentation Dept. *B*	2001–p

[a]*Abbreviations*: **DNRST**, Direction Nationale de la Recherche Scientifique et Technique (National Department of Scientific and Technical Research); **ENATEF**, Ecole Nationale des Agents Techniques des Eaux et Forêts de Mamou (National School for Forest Service Technicians); **ICCRDG**, Institut Central de Coopération pour la Recherche et Documentation de Guinée (Guinean Central Institute of Cooperation for Research and Documentation); **INRDG**, Institut National pour la Recherche et Documentation de Guinée (Guinean Institute for Research and Documentation); **IREB**, Institut de Recherche Environnementale de Bossou (Environmental Research Institute of Bossou); **KUPRI**, Kyoto University Primate Research Institute.

[b]*Abbreviations*: **B**, Bossou; **C**, Conakry; **GC**, Green Corridor; **L**, Liberia (Nimba county); **R**, Déré; **S**, Séringbara; **Y**, Yealé

[c]*Abbreviations*: **Ad**, Adjunct; **Dept,** Department

[d]*Abbreviations*: **p**, present (up to 2010); †, date of death

References

Act for the Establishment of the East Nimba Nature Reserve (2003) Published by Authority, Ministry of Foreign Affairs, 4 November 2003, Monrovia

Adam JG (1971–1983) Flore descriptive des Monts Nimba, vols. 1–6. Mém Mus Natl Hist Nat B20. Publications Scientifiques du Muséum, Paris

Albrecht H, Dunnett SC (1971) Chimpanzees in Western Africa. Piper, Munchen

Allen SD, Emery CL, Lyerly DM (2003) *Clostridium*. In: Murray PR, Baron EJ, Jorgensen JH, Pfaller MA, Yolken RH (eds) Manual of clinical microbiology, 8th edn, vol 1. ASM Press, Washington, pp 835–856

Altman J (1974) Observational study of behavior: sampling methods. Behaviour 49:227–267

Altmann SA (1979) Baboon progressions: order or chaos? A study of one-dimensional group geometry. Anim Behav 27:46–80

Andah BW (1993) Identifying early farming traditions of west Africa. In: Shaw T, Sinclair P, Andah BW, Okpoko A (eds) The Archaeology of Africa: foods, metals and towns. Routledge, London/New York, pp 240–254

Anderson DMW, Morrison NA (1990) Identification of *Albizia* gum exudutes which are not permitted food additives. Food Addit Contam 7:175–180

Anderson JR, Williamson EA, Caster J (1983) Chimpanzees of Sapo Forest, Liberia: density, nests, tools and meat eating. Primates 24:594–601

Anderson JR, Gillies A, Lock LC (2010) Pan thanatology. Curr Biol 20:349–351

Anonymous (1738) London Magazine. 21 September 1738:464–465

Aureli F, de Waal FBM (1997) Inhibition of social behaviour in chimpanzees under high-density conditions. Am J Primatol 41:213–228

Bah M, Thiam A, Keita A, Sylla S, Barry MH, Lauriault J (1997) Monographie nationale sur la diversité biologique, GF-1605-92-74 PNUE. Ministère des Travaux Publics et de L'environnement et Direction Nationale de L'environnement. République de Guinée. 147 pp (unpublished)

Bailey G (2007) Time perspectives, palimpsests and the archaeology of time. J Anthropol Archaeol 26:198–223

Bailey RC, Head G, Jenike M, Owen B, Retchman R, Zechenter E (1989) Hunting and gathering in tropical rain forest: is it possible? Am Anthropol 91:59–82

Bajracharya SB, Furley PA, Newton AC (2005) Effectiveness of community involvement in delivering conservation benefits to the Annapurna Conservation Area, Nepal. Environ Conserv 32:239–247

Banning EB (2002) Archaeological survey. Manuals in archaeological method, theory, and technique. Kluwer, New York

Baratay E, Hardoun-Fugier E (2002) Zoo: a history of zoological gardens in the West. Reaktion Books, London

Barnett AA, Prangley ML (1996) Chimpanzee (*Pan troglodytes*) nest-making behavior in Guinea. African Primates 2:22–23

Barnett AA, Prangley ML (1997) Mammalogy in the Republic of Guinea: an overview of research from 1946 to 1996, a preliminary check-list and a summary of research recommendations for the future. Mammal Rev 27:115–164

Basabose AK (2005) Ranging patterns of chimpanzees in a montane forest of Kahuzi, Democratic Republic of Congo. Int J Primatol 26(1):33–54

Batisse M (1982) The biosphere reserve: a tool for environmental conservation and management. Environ Conserv 9:101–111

Beamont D, Suter J (2004) Outgoing Liberian government passes forest protection laws. Oryx 38(1):13

Beck B, Walkup K, Rodrigues M, Unwin S, Travis D, Stoinski T (2007) Best practice guidelines for the reintroduction of great apes. SSC Primate Specialist Group of the World Conservation Union, Gland

Bermejo M, Rodriguez-Teijeiro JD, Illera G, Barroso A, Vila C, Walsh PD (2006) Ebola outbreak killed 5000 gorillas. Science 314:1564–1564

Bicca-Marques JC, Calegaro-Marques C (1997) Single line progressions in black-and-gold howler monkeys *(Alouatta caraya)*: is there an ordered positioning? (Abstract). Am J Primatol 42:95

BirdLife International (2008) BirdLife International Data Zone. http://www.birdlife.org/datazone/home. Accessed 14 December 2010

Biro D, Inoue-Nakamura N, Tonooka R, Yamakoshi G, Sousa C, Matsuzawa T (2003) Cultural innovation and transmission of tool use in wild chimpanzees: evidence from field experiments. Anim Cogn 6:213–223

Biro D, Sousa C, Matsuzawa T (2006) Ontogeny and cultural propagation of tool use by wild chimpanzees at Bossou, Guinea: case studies in nut cracking and leaf folding. In: Matsuzawa T, Tomonaga M, Tanaka M (eds) Cognitive development in chimpanzees. Springer, Tokyo, pp 476–508

Biro D, Humle T, Koops K, Sousa C, Hayashi M, Matsuzawa T (2010) Chimpanzee mothers at Bossou, Guinea carry the mummified remains of their dead infants. Curr Biol 20: R351–R352

Biro D, Carvalho S, Matsuzawa T (2010) Tools, traditions, and technologies: interdisciplinary approaches to chimpanzee nut-cracking. In: Lonsdorf EV, Ross SR, Matsuzawa T (eds) The mind of the chimpanzee: ecological and experimental perspectives. University of Chicago Press, Chicago, pp 141–155

Black R, Sessay M (1997) Forced migration, land-use change and political economy in the forest region of Guinea. Afr Aff (Lond) 96:587–605

Blunt W (1976) The ark in the park. Hamish Hamilton, London

Boesch C (1991a) The effect of leopard predation on grouping patterns in forest chimpanzees. Behaviour 117:220–242

Boesch C (1991b) Teaching among wild chimpanzees. Anim Behav 41:530–532

Boesch C (1994) Cooperative hunting in wild chimpanzees. Anim Behav 48:653–667

Boesch C (1995) Innovation in wild chimpanzees *(Pan troglodytes)*. Int J Primatol 16:1–16

Boesch C (1996a) The emergence of cultures among wild chimpanzees. Proc Br Acad 88:251–268

Boesch C (1996b) Three approaches for assessing chimpanzee culture. In: Russon AE, Bard KA, Parker ST (eds) Reaching into thought. Cambridge University Press, Cambridge, pp 404–429

Boesch C (1997) Evidence for dominant wild female chimpanzees investing more in sons. Anim Behav 54:811–815

Boesch C, Boesch H (1981) Sex differences in the use of natural hammers by wild chimpanzees: a preliminary report. J Hum Evol 10:585–593

Boesch C, Boesch H (1983) Optimisation of nut cracking with natural hammers by wild chimpanzees. Behaviour 83:265–286

Boesch C, Boesch H (1984a) Possible causes of sex-differences in the use of natural hammers by wild chimpanzees. J Hum Evol 13:415–440

Boesch C, Boesch H (1984b) Mental map in wild chimpanzees: an analysis of hammer transports for nut cracking. Primates 25:160–170

Boesch C, Boesch H (1989) Hunting behavior of wild chimpanzees in the Taï National Park. Am J Phys Anthropol 78: 547–573

Boesch C, Boesch H (1990) Tool use and tool making in wild chimpanzees. Folia Primatol 54:86–99

Boesch C, Boesch-Achermann H (2000) The chimpanzees of the Taï forest. Oxford University Press, Oxford

Boesch C, Marchesi P, Marchesi N, Fruth B, Joulian F (1994) Is nut-cracking in wild chimpanzees a cultural behaviour? J Hum Evol 26:325–338

Bongers RM (2001) An action perspective on tool use and its development. Doctoral Thesis. University of Nijmegen, Nijmegen

Borriello SP (1995) Clostridial disease of the gut. Clin Infect Dis 20(Suppl 2):S242–S250

Borrow N, Demey R (2001) Birds of West Africa: an identification guide. Helm Identification Guide Series, London

Bourque JD, Wilson R (1990) Rapport de l'Étude d'Impact Ecologique d'un Projet Amenagement Forestier Concernant les Forêts Classées de Ziama et de Diécké en République de Guinée. IUCN, Gland, 121 pp

Bown TM, Kraus M (1993) Soils, time, and primate paleoenvironments. Evol Anthropol 2:11–21

Brent L (2004) Solutions for research chimpanzees. Lab Anim 33(1):37–43

Bril B, Dietrich G, Foucart J, Fuwa K, Hirata S (2008) Tool use as a way to assess cognition: how do captive chimpanzees handle the weight of the hammer when cracking a nut? Anim Cogn 12:217–235

Bronson FH (1989) Mammalian reproductive biology. The University of Chicago Press, Chicago

Brugière D, Kormos R (2009) A review of the protected areas network in Guinea, West Africa, with recommendations on the selection of new sites for biodiversity conservation. Biodivers Conserv 18:847–868

Brugière D, Magassouba B (2009) Pattern and sustainability of the bushmeat trade in the Haut Niger National Park, Republic of Guinea. Afr J Ecol 47:630–639

Burbridge B (1928) Gorilla: tracking and capturing the ape-man of Africa. George G. Harrap & Company, London

Busse C (1980) Leopard and lion predation upon chacma baboons living in the Moremi Wildlife Reserve. Botsw Notes Rec 12:15–21

Butynski TM (2003) The robust chimpanzee Pan troglodytes: taxonomy, distribution, abundance and conservation status. In: Kormos R, Boesch C, Bakarr MI, Butynski TM (eds) West African chimpanzees: status survey and conservation action plan. IUCN/SSC Primate Specialist Group. IUCN, Gland and Cambridge, pp 5–12

Caldecott J, Kapos V (2005) Great ape habitats: tropical moist forests of the old world. In: Caldecott J, Miles L (eds) World atlas of great apes and their conservation. University of California Press, Berkeley

Callicott JB (1983) Animal liberation: a triangular affair. In: Scherer D, Attig T (eds) Ethics and the environment. Prentice-Hall, Upper Saddle River, pp 54–67, 72

Camara L (1996) Bossou et ses mystères. Horoya 4310:4

Campbell G, Radley P (2006) Primate and bird diversity in the Fazao-Malfakassa National Park, Togo. University of Calgary, Calgary

Carlsen F, de Jongh T (2006) European studbook for the chimpanzee Pan troglodytes. Copenhagen Zoo, Copenhagen

Carter J (2003) Orphan chimpanzees in West Africa: experiences and prospects for viability in chimpanzee rehabilitation. In: Kormos R, Boesch C, Bakarr MI, Butynski TM (eds) West African chimpanzees: status survey and conservation action plan. IUCN – World Conservation Union, Gland, pp 157–167

Carvalho S (2007) Applying the concept of chaîne opératoire to nut-cracking: an approach based on studying communities of chimpanzees (Pan troglodytes verus) in Bossou and Diecké (Guinea). Masters Thesis, University of Coimbra, Portugal

Carvalho S, Sousa C, Matsuzawa T (2007) New nut-cracking sites in Diecké Forest, Guinea: an overview of the surveys. Pan Africa News 14:11–13

Carvalho S, Cunha E, Sousa C, Matsuzawa T (2008) Chaînes opératoires and resource exploitation strategies in chimpanzee nut-cracking (*Pan troglodytes*). J Hum Evol 55:148–163

Carvalho S, Biro D, McGrew WC, Matsuzawa T (2009) Tool-composite reuse in wild chimpanzees (*Pan troglodytes*): archaeologically invisible steps in the technological evolution of early hominins? Anim Cogn 12:S103–S114

Caspary HU, Koné I, Prout C, de Pauw M (2001) La chasse et la filière viande de brousse dans l'espace Taï, Côte d'Ivoire. Tropenbos Série 2: Programme Tropenbos-Côte d'Ivoire. Abidjan, Côte d'Ivoire

Cavalieri P, Singer P (eds) (1993) The great ape project: equality beyond humanity. Fourth Estate, London

Chapman CA, Peres CA (2001) Primate conservation in the new millennium: the role of scientists. Evol Anthropol 10:16–33

Chapman CA, Gillespie TR, Goldberg TL (2005) Primates and the ecology of their infectious diseases: how will anthropogenic change affect host-parasite interactions? Evol Anthropol 14:134–144

Cherfas J (1989) Pharmaceuticals company "coerced" the press. New Sci 1661:32

Chi F, Leider M, Leendertz F, Bergmann C, Boesch C, Schenk S, Pauli G, Ellerbrok H, Hakenbeck R (2007) New *Streptococcus pneumoniae* clones in deceased wild chimpanzees. J Bacteriol 189:6085–6088

Chimp Haven (2010) http://www.chimphaven.org/about-history.cfm. Accessed 14 December 2010

Chimpanzee Sanctuary Uto (2010) http://cs-uto.org/(in Japanese). Accessed 14 December 2010

Chimpanzee Species Survival Plan (2006) 2006 North American regional chimpanzee studbook (*Pan troglodytes*). Lincoln Park Zoo, Chicago

CIA (2008) The world factbook. The Central Intelligence Agency, Langley

Clark JD (1967) The atlas of African prehistory. University of Chicago Press, Chicago and London

Clark AP (1993) Rank differences in the production of vocalizations by wild chimpanzees as a function of social context. Am J Primatol 31:159–179

Clerici N (2006) Monitoring and assessing fire impacts and land-cover change in tropical and subtropical ecosystems using satellite remote sensing and GIS techniques. PhD Thesis. Official Publications of the European Communities, Luxembourg

Collins DA, McGrew WC (1987) Termite fauna related to differences in tool-use between groups of chimpanzees (*Pan troglodytes*). Primates 28:457–471

Committee on Long-Term Care of Chimpanzees, Institute for Laboratory Animal Research, Commission on Life Sciences, National Research Council (1997) Chimpanzees in research: strategies for their ethical care, management and use. National Academy Press, Washington

Compton RR (1985) Geology in the field. Wiley, New York

Conlee KM, Boysen ST (2005) Chimpanzees in research: past, present, and future. In: Salem DJ, Rowan N (eds) The state of the animals, vol III. Humane Society Press, Washington, pp 119–137

Connolly K, Dalgleish M (1989) The emergence of a tool-using skill in infancy. Dev Psychol 25:894–912

Conservation International (2008) Biodiversity hotspots. http://www.biodiversityhotspots.org/xp/hotspots/west_africa/Pages/default.aspx. Accessed 12 December 2010

Constable JL, Ashley MV, Goodall J, Pusey AE (2001) Noninvasive paternity assignment in Gombe chimpanzees. Mol Ecol 10:1279–1300

Corbey R (2005) The metaphysics of apes: negotiating the animal–human boundary. Cambridge University Press, Cambridge

Corbey R, Theunissen B (eds) (1995) Ape, man, apeman: changing views since 1600. Department of Prehistory, Leiden University, Leiden

Courtenay J (1987) Post-partum amenorrhea, birth intervals and reproductive potential in captive chimpanzees. Primates 27:543–546

Cress D, Rosen N (eds) (2006) Pan African Sanctuary Alliance (PASA) 2006 Management Workshop Report. Pan African Sanctuary Alliance

Cummings JH, Rombeau JL, Sakata T (2004) Physiological and clinical aspects of short-chain fatty acids. Cambridge University Press, Cambridge

Danielsen F, Burgess ND, Balmford A (2005) Monitoring matters: examining the potential of locally-based approaches. Biodivers Conserv 14:2507–2542

de Bournonville D (1967) Contribution à l'étude du chimpanzé en République de Guinée. Bull Inst Fond Afr Noire 29A:1188–1269

de Nijs G (1995) The chimpanzees of the Chimfunshi Wildlife Orphanage. Pan Africa News 2:1

de Waal FBM (1982) Chimpanzee politics: power and sex among apes. Jonathan Cape, London

de Waal FBM (2001) The ape and the sushi master. Basic Books, New York

Debonnet G, Collin G (2007) Mission conjointe de suivi réactif UNESCO/UICN à la Réserve naturelle intégrale des Monts Nimba République de Guinée (WHC-07/31.COM/7A). World Heritage Committee Report. UNESCO, New Zealand

Delagnes A, Roche H (2005) Late Pliocene hominid knapping skills: the case of Lokalalei 2C, West Turkana, Kenya. J Hum Evol 48:435–472

Delorne N (1998) Aménagement forestier en Guinée – Etude de cas, *série FORAFRI,* Document 6, CIRAD-Forêt, Montpellier

Derrick R (1994) Culture in a nutshell: chimp gives lessons in learning. BBC Wildl Mag 12:10

DeVore I, Washburn SL (1963) Baboon ecology and human evolution. In: Howelland FC, Bourliere F (eds) African ecology and human evolution. Aldine, Chicago, pp 335–367

Dux D, Souaré El HS, Kamano P (2002) Atlas Scolaire de la Guinée. PRINT-64, Germany, 60 pp

Ehle FR, Robertson JB, Van Soest PJ (1982) Influence of dietary fibers on fermentation in the human large intestine. J Nutr 112:158–166

Ehmke MD, Shogren JF (2008) Experimental methods for environment and development economics. Environ Dev Econ 14:419–456

Ely J, Dye B, Frels W, Fritz J, Gagneux P, Khun H, Switzer W, Lee D (2005) Subspecies composition and founder contribution of the captive U.S. shimpanzee (*Pan troglodytes*) population. Am J Primatol 67(2):223–241

Emery-Thompson M, Jones J, Pusey A, Brewer-Marsden S, Goodall J, Matsuzawa T, Nishida T, Reynolds V, Sugiyama Y, Wrangham R (2007) Aging and fertility in wild chimpanzees: implication for the evolution of menopause. Curr Biol 17:2150–2156

Fa JE, Funk SM (2007) Global endemicity centres for terrestrial vertebrates: an ecoregions approach. Endanger Species Res 3:31–42

Ferber D (2000) Human diseases threaten great apes. Science 289:1277–1278

Fineg J, Prinne JR, Van Riper DC, Day PW (1967) A new concept in chimpanzee management for research. In: Starck D, Scheneider R, Kuhn H-J (eds) Progress in primatology. Fischer, Stuttgart, pp 345–356

Fleury-Brugière M-C, Brugière D (2002) Estimation de la population et analyse du comportement nidificateur des chimpanzés dans la zone intégralement protégée Mafou du Parc national du Haut-Niger. Parc National du Haut-Niger/AGIR project, Faranah

Fleury-Brugière MC, Brugière D (2010) High population density of *Pan troglodytes verus* in the Haut Niger National Park, Republic of Guinea: implications for local and regional conservation. Int J Primatol 31:383–392

Formentry P, Boesch C, Wyers M, Steiner C, Donati F, Dind F, Walker F, Le Guenno B (1999) Ebola virus outbreak among wild chimpanzees living in a rain forest of Côte d'Ivoire. J Infect Dis 179 (Suppl. 1):S120–S126

Formenty P, Karesh W, Froment JM, Wallis J (2003) Infectious diseases in West Africa: a common threat to chimpanzees and humans. In: Kormos R, Boesch C, Bakarr MI, Butynski TM (eds) West African chimpanzees Status Survey and Conservation Action Plan. IUCN/SSC Primate Specialist Group, Gland, pp 169–174

Fossey D (1982) Reproduction among free-living mountain gorillas. Am J Primatol (Suppl.) 1:97–104

Foucart J, Bril B, Hirata S, Morimura N, Houki C, Ueno Y, Matsuzawa T (2005) A preliminary analysis of nut-cracking movement in a captive chimpanzee: adaptation to the properties of tools and nuts. In: Roux V, Bril B (eds) Stone knapping: the necessary conditions for a uniquely hominid behavior. McDonald Press, Cambridge, pp 147–157

Fouts RS (1997) Next of kin: what chimpanzees have taught me about who we are. Morrow, New York

Fowler A, Sommer V (2007) Subsistence technology of Nigerian chimpanzees. Int J Primatol 28:997–1023

Fragaszy DM, Adams-Curtis LE (1991) Generative aspects of manipulation in tufted capuchin monkeys (*Cebus apella*). J Comp Psychol 105:387–397

Fragaszy DM, Izar P, Visalberghi E, Ottoni EB, Gomes de Oliveira M (2004) Wild capuchin monkeys (*Cebus libidinosus*) use anvils and stone pounding tools. Am J Primatol 64:359–366

Fridman EP, Nadler RD (2002) Medical primatology: history, biological foundations and applications. Taylor and Francis, London

Fruth B, Hohmann G (1994) Comparative analyses of nest building behavior in bonobos and chimpanzees. In: Wrangham RW, McGrew WC, de Waal FBM, Heltne PG (eds) Chimpanzee cultures. Harvard University Press, Cambridge, pp 109–128

Fruth B, Hohmann G (1996) Nest building behavior in the great apes: the great leap forward? In: McGrew WC, Marchant LF, Nishida T (eds) Great ape societies. Cambridge University Press, Cambridge, pp 225–240

Fuentes A, Hockings KJ (2010) The ethnoprimatological approach in primatology. Am J Primatol 72:841–847 (DOI 10. 1002/ajp.20844)

Fujita S, Kageyama, T (2007) Polymerase chain reaction detection of *Clostridium perfringens* in feces from captive and wild chimpanzees, *Pan troglodytes*. J Med Primatol 36:25–32

World Wildlife Fund (2007) Western Guinean lowland forests. In: Cleveland CJ (ed) Encyclopedia of earth. Environmental information coalition. National Council for Science and the Environment, Washington, DC

Furuichi T, Idani G, Ihobe H, Kuroda S, Kitamura K, Mori A, Enomoto T, Okayasu N, Hashimoto C, Kano T (1998) Population dynamics of wild bonobos (*Pan paniscus*) at Wamba. Int J Primatol 19:1029–1043

Fushimi T, Sakura O, Matsuzawa T, Ohno H, Sugiyama Y (1991) Nut-cracking behavior of wild chimpanzees (*Pan troglodytes*) in Bossou, Guinea, (West Africa). In: Ehara A, Kimura T, Takenaka O, Iwamoto M (eds) Primatology today. Elsevier, Amsterdam, pp 695–696

Gagneux P, Woodruff D, Boesch C (1997) Furtive mating in female chimpanzees. Behaviour 130:211–228

Gagneux P, Boesch C, Woodruff DS (1999a) Female reproductive strategies, paternity and community structure in wild West African chimpanzees. Anim Behav 57:19–32

Gagneux P, Wills C, Gerloff U, Tautz D, Morin PA, Boesch C, Fruth B, Hohmann G, Ryder OA, Woodruff DS (1999b) Mitochondrial sequences show diverse evolutionary histories of African hominoids. Proc Natl Acad Sci U S A 96:5077–5082

Gagneux P, Gonder MK, Goldberg TL, Morin PA (2001) Gene flow in wild chimpanzee populations: what genetic data tell us about chimpanzee movement over space and time. Philos Trans R Soc Lond B Biol Sci 356:889–897

Garcia-Quintanilla A, Garcia L, Tudo G, Navarro M, Gonzalez J, Jimenez de Anta MT (2000) Single-tube balanced heminested PCR for detecting *Mycobacterium tuberculosis* in smear-negative samples. J Clin Microbiol 38:1166–1169

Gardner RA, Gardner BT, Van Cantfrot TE (1989) Teaching sign language to chimpanzees. State University of New York Press, New York

Gaubert P, Veron G, Colyn M, Dunham A, Shultz S, Tranier M (2002) A reassessment of the distribution of the rare *Genetta johnstoni* (Viverridae, Carnivora) with some newly discovered specimens. Mammal Rev 32:132–144

Germain J (1984) Guinée: Peuples de la forêt. Académie des sciences d'Outre-Mer, Paris

Ghiglieri MP (1984) The chimpanzees of Kibale Forest. Columbia University Press, New York

Gibson JJ (1977) The theory of affordances. In: Shaw R, Bransford J (eds) Perceiving, acting, and knowing. Erlbaum Associates, Hillsdale, pp 67–82

Gibson JJ (1979) The ecological approach to visual perception. Houghton Mifflin, Boston

Gibson GR, Macfarlane GT (1995) Human colonic bacteria: role in nutrition, physiology, and pathology. CRC Press, London

Gilby IC (2006) Meat sharing among the Gombe chimpanzees: harassment and reciprocal exchange. Anim Behav 71:953–963

Gippoliti S, Dell'Omo G (1996) Primates of the Cantanhez Forest and the Cacine Basin, Guinea-Bissau. Oryx 30:74–80

Gippoliti S, Sousa C (2004) The chimpanzee, *Pan troglodytes*, as an 'umbrella' species for conservation in Guinea-Bissau, West Africa: opportunities and constraints (abstract). Folia Primatol 75:386

Goldberg TL, Ruvolo M (1997a) Molecular phylogenetics and historical biogeography of east African chimpanzees. Biol J Linn Soc Lond 61:301–324

Goldberg TL, Ruvolo M (1997b) The geographic apportionment of mitochondrial genetic diversity in east African chimpanzees, *Pan troglodytes schweinfurthii*. Mol Biol Evol 14:976–984

Goldberg TL, Gillespie TR, Rwego IB, Wheeler E, Estoff EL, Chapman CS (2007) Patterns of gastrointestinal bacterial exchange between chimpanzees and humans involved in research and tourism in western Uganda. Biol Conserv 135:511–517

Gonder MK, Oates JF, Disotell TR, Forstner MRJ, Morales JC, and Melnick DJ (1997) A new west African chimpanzee subspecies? Nature 388(6640):337

Goodall J (1964) Tool-using and aimed throwing in a community of free-living chimpanzees. Nature 201:1264–1266

Goodall J (1968) The behaviour of free-living chimpanzees in the Gombe Stream Reserve. Anim Behav Monogr 1:163–311

Goodall J (1971) In the shadow of man. Collins, London

Goodall J (1983) Population dynamics during a 15 year period in one community of free-living chimpanzees in the Gombe National Park, Tanzania. Z Tierpsychol 61:1–60

Goodall J (1986) The chimpanzees of Gombe: patterns of behavior. Belknap Press of Harvard University Press, Cambridge

Goodall J (1989) Glossary of chimpanzee behaviors. Jane Goodall Institute, Tucson

Goodman M, Porter CA, Czelusniak J, Page SL, Schneider H, Shoshani J, Gunnell G, Groves CP (1998) Toward a phylogenetic classification of Primates based on DNA evidence complemented by fossil evidence. Mol Phylogenet Evol 9: 585–598

Goossens B, Setchell J, Tchidongo E, Dilambaka E, Vidal C, Ancrenaz A, Jamart A (2005) Survival, interactions with conspecifics and reproduction in 37 chimpanzees released into the wild. Biol Conserv 123(4):461–475

Gotwald WH (1972) *Oecophylla longinoda*, an ant predator of *Anomma* driver ants (Hymenoptera: Formicidae). Psyche 79:348–356

Gradstein FM, Ogg JG, Smith AG (2004) A geologic time scale 2004. Cambridge University Press, Cambridge

Granier N, Martinez L (2004) First survey of chimpanzees *Pan troglodytes verus* in the Transboundary Protected Area, between Guinea and Mali (West Africa). Primatologie 6:423–447

Granier N, Huynen MC, Matsuzawa T (2007a) Preliminary surveys of chimpanzees in Gouéla area and Déré forest, the Nimba Mountain Biosphere Reserve, Republic of Guinea. Pan Africa News 14:20–22

Granier N, Kolié CI, Soumah AG, Kpoghomou E (2007b) Inventaire des mammifères du Mont Nimba. In: Kourouma M, Moloumou P, Mahomi CG, Soropogui PE (eds) La diversité biologique de la Réserve de Biosphère des Monts Nimba. Programme de Conservation de la Biodiversité des Monts Nimba. CSE-RBMN Report, Gbakoré, pp 20–47

Greengrass E (2000) The sudden decline of a community of chimpanzees at Gombe National Park: a supplement. Pan Africa News 7:25–26

Guillaumet J, Adjanohoun E (1971) Le milieu naturelle de la Côte d'Ivoire. Mémoire ORSTOM 50:157–264

Ham R (1998) Chimpanzee survey in the Republic of Guinea. Report for the European Union, Conakry

Hamilton WD (1971) Geometry for the selfish herd. J Theor Biol 7:295–311

Hanamura S, Kiyono M, Nakamura M, Sakamaki T, Itoh N, Zamma K, Kitopeni R, Matumula M, Nishida T (2006) A new code of observation employed at Mahale prevention against a flu-like disease. Pan Africa News 13:13–16

Hanamura S, Kiyono M, Lukasik-Braum M, Mlengeya T, Fujimoto M, Nakamura M, Nishida T (2008) Chimpanzee deaths at Mahale caused by flu-like disease. Primates 49:77–80

Hannah AC, McGrew WC (1991) Rehabilitation of captive chimpanzees. In: Box HO (ed) Primate responses to environmental change. Chapman & Hall, London, pp 167–186

Harcourt AH (1987) Options for unwanted or confiscated primates. Primate Conserv 8:111–113

Harris E (1989) Principles of archaeological stratigraphy. Academic, London

Harrison B (1971) Conservation of nonhuman primates in 1970. S. Karger, Basel

Hartley CWS (1988) The Oil Palm, Third edition. Longman, London

Hartley A, Nelson A, Mayaux P, Grégoire JM (2008) The Assessment of African Protected Areas. Scientific and Technical Reports. Office for Official Publications of the European Communities, Luxembourg

Hashimoto C, Furuichi T, Tashiro Y (2000) Ant dipping and meat eating by wild chimpanzees in the Kalinzu forest, Uganda. Primates 41:103–108

Hatheway CL (1990) Toxigenic *Clostridia*. Clin Microbiol Rev 3:66–98

Hawkes K, O'Connell J, Blurton-Jones N, Alvarez H, Charnov E (1990) Grandmothering, menopause, and the evolution of human life histories. Proc Natl Acad Sci U S A 95:1336–1339

Hawthorne WD, Jongkind C (2006) Woody plants of Western African forests: a guide to the forest trees, shrubs and lianes from Senegal to Ghana. Royal Botanic Garden, Kew

Hayakawa S, Takenaka O (1999) Urine as another potential source for template DNA in polymorphic chain reaction (PCR). Am J Primatol 49:299–304

Hayaki H (1990) Social context of pant-grunting in young chimpanzees. In: Nishida T (ed) The chimpanzees of the Mahale Mountains: sexual and life history strategies. University of Tokyo Press, Tokyo, pp 189–206

Hayashi M, Matsuzawa T (2003) Cognitive development in object manipulation by infant chimpanzees. Anim Cogn 6:225–233

Hayashi M (2007) A new notation system of object manipulation in the nesting-cup task for chimpanzees and humans. Cortex 43(3):308–318

Hayashi M, Mizuno Y, Matsuzawa T (2005) How does stone-tool use emerge? Introduction of stones and nuts to naïve chimpanzees in captivity. Primates 46:91–102

Hayashi M, Takeshita H, and Matsuzawa T (2006) Cognitive development in apes and humans assessed by object manipulation. In: Matsuzawa T, Tomonaga M, Tanaka M (eds) Cognitive development in chimpanzees. Springer, New York, pp 395–410

Hernandez-Aguilar AR, Moore J, Pickering TR (2007) Savanna chimpanzees use tools to harvest the underground storage organs of plants. Proc Natl Acad Sci U S A 104:19210–19213

Hill CM (1997) Crop-raiding by wild vertebrates: the farmer's perspective in an agricultural community in western Uganda. Int J Pest Manag 43:77–84

Hillers A, Loua NS, Rödel MO (2008) Assessment of the distribution and conservation of the viviparous toad *Nimbaphrynoides occidentalis* on Monts Nimba, Guinea. Endanger Species Res 5:13–19

Hiraiwa-Hasegawa M (1989) Sex differences in the behavioral development of chimpanzees at Mahale. In: Heltne PG, Marquardt LA (eds) Understanding chimpanzees. Harvard University Press, Cambridge, pp 104–115

Hiraiwa-Hasegawa M, Hasegawa T, Nishida T (1984) Demographic study of a large-sized unit-group of chimpanzees in the Mahale-Mountains, Tanzania: a preliminary report. Primates 25:401–413

Hirata S, Celli M (2003) Role of mothers in the acquisition of tool-use behaviors by captive infant chimpanzees. Anim Cogn 6:235–244

Hirata S, Morimura N, Matsuzawa T (1998a) Green passage plan (tree-planting project) and environmental education using documentary videos at Bossou: a progress report. Pan Africa News 5:18–20

Hirata S, Myowa M, Matsuzawa T (1998b) Use of leaves as cushions to sit on wet ground by wild chimpanzees. Am J Primatol 44:215–220

Hirata S, Yamakoshi G, Fujita S, Ohashi G, Matsuzawa T (2001) Capturing and toying with hyraxes (*Dendrohyrax dorsalis*) by wild chimpanzees (*Pan troglodytes*) at Bossou, Guinea. Am J Primatol 53:93–97

Hirata S, Morimura N, Houki C (2009) How to crack nuts: acquisition process in captive chimpanzees (*Pan troglodytes*) observing a model. Anim Cogn 12:S87–S101

Hoasaka K (1995) Mahale: a single flu epidemic killed at least 11 chimps. Pan Africa News 2:3–4

Hobson PN (1988) The rumen microbial ecosystem. Chapman & Hall, London

Hockings KJ (2007) Human-chimpanzee coexistence at Bossou, the Republic of Guinea: a chimpanzee perspective. PhD Thesis, University of Stirling, Stirling

Hockings KJ, Humle T (2009) Best practice guidelines for the prevention and mitigation of conflict between humans and great apes. IUCN/SSC Primate Specialist Group (PSG), Gland

Hockings KJ, Anderson JR, Matsuzawa T (2006) Road crossing in chimpanzees: a risky business. Curr Biol 16:668–670

Hockings KJ, Humle T, Anderson JR, Biro D, Sousa C, Ohashi G, Matsuzawa T (2007) Chimpanzees share forbidden fruit. PLoS ONE 2(9):e886. doi:10.1371/journal.pone.0000886

Hockings KJ, Anderson JR, Matsuzawa T (2009) Use of wild and cultivated foods by chimpanzees at Bossou, Republic of Guinea: feeding dynamics in a human-influenced environment. Am J Primatol 71:636–646

Hockings KJ, Yamakoshi G, Kabasawa A, Matsuzawa T (2010a) Attacks on local persons by chimpanzees in Bossou, Republic of Guinea: long-term perspectives. Am J Primatol 72:887–896

Hockings KJ, Anderson JR, Matsuzawa T (2010b) Flexible feeding on cultivated underground storage organs by forest-dwelling chimpanzees at Bossou, West Africa. J Human Evol 58:227–233 (DOI: 10.1016/j.jhevol.2009.11.004)

Hohmann G (2001) Association and social interactions between strangers and residents in bonobos (Pan paniscus). Primates 42:91–99

Holas B (1952a) Échantillon du folklore Kono (Haute-Guinée Française). Études Guinéennes 9:3–90

Holas B (1952b) Notes complémentaires sur l'abri sous roche Blandé, fouilles 1951. Bull IFAN T. XIV-4

Holas B (1954) Le culte de Zié. Mém Inst Fr Afr Noire 39:1–275

Holas B (1975) Contes Kono: traditions populaires de la Forêt Guinéenne. Edition G.P. Maisonneuve et Larose, Paris

Homsy J (1999) Ape tourism and human diseases: how close should we get? Report to the International Gorilla Conservation Programme. <http://www.igcp.org/pdf/homsy_rev.pdf>

Hoppe-Dominik B (1991) Distribution and status of chimpanzees (Pan troglodytes verus) on the Ivory Coast. Primate Report 35:45–75

Hopper LM, Spiteri A, Lambeth SP, Schapiro SJ, Horner V, Whiten A (2007) Experimental studies of traditions and underlying transmission processes in chimpanzees. Anim Behav 73:1021–1032

Horai S, Hayasaka K, Kondo R, Tsugane K, Takahata N (1995) Recent African origin of modern humans revealed by complete sequences of hominoid mitochondrial DNAs. Proc Natl Acad Sci U S A 92:532–536

Horner V, Whiten A, Flynn E, de Waal FBM (2006) Faithful replication of foraging techniques along cultural transmission chains by chimpanzees and children. Proc Natl Acad Sci U S A 103:13878–13883

Hortal J, Lobo JM (2005) An ED-based protocol for optimal sampling of biodiversity. Biodivers Conserv 14:2913–2947

Hosaka K (1995) A single flu epidemic killed at least 11 chimps. Pan Africa News 2:3–4

Huffman MA, Wrangham RW (1994) Diversity of medicinal plant use by chimpanzees in the wild. In: Wrangham RW, McGrew WC, de Waal FBM, Heltne PG (eds) Chimpanzee cultures. Harvard University Press, Cambridge, pp 129–148

Huffman MA, Ohigashi H, Kawanaka M, Page JE, Kirby GC, Gasquet M, Murakami A, Koshimizu K (1998) African great ape self-medication: a new paradigm for treating parasite disease with natural medicines? In: Ebizuka Y (ed) Towards natural medicine research in the 21st century. Elsevier, Amsterdam, pp 113–123

Huijbregts B, de Watcher P, Ndong Obiang LS, Akou M (2003) Forte baisse des populations de grands singes dans le massif forestier de Minkebe, au nord-est du Gabon. Canopee 18:12–15

Humle T (1999) New record of fishing for termites (Macrotermes) by the chimpanzees of Bossou (Pan troglodytes verus), Guinea. Pan Africa News 6:3–4

Humle T (2003a) Chimpanzees and crop raiding in West Africa. In: Kormos R, Boesch C, Bakarr MI, Butynski TM (eds) West African Chimpanzees. Status Survey and Conservation Action Plan, IUCN/SSC Primate Specialist Group. IUCN, Gland and Cambridge, pp 147–155

Humle T (2003b) Culture and variation in wild chimpanzee behaviour: a study of three communities in West Africa. PhD Thesis, University of Stirling, Stirling

Humle T, Matsuzawa T, Yamakoshi G (2004) Chimpanzee conservation and environmental education in Bossou and Nimba, Guinea, West Africa. Folia Primatol 75(S1):280–281

Humle T (2006) Ant dipping in chimpanzees: an example of how microecological variables, tool use, and culture reflect the cognitive abilities of chimpanzees. In: Matsuzawa T, Tomonaga M, Tanaka M (eds) Cognitive development in chimpanzees. Springer, Tokyo, pp 452–475

Humle T (2007) Behavioral and ecological monitoring of wild and released chimpanzees (*Pan troglodytes verus*) in the 'Parc National du Haut Niger,' Guinea. Progress Report. Conservational International, Washington

Humle T, Matsuzawa T (2001) Behavioural diversity among the wild chimpanzee populations of Bossou and neighbouring areas, Guinea and Côte d'Ivoire, West Africa. Folia Primatol 72:57–68

Humle T, Matsuzawa T (2002) Ant dipping among the chimpanzees of Bossou, Guinea, and comparisons with other sites. Am J Primatol 58:133–148

Humle T, Matsuzawa T (2004) Oil palm use by adjacent communities of chimpanzees at Bossou and Nimba Mountains, West Africa. Int J Primatol 25:551–581

Humle T, Matsuzawa T (2008) Laterality in hand use across four tool use behaviors among the wild chimpanzees of Bossou, Guinea, West Africa. Am J Primatol 71:40–48

Humle T, Matsuzawa T (2009) Laterality in hand use across four tool-use behaviors among the wild chimpanzees of Bossou, Guinea, West Africa. Am J Primatol 71:40–48

Humle T, Matsuzawa T, Yamakoshi G (2004) Chimpanzee conservation and environmental education in Bossou and Nimba, Guinea, West Africa. Folia Primatol 75(S1):280–281

Humle T, Snowdon CT, Matsuzawa T (2009) Social influences on the acquisition of ant-dipping among the wild chimpanzees (*Pan troglodytes verus*) of Bossou, Guinea, West Africa. Anim Cogn 12:S37–S48

Humle T, Colin C, Laurans M, Raballand E (in press) Group release of sanctuary chimpanzees in the Haut Niger National Park, Guinea, West Africa: ranging patterns and lessons so far. Int J Primatol

Idani G (1991) Social relationships between immigrant and resident bonobos (*Pan paniscus*) females at Wamba. Folia Primatol 57:83–95

Inoue-Nakamura N, Matsuzawa T (1997) Development of stone tool use by wild chimpanzee (*Pan troglodytes*). J Comp Psychol 111:159–173

Irbis C, Garriga R, Kabasawa A, Ushida K (2008) Phylogenetic analysis on *Troglodytella abrassarti* isolated from chimpanzees (*Pan troglodytes verus*) in the wild and in captivity. J Gen Appl Microbiol 54:409–413

Isaac GL (1986) Foundation stones: early artefacts as indicators of activities and abilities. In: Bailey GN, Callow P (eds) Stone age prehistory. Cambridge University Press, Cambridge, pp 221–242

Itani J, Suzuki A (1967) The social unit of chimpanzees. Primates 8:355–381

IUCN (2007) The 2007 IUCN Red List of Threatened Species. IUCN, Gland

IUCN (2009) IUCN Red List of Threatened Species. Version 2009.1. <http://www.iucnredlist.org>

Izawa K, Mizuno A (1977) Palm-fruit cracking behaviour of wild black-capped capuchin (*Cebus apella*). Primates 18:773–792

Jenkins M (2008) Who murdered the Virunga gorillas? National Geographic. http://ngm.nationalgeographic.com/2008/07/virunga/jenkin-text. Accessed 14 December 2010

Johanson D, Edgar B (1996) From Lucy to language. Simon & Schuster, New York

Johns BG (1996) Responses of chimpanzees to habituation and tourism in the Kibale Forest, Uganda. Biol Conserv 78:257–262

Johnson DL (2002) Darwin would be proud: bioturbation, dynamic denudation, and the power of theory in science. Geoarcheology 17:7–40

Joulian F (1994) Culture and material culture in chimpanzees and early hominids. In: Roeder JJ, Thierry B, Anderson JR, Herrenschmidt N (eds) Current primatology, vol II. Social development, learning and behaviour. Université Louis Pasteur, Strasbourg, pp 397–404

Joulian F (1996) Comparing chimpanzee and early hominid techniques: some contributions to cultural and cognitive questions. In: Mellars P, Gibson K (eds) Modelling the early human mind. McDonald Institute Monographs, Cambridge, pp 173–189

Kabasawa A (2009) Current situation of the chimpanzee pet trade in Sierra Leone. Afr Stud Monogr 30(1):37–54

Kabasawa A, Garriga RM, Amarasekaran B (2008) Human fatality by escaped *Pan troglodytes* in Sierra Leone. Int J Primatol 29:1671–1685 (DOI: 10.1007/s10764-008-9323-0)

Kalema-Zikusoka G, Kock RA, Macfie EJ (2002) Scabies in free-ranging mountain gorillas (*Gorilla beringei beringei*) in Bwindi Impenetrable National Park, Uganda. Vet Rec 150:12–15

Kano T (1992) The last ape: pygmy chimpanzee behavior and ecology. Stanford University Press, Menlo Park

Karesh WB (1995) Wildlife rehabilitation: additional considerations for developing countries. J Zoo Wildl Med 26(1):2–9

Kaur T, Huffman MA (2004) Descriptive urological record of chimpanzees (*Pan troglodytes*) in the wild and limitations associated with using multi-reagent dipstick test strips. J Med Primatol 33:187–196

Kaur T, Singh J, Tong S, Humphrey C, Clevenger D, Tan W, Szekely B, Wang Y, Li Y, Muse EA, Kiyono M, Hanamura S, Inoue E, Nakamura M, Huffman MA, Jiang B, Nishida T (2008) Descriptive epidemiology of fatal respiratory outbreaks and detection of a human-related metapneumovirus in wild chimpanzees (*Pan troglodytes*) at Mahale Mountains National Park, Western Tanzania. Am J Primatol 70:755–765

Kirschofer R (1987) International studbook of the gorilla, *Gorilla gorilla*, 1985. Frankfurt Zoological Garden, Frankfurt

Kishimoto A, Ushida K, Phillips GO, Ogasawara T, Sasaki Y (2006) Identification of intestinal bacteria responsible for fermentation of gum arabic in pig model. Curr Microbiol 53:173–177

Kitamura K (1989) Genito-genital contacts in the pygmy chimpanzee (*Pan paniscus*). Afr Study Monogr 10:49–67

Knott CD (1996) Monitoring health status of wild orangutans through field analysis of urine. Am J Phys Anthropol (Suppl) 22:139–140

Knott CD (1998) Changes in orangutan caloric intake, energy balance, and ketones in response to fluctuating fruit availability. Int J Primatol 19:1061–1079

Köhler W (1925) The mentalities of apes. Harcourt, New York

Köndgen S, Kühl H, N'Goran PK, Walsh PD, Schenk S, Ernst N, Biek R, Formenty P, Mätz-Rensing K, Schweiger B, Junglen S, Ellerbrok H, Nitsche A, Briese T, Lipkin WI, Pauli G, Boesch C, Leendertz FH (2008) Pandemic human viruses cause decline of endangered great apes. Curr Biol 18:260–264

Koops K (2005) Nesting patterns and characteristics of the chimpanzees in the Nimba Mountains, Guinea, West Africa. Department of Behavioural Biology, University of Utrecht, Utrecht

Koops K, Matsuzawa T (2006) Hand clapping by a chimpanzee in the Nimba Mountains, Guinea, West Africa. Pan Africa News 13:19–21

Koops K, Humle T, Sterck EH, Matsuzawa T (2007) Ground-nesting by the chimpanzees of the Nimba Mountains, Guinea: environmentally or socially determined? Am J Primatol 69:407–419

Koops K, McGrew WC, Matsuzawa T (2010) Do chimpanzees (*Pan troglodytes*) use cleavers and anvils to fracture *Treculia africana* fruits? Preliminary data on a new form of percussive technology. Primates 51:175–178

Kormos R, Boesch C, Bakarr MI, Butynski TM (2003a) West African Chimpanzees: Status Survey and Conservation Action Plan. IUCN/SSC Primate Specialist Group IUCN, Gland and Cambridge

Kormos R, Humle T, Brugière D, Fleury-Brugière M-C, Matsuzawa T, Sugiyama Y, Carter J, Diallo MS, Sagno C, Tounkara EO (2003b) Status surveys and recommendations: country reports: The Republic of Guinea. In: Kormos R, Boesch C, Bakarr MI, Butynski TM (eds) Status survey and conservation action plan: West African Chimpanzees. IUCN/SSC Primate Specialist Group, Gland and Cambridge, pp 63–76

Kortlandt A (1967) Experimentation with chimpanzees in the wild. In: Starck D, Schneider R, Kuhn HJ (eds) Neue ergebnisse der primatologie (Progress in primatology). Fischer, Stuttgart, pp 208–224

Kortlandt A (1968) Die schlacht der schimpansen gegen ihren erbfeind. Das Tier 8(12):10–15

Kortlandt A (1972) New perspectives on ape and human evolution. University of Amsterdam, Amsterdam

Kortlandt A (1986) The use of stone tools by wild-living chimpanzees and earliest hominids. J Hum Evol 15:77–132

Kortlandt A (1989) The use of stone tools by wild-living chimpanzees. In: Heltne PG, Marquardt LG (eds) Understanding chimpanzees. Harvard University Press, Cambridge, pp 146–147

Kortlandt A, Holzhaus E (1987) New data on the use of stone tools by chimpanzees in Guinea and Liberia. Primates 28:473–496

Koltlandt A, van Orshoven J, Pfeijffers R, van Zon JCJ (1981) Chimpanzees in the wild, Guinea 1966–1967: Sixth Netherlands chimpanzee expedition (transcripts of film/video texts). University of Amsterdam, Amsterdam

Kourouma M, Moloumou P, Mahomi CG, Soropogui PE (2008) La diversité biologique de la Réserve de Biosphère des Monts Nimba. Programme de Conservation de la Biodiversité des Monts Nimba, CSE-RBMN Report, Gbakoré

Krief S, Huffman MA, Sévenet T, Guillot J, Bories C, Hladik CM, Wrangham RW (2005) Noninvasive monitoring of the health of *Pan troglodytes schweinfurthii* in the Kibake National Park, Uganda. Int J Primatol 26:467–490

Kühl H, Maisels F, Ancrenaz M, Williamson EA (2008) Best practice guidelines for surveys and monitoring of great ape populations. IUCN SSC Primate Specialist Group (PSG), Gland

Laden G, Wrangham R (2005) The rise of the hominids as an adaptive shift in fallback foods: plant underground storage organs (USOs) and australopith origins. J Hum Evol 49:482–498

Lahm SA, Kombila M, Swanepoel R, Barnes RFW (2007) Morbidity and mortality of wild animals in relation to outbreaks of Ebola haemorrhagic fever in Gabon, 1994–2003. Trans R Soc Trop Med Hyg 101:64–78

Lamotte M (1942) La faune mammalogique du Mont Nimba (Haute Guinée). Mammalia 6:114–119

Lamotte M (1998a) Le Climat du Nimba. In: Lamotte M (ed) Le Mont Nimba, Réserve de la Biosphère et site du Patrimoine Mondial-Initiation à la géomorphologie et à la biogéographie. UNESCO, Paris, pp 37–54

Lamotte M (1998b) Le Mont Nimba, Réserve de Biosphère et Site du Patrimoine Mondial (Guinée - Côte d'Ivoire) – Initiation à la Géomorphologie et Biogéographie. UNESCO, Paris

Lamotte M, Rougerie G (1998) Les traits principaux de la géologie et de la géomorphologie du Nimba. In: Lamotte M (ed) Le Mont Nimba, Réserve de la Biosphère et site du Patrimoine Mondial-Initiation à la géomorphologie et à la biogéographie. UNESCO, Paris, pp 13–36

Lamotte M, Roy R (2003) Le peuplement animal du mont Nimba (Guinée, Côte d'Ivoire, Liberia). Mémoires du Muséum National d'Histoire Naturelle, series No. 190. Publications Scientifiques du Muséum, Paris

Lamotte M, Roy R, Xavier F (2003) Les premiers temps de l'étude scientifique et de la protection du Nimba (1942–1978). In: Lamotte M, Roy R (eds) Le peuplement animal du Mont Nimba (Guinée, Côte d'Ivoire, Liberia). Publications Scientifiques du Muséum, Paris, pp 11–27

Langouw A (2002) Behavioural adaptations to water scarcity in Tongo chimpanzees. In: Boesch C, Hohmann G, Marchant, LF (eds) Behavioural diversity in chimpanzees and bonobos. Cambridge University Press, Cambridge, pp 52–60

Leblan V (2008) Analyse spatiale des relations entre les hommes te les chimpanzés dans la région de Boké (Guinée). Thèse de Doctorat, École des Hautes Études en Sciences Sociales, Paris

Leciak E, Hladik A, Hladik CM (2005) Le Palmier à huile (*Elaeis guineensis*) et les noyaux de biodiversité des forêts-galeries de Guinée Maritime: à propos du commensalisme de l'homme et du chimpanzé. Rev Écol (Terre Vie) 60:179–184

Leendertz FH, Ellerbrok H, Boesch C, Couacy-Hymann E, Matz-Rensing K, Hakenbeck R, Bergmann C, Abaza P, Junglen S, Moebius Y, Vigilant L, Formenty P, Pauli G (2004) Anthrax kills wild chimpanzees in a tropical rainforest. Nature 430:451–452

Leendertz FH, Lankester F, Guislain P, Néel C, Drori O, Dupain J, Speede S, Reed P, Wolfe N, Loul S, Mpoudi-Noile E, Peeters M, Boesch C, Pauli G, Ellerbrok H, Leory EM (2006a) Anthrax in Western and Central African great apes. Am J Primatol 68:928–933

Leendertz FH, Pauli G, Maetz-Rensing K, Boardman W, Nunn C, Ellerbrok H, Jensen SA, Junglen S (2006b) Pathogens as drivers of population declines: the importance of systematic monitoring in great apes and other threatened mammals. Biol Conserv 131:325–337

Leory EM, Rouquet P, Formentry P, Souquière S, Kibourne A, Froment J-M, Bermejo M, Smit S, Karesh W, Swanepoel R, Zaki SR, Rollin PE (2004) Multiple ebola virus transmission events and rapid decline of central African wildlife. Science 303:387–390

Littleton J (2005) Fifty years of chimpanzee demography at Taronga Park Zoo. Am J Primatol 67:281–298

Lonsdorf EV (2005) Sex differences in the development of termite-fishing skills in the wild chimpanzees, *Pan troglodytes schweinfurthii*, of Gombe National Park, Tanzania. Anim Behav 70:673–683

Lonsdorf EV (2006) What is the role of mothers in the acquisition of termite-fishing behaviors in wild chimpanzees (*Pan troglodytes schweinfurthii*)? Anim Cogn 9:36–46

Lonsdorf EV, Eberly LE, Pusey AE (2004) Sex differences in learning in chimpanzees. Nature 428 (6984):715–716

Lonsdorf EV, Travis D, Pusey AE, Goodall J (2006) Using retrospective health data from the Gombe chimpanzee study to inform future monitoring efforts. Am J Primatol 68:897–908

Lycett SJ, Collard M, McGrew WC (2007) Phylogenetic analyses of behavior support existence of culture among wild chimpanzees. Proc Natl Acad Sci U S A 104:17588–17592

Macfie LJ, Williamson EA (2010) Best practice guidelines for great ape tourism. Gland, Switzerland: IUCN/SSC Primate Specialist Group (PSG)

Maple TL (1979) Great apes in captivity: the good, the bad, and the ugly. In: Erwin J, Maple TL, Mitchell G (eds) Captivity and behavior. Van Nostrand Reinhold, New York, pp 239–272

Marchant LF, McGrew WC (2005) Percussive technology: chimpanzee baobab smashing and the evolutionary modelling of hominin knapping. In: Roux V, Bril B (eds) Stone knapping: the necessary conditions for a uniquely hominin behavior. McDonald Institute for Archaeological Research, Cambridge, pp 341–350

Marchant LF, McGrew WC (2007) Ant fishing by wild chimpanzees is not lateralised. Primates 48:22–26

Marshall AJ, Wrangham RW (2007) Evolutionary consequences of fallback foods. Int J Primatol 28:1219–1235

Martin P, Bateson P (2007) Measuring behavior: an introductory guide, 3rd edn. Cambridge University Press, Cambridge

Matsusaka T, Nishie H, Shimada M, Kutsukake N, Zamma K, Nakamura M, Nishida T (2006). Tool-use for drinking water by immature chimpanzees of Mahale: prevalence of an unessential behavior. Primates 47:113–122

Matsuzawa T (1991) Nesting cups and metatools in chimpanzees. Behav Brain Sci 14:570–571

Matsuzawa T (1992) Death of a chimpanzee and care of the dead infant. Hattatsu 50:95–104 (in Japanese)

Matsuzawa T (1994) Field experiments on use of stone tools by chimpanzees in the wild. In: Wrangham RW, McGrew WC, de Waal FBM, Heltne PG (eds) Chimpanzee cultures. Harvard University Press, Cambridge, pp 351–370

Matsuzawa T (1995) Chimpanzee being (in Japanese). Iwanami-Shoten, Tokyo, pp 121–126

Matsuzawa T (1996) Chimpanzee intelligence in nature and in captivity: isomorphism of symbol use and tool use. In: McGrew WC, Marchant LF, Nishida T (eds) Great ape societies. Cambridge University Press, Cambridge, pp 196–209

Matsuzawa T (1997a) Phylogeny of intelligence: a view from cognitive behavior of chimpanzees. IIAS Rep 1997:17–26

Matsuzawa T (1997b) The death of an infant chimpanzee at Bossou, Guinea. Pan Africa News
    4:4–6
Matsuzawa T (1998) Chimpanzee behavior: a comparative cognitive perspective. In: Greenberg
    G, Haraway M (eds) Comparative psychology: a handbook. Garland, New York, pp 360–375
Matsuzawa T (1999) Communication and tool use in chimpanzees: cultural and social context. In:
    Hauser M, Konishi M (eds) The design of animal communication. MIT Press, Cambridge, pp
    645–671
Matsuzawa T (2001a) Primate foundations of human intelligence: a view of tool use in nonhuman
    primates and fossil hominids. In: Matsuzawa T (ed) Primate origins of human cognition and
    behavior. Springer, Tokyo, pp 3–25
Matsuzawa T (ed) (2001b) Primate origins of human cognition and behavior. Springer, Tokyo
Matsuzawa T (2005) Book review of the cultured chimpanzee: reflections on cultural primatology.
    Nature 434:21–22
Matsuzawa T (2006a) Bossou: 30 years. Pan Africa News 13:16–19
Matsuzawa T (2006b) Sociocognitive development in chimpanzees: a synthesis of laboratory
    work and fieldwork. In: Matsuzawa T, Tomonaga M, Tanaka M (eds) Cognitive development
    in chimpanzees. Springer, Tokyo, pp 3–33
Matsuzawa T (2006c) Comparative cognitive development. Dev Sci 10:97–103
Matsuzawa T (2007) Assessment of the planted trees in Green Corridor Project. Pan Africa News
    14:27–29
Matsuzawa T (2011) What is uniquely human? A view from comparative cognitive development
    of humans and chimpanzees. In: de Waal FBM, Ferrari P (eds) Primate social mind. Cambridge
    University Press, Cambridge (in press)
Matsuzawa T, Kourouma M (2008) The Green corridor project: long-term research and conserva-
    tion in Bossou, Guinea. In: Ross E, Wrangham R (eds) Long-term research leads to conserva-
    tion: examples from African forests. Chicago University Press, Chicago, pp 201–212
Matsuzawa T, Sakura O (1988) Choice of foraging sites in wild chimpanzees: Analysis by observ-
    ing progressions and foot print identification. Reichoru Kenkyu/Primate Res 4:155
Matsuzawa T, Yamakoshi G (1996) Comparison of chimpanzee material culture between Bossou
    and Nimba, West Africa. In: Russon AE, Bard KA, Parker S (eds) Reaching into thought: the
    mind of the great apes. Cambridge University Press, Cambridge, pp 211–232
Matsuzawa T, Sakura O, Kimura T, Hamada Y, Sugiyama Y (1990) Case report on the death of a
    wild chimpanzee (*Pan troglodytes verus*). Primates 31:635–641
Matsuzawa T, Yamakoshi G, Humle T (1996) Newly found tool use by wild chimpanzees: algae
    scooping (in Japanese). Primate Res 12:283
Matsuzawa T, Takemoto H, Hayakawa S, Shimada M (1999) Diécké forest in Guinea. Pan Africa
    News 6:10–11
Matsuzawa T, Biro D, Humle T, Inoue-Nakamura N, Tonooka R, Yamakoshi G (2001) Emergence
    of culture in wild chimpanzees: education by master-apprenticeship. In: Matsuzawa T (ed)
    Primate origins of human cognition and behavior. Springer, Tokyo, pp 557–574
Matsuzawa T, Humle T, Koops K, Biro D, Hayashi M, Sousa C, Mizuno Y, Kato A, Yamakoshi
    G, Ohashi G, Sugiyama Y, Kourouma M (2004) Wild chimpanzees at Bossou-Nimba: deaths
    through a flu-like epidemic in 2003 and the green-corridor project. Primate Res 20:45–55 (in
    Japanese)
Matsuzawa T, Tomonaga M, Tanaka M (eds) (2006) Cognitive development in chimpanzees.
    Springer, Tokyo
Mauny R, Holas B (1953) Nouvelles fouilles à l'abri sous roche Blandé (Guinée). Bull
    IFAN T XV-4
McCullough J, Wright HE, Alonso LE, Diallo MS (2006) A rapid biological assessment of three
    classified forests in south-eastern Guinea. RAP Bulletin of Biological Assessment 40.
    Conservation International, Washington
McDonel JL (1980) *Clostridium perfringens* toxins (type A, B, C, D, E). Pharmacol Ther 10:617–655
McGreal S (1983) Letter to editor: a project with potential to spread non-A, non-B hepatitis in
    west Africa. J Med Primatol 12:280–281

McGrew WC (1974) Tool use by wild chimpanzees in feeding upon driver ants. J Hum Evol 3:501–508

McGrew WC (1977) Socialization and object manipulation of wild chimpanzees. In: Chevalier-Skolnikoff S, Poirier F (eds) Primate biosocial development. Garland, New York

McGrew WC (1979) Evolutionary implications of sex differences in chimpanzee predation and tool use. In: Hamburg DA, McCrown ER (eds) The great apes. Benjamin Cummings, London, pp 441–463

McGrew WC (1983) Animal foods in the diets of wild chimpanzees (*Pan troglodytes*): why cross-cultural variation? J Ethol 1:46–61

McGrew WC (1992) Chimpanzee material culture: implications for human evolution. Cambridge University Press, Cambridge

McGrew WC (2004) The cultured chimpanzee: reflections on cultural primatology. Cambridge University Press, Cambridge

McGrew WC, Marchant LF (1999) Laterality of hand use pays off in foraging success for wild chimpanzees. Primates 40:509–513

McGrew WC, Collins DA (1985) Tool use by wild chimpanzees (*Pan troglodytes*) to obtain termites (*Macrotermes herus*) in the Mahale Mountains, Tanzania. Am J Primatol 9:47–62

McGrew WC, Tutin CEG (1978) Evidence for a social custom in wild chimpanzees? Man 13:234–251

McGrew WC, Ham RM, White LJT, Tutin CEG, Fernandez M (1997) Why don't chimpanzees in Gabon crack nuts? Int J Primatol 18:353–374

McGrew WC, Ensminger AL, Marchant LF, Pruetz JD, Vigilant L (2004) Genotyping aids field study of unhabituated wild chimpanzees. Am J Primatol 63:87–93

McNeal NI (1984) The contribution of the large intestine to energy supplies in man. Am J Clin Nutr 39:338–342

McRitchei R (1967) Chimpanzees in bio-medical research at the 6571st aeromedical research laboratory, U.S. Air Force. In: Starck D, Scheneider R, Kuhn H-J (eds) Progress in primatology. Gustav Fischer, Stuttgart, pp 363–372

Mercader J (ed) (2003) Under the canopy. The archaeology of tropical rain forests. Rutgers University Press, New Brunswick

Mercader J, Panger M, Boesch C (2002) Excavation of a chimpanzee stone tool site in the African rainforest. Science 296:1452–1455

Mercader J, Barton H, Gillespie J, Harris J, Kuhn S, Tyler R, Boesch C (2007) 4,300-year-old chimpanzee sites and the origins of percussive stone technology. Proc Natl Acad Sci U S A 104:1–7

Miller LE, Treves A (2006). Predation on primates. In: Campbell C, Fuentus A, MacKinnon K, Panger K, Bearder SK (eds) Primates in perspective. Oxford University Press, New York, pp 525–536

Mills W, Cress D, Rosen N (2005) Pan African Sanctuary Alliance (PASA) 2005 workshop report. Conservation Breeding Specialist Group (SSC/IUCN), Gland

Milton K, Demment MW (1988) Digestion and passage kinetics of chimpanzees fed high and low fiber diets and comparison with human data. J Nutr 118:1082–1088

Mitani JC, Nishida T (1993) Contexts and social correlates of long-distance calling by male chimpanzees. Anim Behav 45:735–746

Mitani JC, Watts D (2001) Why do chimpanzees hunt and share meat? Anim Behav 61:915–924

Mitani JC, Merriwether DA, Zhang C (2000) Male affiliation, cooperation and kinship in wild chimpanzees. Anim Behav 59:885–893

Mitsuoka T (1982) Recent trends in research on intestinal flora. Bifidobacteria Microflora 1:3–24

Mittermeier RA, Myers N, Mittermeier CG (1999) Hotspots: earth's biologically richest and most endangered terrestrial ecoregions. CEMEX, Mexico City

Mizuno Y, Takeshita H, Matsuzawa T (2006) Behavior of infant chimpanzees during the night in the first 4 months of life: smiling and suckling in relation to behavioral state. Infancy 9(2):221–240

Möbius Y, Schöning C, Koops K, Matsuzawa T, Boesch C, Humle T (2008) Comparative study of chimpanzee predation on army ants between Bossou, Guinea and Taï, Côte d'Ivoire, West Africa: is ant-dipping cultural? Anim Behav 76:37–45

Morgan BJ, Abwe EE (2006) Chimpanzees use stone hammers in Cameroon. Curr Biol 16:R632–R633

Morgan D, Sanz C (2003) Naïve encounters with chimpanzees in the Goualougo Triangle, Republic of Congo. Int J Primatol 24:369–381

Morgan D, Sanz C (2007) Best practice guidelines for reducing the impact of commercial logging on Great Apes in Western Equatorial Africa. IUCN SSC Primate Specialist Group (PSG), Gland

Morin PA, Moore JJ, Chakraborty R, Jin L, Goodall J, Woodruff DS (1994a) Kin selection, social structure, gene flow, and the evolution of chimpanzees. Science 265:1193–1201

Morin PA, Wallis J, Moore JJ, Woodruff DS (1994b) Paternity exclusion in a community of wild chimpanzees using hypervariable simple sequence repeats. Mol Ecol 3:469–477

Morin PA, Chambers KE, Boesch C, Vigilant L (2001) Quantitative polymerase chain reaction analysis of DNA from noninvasive samples for accurate microsatellite genotyping of wild chimpanzees. Mol Ecol 10:1835–1844

Morris R, Morris D (1966) Men and apes. McGraw-Hill, New York

Muroyama Y, Sugiyama Y (1994) Grooming relationships in two species of chimpanzees. In: Wrangham RW, McGrew WC, de Waal FBM, Heltne PG (eds) Chimpanzee cultures. Harvard University Press, Cambridge, pp 169–180

Myers N, Mittermeier RA, Mittermeier CG, da Fonseca GAB, Kent J (2000) Biodiversity hotspots for conservation priorities. Nature 403: 853–858

Nakamichi M, Koyama N, Jolly A (1996) Maternal responses to dead and dying infants in wild troops of ring-tailed lemurs at the Berenty Reserve, Madagascar. Int J Primatol 17:505–523

Nakamura M (2000) Is human conversation more efficient than chimpanzee grooming? Comparison of clique sizes. Hum Nat 11:281–297

Nakamura M (2002) Grooming-hand-clasp in Mahale M group chimpanzees: implication for culture in social behaviors. In: Boesch C, Hohmann G, Marchant LF (eds) Behavioural diversity in chimpanzees and bonobos. Cambridge University Press, Cambridge, pp 71–83

Nakamura M (2003a) 'Gatherings' of social grooming among wild chimpanzees: implications for evolution of sociality. J Hum Evol 44:59–71

Nakamura M (2003b) Questions about chimpanzee culture studies (in Japanese). Ecosophia 12:55–61

Nakamura M (2010) Ubiquity of culture and possible social inheritance of sociality among wild chimpanzees. In: Lonsdorf EV, Ross SR, Matsuzawa T (eds) The mind of the chimpanzee: ecological and experimental perspectives, University of Chicago Press, Chicago, pp 156–167

Nakamura M, Nishida T (2006) Subtle behavioral variation in wild chimpanzees, with special reference to Imanishi's concept of kaluchua. Primates 47:35–42

Nakamura M, Ohashi G (2003) Eleven-year old male chimpanzee outranks ex-alpha adult male at Bossou. Pan Africa News 10:9–11

Nakamura M, McGrew WC, Marchant LF, Nishida T (2000) Social scratch: another custom in wild chimpanzees? Primates 41:237–248

Naughton-Treves L, Treves A, Chapman C, Wrangham R (1998) Temporal patterns of crop-raiding by primates: linking food availability in croplands and adjacent forest. J Appl Ecol 35:596–606

Newton-Fisher NE (1999) Association by male chimpanzees: a social tactic? Behaviour 136:705–730

Niilo L (1980) Clostridium perfringens in animal diseases: a review of current knowledge. Can Vet J 21:141–148

Nishida T (1973) The ant-gathering behavior by the use of tools among wild chimpanzees of the Mahale Mountains. J Hum Evol 2:357–370

Nishida T (1980) The leaf-clipping display: a newly-discovered expressive gesture in wild chimpanzees. J Hum Evol 9:117–128

Nishida T (1989) Social interaction between resident and immigrant female chimpanzees. In: Heltne PG, Marquardt LA (eds) Understanding chimpanzees. Harvard University Press, Cambridge, pp 68–89

Nishida T (1990a) A quarter century of research in the Mahale mountains: an overview. In: Nishida T (ed) The chimpanzees of the Mahale mountains: sexual and life history strategies. University of Tokyo Press, Tokyo, pp 3–35

Nishida T (ed) (1990b) The chimpanzees of the Mahale Mountains. University of Tokyo Press, Tokyo

Nishida T (1994) Review of recent findings on Mahale chimpanzees: implications and future research directions. In: Wrangham RW, McGrew WC, de Waal FBM, Heltne PG (eds) Chimpanzee cultures. Harvard University Press, Cambridge, pp 373–396

Nishida T (1997) Sexual behavior of adult male chimpanzees of the Mahale Mountains National Park, Tanzania. Primates 38:379–398

Nishida T (2008) Why were guava trees cut down in Mahale Park? The question of exterminating all introduced plants. Pan Africa News 15:12–14

Nishida T, Hosaka K (1996) Coalition strategies among adult male chimpanzees of the Mahale Mountains, Tanzania. In: McGrew WC, Marchant LF, Nishida T (eds) Great ape societies. Cambridge University Press, Cambridge, pp 114–134

Nishida T, Uehara S (1980) Chimpanzees, tools, and termites: another example from Tanzania. Curr Anthropol 21:671–672

Nishida T, Uehara S (1983) Natural diet of chimpanzees (*Pan troglodytes schweinfurthii*): long term record from the Mahale Mountains, Tanzania. Afr Stud Monogr 3:109–130

Nishida T, Wallauer W (2003) Leaf-pile pulling: an unusual play pattern in wild chimpanzees. Am J Primatol 60:167–173

Nishida T, Wrangham RW, Goodall J, Uehara S (1983) Local differences in plant-feeding habits of chimpanzees between the Mahale Mountains and Gombe National Park. J Hum Evol 12:467–480

Nishida T, Hiraiwa-Hasegawa M, Hasegawa T, Takahata Y (1985) Group extinction and female transfer in wild chimpanzees in the Mahale National Park, Tanzania. Z Tierpsychol 67:284–301

Nishida T, Takasaki H, Takahata Y (1990) Demography and reproductive profiles. In: Nishida T (ed) The chimpanzees of the Mahale Mountains. University of Tokyo Press, Tokyo, pp 9–10

Nishida T, Kano T, Goodall J, McGrew WC, Nakamura M (1999) Ethogram and ethnography of Mahale chimpanzees. Anthropol Sci 107:141–188

Nishida T, Uehara S, Kawanaka K (eds) (2002) The Mahale chimpanzees (in Japanese). Kyoto University Press, Kyoto

Nishida T, Corp N, Hamai M, Hasegawa T, Hiraiwa-Hasegawa M, Hosaka K, Hunt KV, Itoh N, Kawanaka K, Matsumono-Oda A, Mitani JC, Nakamura M, Norikoshi K, Sakamaki T, Turner L, Uehara S, Zamma K (2003) Demography, female life history, and reproductive profiles among the chimpanzees of Mahale. Am J Primatol 59:99–121

Nishida T, Mitani JC, Watts DP (2004) Variable grooming behaviours in wild chimpanzees. Folia Primatol 75:31–36

Nissen HW (1932) A field study of the chimpanzee: observations of chimpanzee behavior and environment in western French Guinea. Comp Psychol Monogr 8:1–122

Ohashi G (2006a) Behavioral repertoire of tool use in the wild chimpanzees at Bossou. In: Matsuzawa T, Tomonaga M, Tanaka M (eds) Cognitive development in chimpanzees. Springer, Tokyo, pp 439–451

Ohashi G (2006b) Bossou chimpanzees crossed the national border of Guinea into Liberia. Pan Africa News 13:10–12

Ohashi G (2007) Papaya fruit sharing in wild chimpanzees at Bossou, Guinea. Pan Africa News 14:14–16

Ott-Joslin JE (1993) Zoonotic diseases of nonhuman primates. In: Fowler ME (ed) Zoo and wild animal medicine. Saunders, Philadelphia, pp 358–373

Pan African Sanctuary Alliance (2008) Guinea chimpanzee reintroduction brings surprises, success. PASA Newsl 5 September 2008

Panger MA, Brooks AS, Richmond BG, Wood B (2002) Older than the Oldowan? Rethinking the emergence of hominin tool use. Evol Anthropol 11:235–245

Pascual JF (1988) Les sols actuels et les formations superficielles des crêtes nord-est du Nimba (Guinée). Contribution à l'étude morphologique du Quaternaire de la chaîne. Cuk. ORSTOM Sér Pédol 24(2):137–162

Pellant C (1992) Rocks and Minerals. Dorling Kindersley, New York

Peterson D, Goodall J (1993) Visions of Caliban: on chimpanzees and people. Houghton Mifflin, Boston

Plumptre AJ, Johns AG (2001) Changes in primate communities following logging disturbance. In: Fimbel RA, Grajal A, Robinson JG (eds) The cutting edge: conserving wildlife in logged tropical forest. Columbia University Press, New York, pp 71–92

Pruetz JD (2002) Competition between savanna chimpanzees and humans in southeastern Senegal (abstract). Am J Phys Anthropol 34:128

Pruetz JD, Bertolani P (2007) Savanna chimpanzees, *Pan troglodytes verus*, hunt with tools. Curr Biol 17:412–417

Pusey A, Williams J, Goodall J (1997) The influence of dominance rank on the reproductive success of female chimpanzees. Science 277:828–830

Quiatt D, Kiwede ZT (1994) Leaf sponge drinking by the Budongo forest chimpanzees (abstract). Am J Primatol 33:236

Reclus E (1892) The earth and its inhabitants: Africa. Volume III. West Africa. D. Appleton and Company, New York

Reed ES (1996) Encountering the world. Toward an ecological psychology. Oxford University Press, New York

Rémy M (1999) La Guinée Aujourd'hui. Les Editions du Jaguar, Paris

Renfrew C, Bahn P (1998) Arqueología, Teorías, métodos y práctica. 2ª edicíon. Ediciones Akal, Madrid

Resende BD, Ottoni EB, Fragaszy DM (2008) Ontogeny of manipulative behavior and nutcracking in young tufted capuchin monkeys (*Cebus apella*): a perception-action perspective. Dev Sci 11:828–840

Reynolds V (1967) The apes: the gorilla, chimpanzee, orangutan, and gibbon—their history and their world. Dutton, New York

Reynolds V (2005a) The chimpanzees of the Budongo Forest: ecology, behaviour, and conservation. Oxford University Press, Oxford

Reynolds V (2005b) The problem of snares. In Reynolds V (ed) The chimpanzees of the Budongo Forest: ecology, behaviour and conservation. Oxford University Press, New York, pp 164–190

Rhine RJ (1975) The order of movement of yellow baboons *(Papio cynocephalus)*. Folia Primatol 23:72–104

Rhine RJ, Tilson R (1987) Reactions to fear as a proximate factor in the sociospatial organization of baboon progressions. Am J Primatol 13:119–128

Rhine RJ, Westlund BJ (1981) Adult male positioning in baboon progressions: order and chaos revisited. Folia Primatol 35:77–116

Richards PW (1996) The tropical rain forest, 2nd edn. Cambridge University Press, Cambridge

Richards P (2000) Chimpanzees as political animals in Sierra Leone. In: Knight J (ed) Natural enemies: people–wildlife conflicts in anthropological perspective. Routledge, London, pp 78–103

Rochat P (2001) The infant's world. Harvard University Press, Cambridge

Rondeau G, Lebbie A (2007) Faunal species of conservation concern at Mount Nimba (literature review). Environmental impact study. Société des Mines de Fer de Guinée (SMFG), Gbakoré

Roskam S (2001) Excavation. Cambridge University Press, Cambridge

Ross SR, Lukas KE, Lonsdorf EV, Stoinski TS, Hare B, Shumker R, Goodall J (2008) Inappropriate use and portrayal of chimpanzees. Science 319:1487

Rossianov K (2002) Beyond species: Il'ya Ivanov and his experiments on cross-breeding humans with anthropoid apes. Sci Context 15(2):277–316

Rothfels N (2002) Savages and beasts: the birth of the modern zoo. The John Hopkins University Press, Baltimore

Rouquet P, Friment J-M, Bermejo M, Kibourn A, Karesh W, Reed P, Kumulungui B, Yaba P, Délicat A, Rollin PE, Leroy EM (2005) Wild animal mortality monitoring and human ebola outbreaks, Gabon and Republic of Congo, 2001-2003. Emerg Infect Dis 11:283–290

Rumbaugh DM (1977) Language learning by a chimpanzee. The Lana project. Academic, New York

Saito M (1990) Production of enterotoxin by *Clostridium perfringens* derived from human, animals, foods, and the natural environment in Japan. J Food Prot 53:115–118

Sakamaki T (1998) First record of algae-feeding by a female chimpanzee at Mahale. Pan African News 5:1–3

Sakho D (1997) Elaboration d'un plan d'aménagement pastoral dans la Réserve de Biosphère des Monts Nimba. http://www.unesco.org/mab/doc/mys/95/sakho/95_sakho.pdf. Accessed 14 December 2010

Sakura O (1991) On the concept of 'group' in primates, with special reference to fission-fusion of chimpanzees. In: Ehara A, Kimura T, Takenaka O, Iwamoto M (eds) Primatology today. Elsevier, Amsterdam, pp 247–250

Sakura O (1994) Factors affecting party size and composition of chimpanzees *(Pan troglodytes verus)* at Bossou, Guinea. Int J Primatol 15:167–181

Sakura O (1995) What is this thing called "group of animals"? A proposal of the pluralistic terminology implicated from non-human primate ecology. Ann Jpn Assoc Philos Sci 8:237–252

Sakura O, Matsuzawa T (1991) Flexibility of wild chimpanzee nut-cracking behavior using stone hammers and anvils: an experimental analysis. Ethology 87:237–248

Sakura O, Fushimi T, Matsuzawa T, Ohno H, Sugiyama Y (1991) Social behavior of wild chimpanzees in Bossou, Guinea, West Africa. In: Ehara A, Kimura T, Takenaka O, Iwamoto M (eds) Primatology today. Elsevier, Amsterdam, pp 713–714

Sanz CM, Morgan DB (2007) Chimpanzee tool technology in the Goualougo Triangle, Republic of Congo. J Hum Evol 52:420–433

Sanz CM, Morgan DB (2010) The complexity of chimpanzee tool-use behaviors. In: Lonsdorf EV, Ross SR, Matsuzawa T (eds) The mind of the chimpanzee. The University of Chicago Press, Chicago, pp 127–140

Sanz C, Morgan D, Gulick S (2004) New insights into chimpanzees, tools, and termites from the Congo Basin. Am Nat 164:567–581

Savage T, Wyman J (1844) Observations on the external characters and habits of *Troglodytes niger* Geoff. and on its organization. Boston J Nat Hist 4:362–386

Sayer JA, Harcourt CS, Collins MN (eds) (1992) The conservation atlas of tropical forests: Africa. MacMillan, London

Schiffer MB (1985) Is there a "Pompeii premise" in archaeology? J Anthropol Archaeol 41:18–41

Schlüter T (2006) Geological atlas of Africa: with notes on stratigraphy, tectonics, economic geology, geohazards and geosites of each country. Birkhäuser, Boston

Schnell R (1998) Le Mont Nimba, carrefour West Africain des flores. In: Lamotte M (ed) Le Mont Nimba, Réserve de Biosphère et site du Patrimoine Mondial (Guinée – Côte d'Ivoire) – Initiation à la Géomorphologie et Biogéographie. UNESCO, Paris, pp 55–76

Schöning C, Ellis D, Fowler A, Sommer V (2007) Army ant prey availability and consumption by chimpanzees *(Pan troglodytes vellerosus)* at Gashaka (Nigeria). J Zool 271:125–133

Schöning C, Humle T, Möbius Y, McGrew WC (2008) The nature of culture: technological variation in chimpanzee predation on army ants. J Hum Evol 55(1):48–59

Semaw S, Rogers MJ, Quade J, Renne PR, Butler RF, Dominguez-Rodrigo M, Stout D, Hart WS, Pickering T, Simpson SW (2003) 2.6-Million-year-old stone tools and associated bones from OGS-6 and OGS-7, Gona, Afar, Ethiopia. J Hum Evol 45:169–177

Sept JM (1992) Was there no place like home? A new perspective on early hominid archeological sites from the mapping of chimpanzees nests. Curr Anthropol 33:187–207

Sept JM, Brooks GE (1994) Reports of chimpanzee natural history, including tool use, in 16th- and 17th-century. Int J Primatol 15(6):867–878

Shimada MK (2000) A Survey of Nimba Mountain, West Africa from three routes: confirmed new habitat and ant-catching wands use of chimpanzees. Pan Africa News 7:7–10

Shimada MK, Hayakawa S, Humle T, Fujita S, Hirata S, Sugiyama Y, Saitou N (2004) Mitochondrial DNA genealogy of chimpanzees in the Nimba Mountains and Bossou, West Africa. Am J Primatol 64:261–275

Shimada MK, Hayakawa S, Fujita S, Sugiyama Y, Saitou N (2009) Skewed matrilineal genetic composition in a small wild chimpanzee community. Folia Primatol 80:19–32

Shinoda K, Udono T, Yoshihara K, Shimada M, Takenaka O (2003) Molecular identification of subspecies of captive chimpanzees reared in Japan using mitochondrial DNA. Primate Res 19:145–155

Sholley C, Hasting B (1989) Outbreak of illness among Rwanda's gorillas. Gorilla Conserv News 3:7

Sivert H, Karesh WB, Sunde V (1991) Reproductive intervals in captive female western lowland gorillas with a comparison to wild mountain gorillas. Am J Primatol 24:227–234

Smedley JG III, Fisher DJ, Sayeed S, Chakrabarti G, McClane BA (2004) The enteric toxins of *Clostridium perfringens*. Rev Physiol Biochem Pharmacol 152:183–204

Smith AH, Butler TM, Pace N (1975) Weight growth of colony reared chimpanzees. Folia Primatol 24:29–59

Smitsman AW (1997) The development of tool-use: changing boundaries between organism and environment. In: Dent-Read C, Zukow-Goldring P (eds) Evolving explanations of development: ecological approaches to organism–environment systems. American Psychological Association, Washington, pp 301–333

Soave O (1982) The rehabilitation of chimpanzees and other apes. Lab Primate Newsl 21(4):3–8

Songer JG (1996) Clostridial enteric diseases of domestic animals. Clin Microbiol Rev 9:216–234

Soropogui PE, Gbamy KR, Diakité A (2008) Le climat. In: Kourouma M, Moloumou P, Mahomi CG, Soropogui PE (eds) La diversité biologique de la Réserve de Biosphère des Monts Nimba. Programme de Conservation de la Biodiversité des Monts Nimba. CSE-RBMN Report, Gbakoré, pp 308–318

Sousa C, Biro D, Matsuzawa T (2009) Leaf-tool use for drinking water by wild chimpanzees (*Pan troglodytes*): acquisition patterns and handedness. Anim Cogn 12:S115–S125

Species Survival Network (2007) Sierra Leone outlaws the capture and killing of chimpanzees. CITES Afr 1(2):3

Stevens CE, Hume ID (1995) Comparative physiology of the vertebrate digestive system. Cambridge University Press, Cambridge

Stewart KJ (1988) Suckling and lactational anoestrus in wild gorillas (*Gorilla gorilla*). J Reprod Fertil 83:627–634

Sugiyama Y (1978) Bossou-mura no hito to chinpanji: Nishi-afurika hekichi no seitai (in Japanese). Kinokuniya, Tokyo

Sugiyama Y (1981a) Observations on the population dynamics and behavior of wild chimpanzees at Bossou, Guinea, in 1979–1980. Primates 22:435–444

Sugiyama Y (1981b) Yasei chinpanji no sekai: Jinrui shinka heno michisuji (in Japanese). Kodansha, Tokyo

Sugiyama Y (1984) Population dynamics of wild chimpanzees at Bossou, Guinea, between 1976–1983. Primates 25:391–400

Sugiyama Y (1985) Chimpanzee importation and animal experiment research (original article in Japanese). Kagaku 55:127–130

Sugiyama Y (1987) A preliminary list of chimpanzees alimentation at Bossou, Guinea. Primates 28:133–147

Sugiyama Y (1988) Grooming interactions among adult chimpanzees at Bossou, Guinea, with special reference to social structure. Int J Primatol 9:393–407

Sugiyama Y (1989a) Description of some characteristic behaviors and discussion on their propagation process among chimpanzees of Bossou, Guinea. In: Sugiyama Y (ed) Behavioral studies of wild chimpanzees at Bossou, Guinea. Kyoto University Primate Research Institute, Inuyama, pp 43–76

Sugiyama Y (1989b) Population dynamics of chimpanzees at Bossou, Guinea. In: Heltne PG, Marquardt LG (eds) Understanding chimpanzees. Harvard University Press, Cambridge, pp 134–145

Sugiyama Y (1990) Nut-cracking culture in chimpanzees (in Japanese). In: Kawai M (ed) Jinrui Izen no Shakai Gaku. Kyoikusha, Tokyo, pp 435–453

Sugiyama Y (1991) Habitat isolation and population structure of wild chimpanzees in and around Bossou, West Africa. In: Maruyama N, Bobek B, Ono Y, Regelin W, Bartos L, Ratcliffe PR (eds) Wildlife conservation: present trends and perspectives for the 21th century. Japan Wildlife Research Center, Tokyo, pp 32–35

Sugiyama Y (1993) Local variation of tools and tool use among wild chimpanzee populations. In: Berthelet A, Chavaillon J (eds) The use of tools by human and non-human primates. Clarendon Press, Oxford, pp 175–187

Sugiyama Y (1994a) Age-specific birth-rate and lifetime reproductive success of chimpanzees at Bossou, Guinea. Am J Primatol 32:311–318

Sugiyama Y (1994b) Research at Bossou. Pan Africa News 1:2–3

Sugiyama Y (1994c) Tool use by wild chimpanzees. Nature 367:327

Sugiyama Y (1995a) Drinking tools of wild chimpanzees at Bossou. Am J Primatol 37:263–269

Sugiyama Y (1995b) Tool-use for catching ants by chimpanzees at Bossu and Monts-Nimba, West-Africa. Primates 36:193–205

Sugiyama Y (1997) Social tradition and the use of tool-composites by wild chimpanzees. Evol Anthropol 6:23–27

Sugiyama Y (1998) Local variation of tool-using repertoire in wild chimpanzees. In: Nishida T (ed) Comparative study of the behavior of the genus *Pan* by compiling video ethogram. Nissindo, Kyoto, pp 82–91

Sugiyama Y (1999) Socioecological factors of male chimpanzee migration at Bossou, Guinea. Primates 40:61–68

Sugiyama Y (2004) Demographic parameters and life history of chimpanzees at Bossou, Guinea. Am J Phys Anthropol 124:154–165

Sugiyama Y, Koman J (1979a) Social structure and dynamics of wild chimpanzees at Bossou, Guinea. Primates 20:323–339

Sugiyama Y, Koman J (1979b) Tool-using and making behavior in wild chimpanzees at Bossou, Guinea. Primates 20:513–524

Sugiyama Y, Koman J (1987) A preliminary list of chimpanzees' alimentation at Bossou, Guinea. Primates 28:133–147

Sugiyama Y, Koman J (1992) The flora of Bossou: its utilization by chimpanzees and humans. Afr Study Monogr 13:127–169

Sugiyama Y, Soumah AG (1988) Preliminary survey of the distribution and population of chimpanzees in the Republic of Guinea. Primates 29:569–574

Sugiyama Y, Koman J, Sow MB (1988) Ant-catching wands of wild chimpanzees at Bossou, Guinea. Folia Primatol 51:56–60

Sugiyama Y, Fushimi T, Sakura O, Matsuzawa T (1993a) Hand preference and tool use in wild chimpanzees. Primates 34:151–159

Sugiyama Y, Kawamoto S, Takenaka O, Kumazaki K, Miwa N (1993b) Paternity discrimination and intergroup relationships of chimpanzees at Bossou. Primates 34:545–552

Sugiyama Y, Kurita H, Matsui T, Kimoto S, Shimomura T (2009) Carrying of dead infants by Japanese macaque (*Macaca fuscata*) mothers. Anthropol Sci 117:113–119

Swaine MD (1992) Characteristics of dry forest in West Africa and the influence of fire. J Veget Sci 3:365–374

Takahata Y (1990) Social relationships among adult males. In: Nishida T (ed) The chimpanzees of the Mahale Mountains: sexual and life history strategies. University of Tokyo Press, Tokyo, pp 149–170

Takahata Y, Hiraiwa-Hasegawa M, Takasaki H, Nyundo R (1985) Newly acquired feeding habits among the chimpanzees of the Mahale Mountains National Park, Tanzania. Hum Evol 1:277–284

Takasaki H (1983) Mahale chimpanzees taste mangoes—toward acquisition of a new food item? Primates 24: 273–275

Takemoto H (2002) Feeding ecology of chimpanzees in Bossou, Guinea: coping with the seasonal fluctuation of food supply and micrometeorology in the tropical forest. Primate Research Institute, Kyoto University, Kyoto

Takemoto H (2003) Phytochemical determination for leaf food choice by wild chimpanzees in Guinea, Bossou. J Chem Ecol 29:2551–2573

Takemoto H (2004) Seasonal change in terrestriality of chimpanzees in relation to microclimate in the tropical forest. Am J Phys Anthropol 124:81–92

Takemoto H, Hirata S, Sugiyama Y (2005) The formation of the brush-sticks: modification of chimpanzees or the by-product of folding? Primates 46:183–189

Takeshita H (2001) Development of combinatory manipulation in chimpanzee infants (*Pan troglodytes*). Anim Cogn 4:335–345

Takeshita H, Myowa-Yamakoshi M, Hirata S (2009) The supine position of postnatal human infants: implications for the development of cognitive intelligence. Interaction Stud 10:252–268

Teleki G (1973) The predatory behaviour of wild chimpanzees. Bucknell University Press, Lewisburg

Teleki G (1980) Hunting and trapping wildlife in Sierra Leone: aspects of exploitation and exportation. Report submitted to Office of President, Sierra Leone; World Wildlife Fund–U.S., World Wildlife Fund-International

Teleki G (2001) Sanctuaries for ape refugees. In: Beck B, Stoinski T, Hutchins M, Maple T, Norton B, Rowan A, Stevens E, Arluke A (eds) Great apes and humans: the ethics of coexistence. Smithsonian Institution Press, Washington, pp 133–149

The Chimpanzee Sequencing and Analysis Consortium (2005) Initial sequence of the chimpanzee genome and comparison with the human genome. Nature 437(7055):69–87

Tilley JMA, Terry RA (1963) A two stage technique for the in vitro digestion of forage crops. Grass Forage Sci 18:104–111

Tixier JP (2000) À propos des processus de formation des sites préhistoriques. Paleo 12:379–385

Tomasello M (1999) The cultural origins of human cognition. Harvard University Press, Cambridge

Tomasello M, Call J (1997) Primate cognition. Oxford University Press, New York

Tomonaga M, Tanaka M, Matsuzawa T, Myowa-Yamakoshi M, Kosugi D, Mizuno Y, Okamoto S, Yamaguchi MK, Bard KA (2004) Development of social cognition in chimpanzees (*Pan troglodytes*): face recognition, smiling, mutual gaze, gaze following and the lack of triadic interactions. Jpn Psychol Res 46:227–235

Tonooka R (2001) Leaf-folding behavior for drinking water by wild chimpanzees (*Pan troglodytes verus*) at Bossou, Guinea. Anim Cogn 4:325–334

Tonooka R, Inoue N, Matsuzawa T (1994) Leaf-folding behavior for drinking water by wild chimpanzees at Bossou, Guinea: a field experiment and leaf selectivity (in Japanese with English summary). Primate Res 10:307–313

Tonooka R, Tomonaga M, Matsuzawa T (1997) Acquisition and transmission of tool making and use for drinking juice in a group of captive chimpanzees (*Pan troglodytes*). Jpn Psychol Res 39:253–265

Touré M, Suter J (2002) Initiation d'un programme tri national pour la conservation intégrée des Monts Nimba. 2ème Réunion Tri nationale Côte d'Ivoire, Guinée, Libéria, Février 2002 à N'Zérékoré. FFI-CI-BirdLife Report, Guinea

Trefflich H, Anthony E (1967) Jungle for sale. Hawthorn, New York

Trevarthen C, Hubley P (1978) Secondary intersubjectivity: confidence, confiding and acts of meaning in the first year. In: Lock A (ed) Action, gesture and symbol. Academic, London, pp 183–229

Tsukahara T, Koyama K, Okada A, Ushida K (2002) Stimulation of butyrate production by gluconic acid in pig cecal digesta and identification of butyrate producing bacteria. J Nutr 132:2229–2234

Turvey MT (1992) Affordances and prospective control: an outline of the ontology. Ecol Psychol 4(3):173–187

Tutin CEG (1994) Reproductive success story: variability among chimpanzees and comparisons with gorillas. In: Wrangham RW, McGrew WC, de Waal FBM, Heltne PG (eds) Chimpanzee cultures. Harvard University Press, Cambridge, pp 181–193

Tutin CEG, Fernandez M (1991) Responses of wild chimpanzees and gorillas to the arrival of primatologists: behaviour observed during habituation. In: Box HO (ed) Primate responses to environmental change. Chapman & Hall, London, pp 187–197

Tutin CEG, Ancrenaz M, Paredes J, Vacher-Vallas M, Vidal C, Goosens B, Bruford MW, Jamart A (2001) Conservation biology framework for the release of wild-born orphaned chimpanzees into the Conkouati reserve, Congo. Conserv Biol 15(5):1247–1257

Tutin CEG, Stokes E, Boesch C, Morgan D, Sanz C, Reed T, Blom A, Walsh P, Blake S, Kormos R (2005) Regional Action Plan for the conservation of chimpanzees and gorillas in western equatorial Africa. Conservation International, Washington

Tweheyo M, Lye KA (2003) Phenology of figs in Budongo Forest Uganda and its importance for the chimpanzee diet. Afr J Ecol 41:306–316

Uehara S (1982) Seasonal changes in the techniques employed by wild chimpanzees in the Mahale Mountains, Tanzania, to feed on termites (*Pseudocanthotermes spiniger*). Folia Primatol 37:44–76

Uehara S (1997) Predation on mammals by the chimpanzee (*Pan troglodytes*). Primates 38:193–214

Uenishi G, Fujita S, Ohashi G, Kato A, Yamauchi S, Matsuzawa T, Ushida K (2007) Molecular analyses of the intestinal microbiota of chimpanzees in the wild and in captivity. Am J Primatol 69:367–376

Ueno A, Matsuzawa T (2005) Response to novel food in infant chimpanzees: do infants refer to mothers before ingesting food on their own? Behav Process 68:85–90

Unti B (2006) Chimpanzee protection in Sierra Leone: a law enforcement and legislative review. Chimpanzee Conservation and Sensitization Program, the Jane Goodall Institute, Arlington

Ushida K (2009) Intestinal bacteria of chimpanzees in the wild and in captivity: an application of molecular ecological methodologies. In: Huffman M, Chapman C (eds) Primate parasite ecology. Cambridge University Press, Cambridge, pp 283–295

Ushida K, Sakata T (1998) Effect of pH on oligosaccharides fermentation by porcine cecal digesta. Anim Sci J 69:20–27

Ushida K, Kayouli C, DeSMet S, Jouany JP (1990) Effect of defaunation on protein and fibre digestion in sheep fed ammonia-treated straw based diets with or without maize. Br J Nutr 64:765–775

Ushida K, Fujita S, Ohashi G (2006) Nutritional significance of the selective ingestion of *Albizia zygia* gum exudate by wild chimpanzees in Bossou, Guinea. Am J Primatol 68:143–151

Ushida K, Uwatoko Y, Adachi Y, Soumah AG, Matsuzawa T (2010) Isolation of Bifidobacteria from feces of chimpanzees in the wild. J Gen Appl Microbiol 56:57–60

Van den Ende MC, Brotman B, Prince AM (1980) An open air holding system for chimpanzees in medical experiments. Dev Biol Stand 45:95–98

van den Hoek C, Mann D, Jahns H (1995) Algae. Cambridge University Press, Cmabridge

van Lawick-Goodall J (1972) A preliminary report on expressive movements and communication in the Gombe Stream chimpanzees. In: Dolhinow P (ed) Primate patterns. Holt, Rinehart and Winston, New York, pp 25–84

van Leeuwen L, Smitsman A, van Leeuwen C (1994) Affordance, perceptual complexity, and the development of tool use. J Exp Psychol 20:174–191

van Schaik CP, Ancrenaz M, Borgen G, Galdikas B, Knott CD, Singleton I, Suzuki A, Utami SS, Merrill M (2003) Orangutan cultures and the evolution of material culture. Science 299:102–105

Van Soest PJ (1982) Nutritional ecology of the ruminant. O & B Books, Corvallis

van Zon JCJ, van Orshoven J (1967) Enkele Resultaten van de Zesde Nederlandse Chimpansee-Expeditie. Vakbl Biol 47:161–166

VandeBerg JL, Zola SM (2005) A unique biomedical resource at risk. Nature 437:30–32

Verschuren J (1983) Conservation of tropical rain forest of Liberia. Recommendations for wildlife conservation and national parks. WWF/IUCN Report, World Conservation Center, Switzerland

Visalberghi E (1987) Acquisition of nut-cracking behavior by 2 capuchin monkeys (*Cebus apella*). Folia Primatol 49:168–181

Vogel G (1999) Chimps in the wild show stirrings of culture. Science 284:2070–2073

von Linstow ML, Eugen-Olsen J, Koch A, Winther TM, Westh H, Hogh B (2006) Excretion patterns of human metapneumovirus and respiratory syncytial virus among young children. Eur J Med Res 11:329–335

Wallis J (1997) A survey of reproductive parameters in the free-ranging chimpanzees of Gombe National Park. J Reprod Fertil 109:297–307

Wallis J (2002) Seasonal aspects of reproduction and sexual behavior in two chimpanzee populations: a comparison of Gombe (Tanzania) and Budongo (Uganda). In: Boesch C, Hohmann G, Marchant LF (eds) Behavioral diversity in chimpanzees and bonobos. Cambridge University Press, Cambridge, pp 181–191

Wallis J, Lee DR (1999) Primate conservation: the prevention of disease transmission. Int J Primatol 20:803–826

Walsh PD, Abernethy KA, Bermejo M, Beyers R, De Wachter P, Akou ME, Huijbregts B, Mambounga DI, Toham AK, Kilbourn AM, Lahm SA, Latour S, Maisels F, Mbina C, Mihindou Y, Obiang SN, Effa EN, Starkey MP, Telfer P, Thibault M, Tutin CEG, White LJT, Wilkie DS (2003) Catastrophic ape decline in western equatorial Africa. Nature 422:611–614

Walsh PD, Biek R, Real LA (2005) Wave-like spread of Ebola Zaire. PLOS Biol 3:1946–1953

Warren Y, Williamson EA (2004) Transport of dead infant mountain gorillas by mothers and unrelated females. Zoo Biol 23:375–378

Waser PM (1985) Spatial structure in mangabey groups. Int J Primatol 6:569–580

Watts DP (2000) Causes and consequences of variation in male mountain gorilla life histories and group membership. In: Kappeler PM (ed) Primate males: causes and consequences of variation in group composition. Cambridge University Press, Cambridge, pp 169–179

Watts DP, Mitani JC (2001) Boundary patrols and intergroup encounters in wild chimpanzees. Behaviour 138:299–327

WCMC (1992) Guinea/Côte d'Ivoire-Réserve Naturelle Intégrale des Mont Nimba. UNEP-WCMC (World Conservation Monitoring Centre). Infobase Report

White F (1983) The vegetation of Africa. UNESCO, Paris

White LJT, Abernethy K (1996) Guide de la vegetation de la reserve de la Lopé. Écofac Gabon/Multipress Gabon, Gabon

White LJT, Tutin CEG (2001) Why chimpanzees and gorillas respond differently to logging: a cautionary tale from Gabon. In: Weber W, White LJT, Vedder A, Naughton-Treves L (eds) African rain forest ecology and conservation. Yale University Press, New Haven, pp 449–462

White TD, Suwa G, Asfaw B (1995) Ardipithecus ramidus, a new species of early hominid from Aramis, Ethiopia. Nature 375:88

Whiten A (2010) A Coming of age for cultural panthropology. In: Lonsdorf EV, Ross SR, Matsuzawa T (eds) The mind of the chimpanzee: ecological and experimental perspectives. University of Chicago Press, Chicago, pp 87–100

Whiten A, Ham R (1992) On the nature and evolution of imitation in the animal kingdom: reappraisal of a century of research. Adv Stud Behav 21:239–283

Whiten A, Goodall J, McGrew WC, Nishida T, Reynolds V, Sugiyama Y, Tutin CEG, Wrangham RW, Boesch C (1999) Cultures in chimpanzees. Nature 399:682–685

Whiten A, Goodall J, McGrew WC, Nishida T, Reynolds V, Sugiyama Y, Tutin CEG, Wrangham R, Boesch C (2001) Charting cultural variation in chimpanzees. Behaviour 138:1481–1516

Whiten A, Horner H, de Waal FBM (2005) Conformity to cultural norms of tool use in chimpanzees. Nature 437:737–740

Whiten A, Spiteri A, Horner V, Bonnie KE, Lambeth SP, Schapiro SJ, de Waal FBM (2007) Transmission of multiple traditions within and between chimpanzee groups. Curr Biol 17:1038–1043

Whitesides GH (1985) Nut cracking by wild chimpanzees in Sierra Leone, West Africa. Primates 26:91–94

WHS-UNESCO (2008) World Heritage List. Mount Nimba Strict Nature Reserve. http://whc.unesco.org/en/list/155. Accessed 14 December 2010

Wilkerson JE, Raven PB, Horvath SM (1972) Critical temperature of unacclimatized male Caucasians. J Appl Physiol 33:451–455

Wilkie DS, Carpenter JF (1999) Bushmeat hunting in the Congo Basin. An assessment of impact and options for mitigation. Biodivers Conserv 8:927–945

Wilkinson TJ (2006). The archaeology of landscae. In: Bintliff J (ed) A companion to archaeology. Blackwell, London, pp 334–356

Wilson E (1975) Sociobiology. Harvard University Press, Cambridge

Wilson R (1992) Guinea. In: Sayer JA, Harcourt CA, Collins NM, Billington C, Adam M (eds) The conservation atlas of tropical forests: Africa. Macmillan, London, pp. 193–199

Wilson ML, Hauser MD, Wrangham RW (2001) Does participation in intergroup conflict depend on numerical assessment, range location, or rank for wild chimpanzees? Anim Behav 61:1203–1216

Wolfe ND, Escalante AA, Karesh WB, Kilbourn A, Spielman A, Lala AA (1998) Wild primates populations in emerging infectious disease research: the missing link? Emerg Infect Dis 4:451–457

Woodford MH, Butynski TM, Karesh WB (2002) Habituating the great apes: the disease risks. Oryx 36:153–160

Woodruff DS (2004) Noninvasive genotyping and field studies of free-ranging nonhuman primates. In: Chapais B, Berman CM (eds) Kinship and behavior in primates. Oxford University Press, Oxford, pp 46–68

Wrangham RW (1992) Living naturally: aspects of wild chimpanzee management. In: Erwin J, Landon JC (eds) Chimpanzee observation and public health. Diagnon, Rockville, pp 71–81

Wrangham RW (2006) Chimpanzees: the culture-zone concept becomes untidy. Curr Biol 16:R634–R635

Wrangham RW, Clark AP, Isabirye-Basuta G (1992) Female social relationships and social organization of Kibale Forest chimpanzees. In: Nishida T, McGrew WC, Marler P, Pickford M, de Waal FBM (eds) Topics in primatology, vol 1. Human origins. University of Tokyo Press, Tokyo, pp 81–98

Wrangham R, Chapman C, Clark-Arcadi A, Isabirye-Basuta G (1996) Social ecology of Kanyawara chimpanzee: implications for understanding the costs of great ape groups. In: McGrew WC, Marchant LF, Nishida T (eds) Great ape societies. Cambridge University Press, Cambridge, pp 45–57

Wright JB (1985) Geology and mineral resources of West Africa. Springer, London

Wynn T, McGrew WC (1989) An ape's view of the Oldowan. Man 24:383–398

Yamagiwa J, Kahekwa J (2001) Dispersal patterns, group structure and reproductive parameters of eastern lowland gorillas at Kahuzi in the absence of infanticide. In: Robbins M, Sicotte P, Stewart KJ (eds) Mountain gorillas. Cambridge University Press, Cambridge, pp 89–122

Yamakoshi G (1998) Dietary responses to fruit scarcity of wild chimpanzees at Bossou, Guinea: possible implications for ecological importance of tool use. Am J Phys Anthropol 106:283–295

Yamakoshi G (1999) Chimpanzees in a "sacred grove": coexistence with the people at Bossou, Guinea (in Japanese). Ecosophia 3:106–117

Yamakoshi G (2001) Ecology of tool use in wild chimpanzees: toward reconstruction of early hominid evolution. In: Matsuzawa T (ed) Primate origins of human cognition and behavior. Springer, Tokyo, pp 537–556

Yamakoshi G (2004a) Evolution of complex feeding techniques in primates: is this the origin of great-ape intelligence? In: Russon AE, Begun DR (eds) The evolution of thought: evolutionary origins of great ape intelligence. Cambridge University Press, Cambridge, pp 140–171

Yamakoshi G (2004b) Food seasonality and socioecology in Pan: are West African chimpanzees another bonobo? Afr Study Monogr 25:45–60

Yamakoshi G (2005) What is happening on the border between humans and chimpanzees? Wildlife conservation in West African rural landscapes. In: Hiramatsu K (ed) Coexistence with nature in a 'Globalising' World: field science perspectives. Proceedings of the 7th Kyoto University International Symposium, 2005. Kyoto University, Kyoto, pp 91–97

Yamakoshi G (2006a) An indigenous concept of landscape management for chimpanzee conservation at Bossou, Guinea. In: Maruyama J, Wang L, Fujikura T, Ito M (eds) Proceedings of Kyoto Symposium 2006, Crossing Disciplinary Boundaries and Re-visioning Area Studies: Perspectives from Asia and Africa. Kyoto University, Kyoto, pp 3–10

Yamakoshi G (2006b) The indigenous concept of landscape management for chimpanzee conservation at Bossou, Guinea: the examination of a recent conflict and its subsequent change over autonomy in natural resource management (in Japanese). J Environ Sociol 12:120–135

Yamakoshi G, Koops K (2008) Life history profiles of female chimpanzees in Bossou over 40 years (abstract). Primate Eye 96:305

Yamakoshi G, Myowa-Yamakoshi M (2004) New observations of ant-dipping techniques in wild chimpanzees at Bossou, Guinea. Primates 45:25–32

Yamakoshi G, Sugiyama Y (1995) Pestle-pounding behavior of wild chimpanzees at Bossou, Guinea: a newly observed tool-using behavior. Primates 36:489–500

Yamamoto S, Yamakoshi G, Humle T, Matsuzawa T (2008). Invention and modification of a new tool use behavior: ant-fishing in trees by a wild chimpanzee (*Pan troglodytes verus*) at Bossou, Guinea. Am J Primatol 70:699–702

Yeager CP, Silver SC (1999) Translocation and rehabilitation as primate conservation tools: are they worth the cost? In: Dolhinow P, Fuentes A (eds) The nonhuman primates. Maryfield, Mountain View, pp 164–169

Yerkes RM, Yerkes AW (1929) The great apes: a study of anthropoid life. Yale University Press, New Haven

Yoshihara M (1985) The data of pregnancy, nursing and growth of chimpanzees, *Pan troglodytes*, at Tama Zoological Park (in Japanese). Dohsuishi 27:58–61

Young W, Yerkes R (1943) Factors influencing the reproductive cycle in the chimpanzees: the period of adolescent sterility and related problems. Endocrinology 33:121–154

Zamma K (2002) Leaf-grooming by a wild chimpanzee in Mahale. Primates 43:87–90

Zamma K, Fujita S (2004) Genito-genital rubbing among the chimpanzees of Mahale and Bossou. Pan Africa News 11:5–8

Index

Printed in Japan

Springer